QA351 SLA

OXFORD MATHEMATICAL MONOGRAPHS

Series Editors

J. M. BALL   E. M. FRIEDLANDER   I. G. MACDONALD
L. NIRENBERG   R. PENROSE   J. T. STUART

OXFORD MATHEMATICAL MONOGRAPHS

Special Functions
A Unified Theory Based on Singularities

SERGEI YU. SLAVYANOV
St. Petersburg State University

and

WOLFGANG LAY
Universität Stuttgart

with a Foreword by

ALFRED SEEGER
Max-Planck-Institut für Metallforschung Stuttgart

OXFORD
UNIVERSITY PRESS

OXFORD

UNIVERSITY PRESS

Great Clarendon Street, Oxford OX2 6DP

Oxford University Press is a department of the University of Oxford.
It furthers the University's objective of excellence in research, scholarship,
and education by publishing worldwide in

Oxford New York

Athens Auckland Bangkok Bogotá Buenos Aires Calcutta
Cape Town Chennai Dar es Salaam Delhi Florence Hong Kong Istanbul
Karachi Kuala Lumpur Madrid Melbourne Mexico City Mumbai
Nairobi Paris São Paulo Singapore Taipei Tokyo Toronto Warsaw

with associated companies in Berlin Ibadan

Oxford is a registered trade mark of Oxford University Press
in the UK and in certain other countries

Published in the United States
by Oxford University Press, Inc., New York

A catalogue record for this book is available from the British Library

Library of Congress Cataloging in Publication Data

ISBN 0 19 850573 6

Typeset using the authors' LaTeX files by
Newgen Imaging Systems (P) Ltd., Chennai, India
Printed in Great Britain
on acid-free paper by
Biddles Ltd, Guildford & King's Lynn

FOREWORD

There are several ways in which the mathematical functions encountered in theoretical physics and applied mathematics may be classified. As research students in the late nineteen forties and the early nineteen fifties we used to distinguish between *Johnke-Emde functions* and *non-Jahnke-Emde functions*. This – admittedly somewhat facetious – distinction came from the *Tables of Functions with Formulae and Curves* by Eugen Jahnke and Fritz Emde, the second edition of which had been published by B. G. Teubner (Leipzig and Berlin) in 1933. The "Jahnke–Emde", as it was affectionately called, was a fascinating and quite unusual book. The text was in German and English. It contained tables and diagrams that facilitated computations involving complex numbers and solving cubic equations. In those days, when slide rules and tables of the so-called *elementary transcendents* (logarithms, trigonometric and hyperbolic functions) were the main paraphernalia of numerical computing and when hand-operated desk calculators were highly treasured rarities, the material provided in the book was an immense help in numerical work. The practical mind of Fritz Emde, the sole author of the second edition after Eugen Jahnke had died in 1921 twelve years after the first edition had appeared, may be gathered from the fact that the extensive listing of tables of the elementary transcendental functions provided not only comprehensive characterizations of the contents of the tables but also stated which of them were out of print and which were available, and if so, at what price!

The real treasures of the book were the chapters on the non-elementary functions. They contained collections of formulae, including approximations facilitating numerical computations, tables of numerical values and graphic illustrations, often as function of *complex* arguments. If in our research work we came across a non-standard differential equation or a parametric definite integral, the first question to ask was whether the solution was a *Jahnke–Emde function*, i.e., a function for which numerical results could be obtained using the information provided in the book. Having encountered a *non-Jahnke–Emde function* usually meant either hard numerical work with self-styled approximation methods or, more often than not, having to look out for a simpler approach.

Although I was fortunate in having inherited, in my father's bookcase, the 1933 edition of Jahnke–Emde, from one of my first salaries I purchased the 1952 edition of Jahnke–Emde, which contained much more material than the 1933 edition on what historically have been known as "special functions of mathematical physics", e.g. chapters on the confluent hypergeometric function and on Mathieu functions of integral order, plus more detailed results on Bessel functions. At about the same time, I was able to contribute to the decision of the Stuttgart Institute of Technology to acquire its first commercial electronic computer, a Zuse machine. One of my incentives was that it was clearly unsatisfactory to shun all **non-Jahnke–Emde functions**, which in my research work was tantamount to either avoiding or treating by phase-integral or perturbation approximations all those second-order linear ordinary differential equations of mathematical physics that could not be reduced to the hypergeometric equation or to one of its special or degenerate cases.

I quickly learned that having access to an electronic computer did not do too much for solving the problem of the "non-Jahnke–Emde" functions. The mathematical description of physical problems always requires simplifications, and these often introduce *singularities*. The essentials of a physical problem are, in fact, usually contained in the location and the character of the singularities. However, numerical handling of singularities, be they in the equations or in the solutions, is always a delicate matter, for which still no general recipe is available. The revelation came with Josef Meixner's and Friedrich Wilhelm Schäfke's book *Mathieusche Funktionen und Sphäroidfunktionen mit Anwendungen auf physikalische und technische Probleme*, published by Springer (Berlin–Göttingen–Heidelberg) in 1954. Here, exact methods were used to derive far-reaching analytical results on functions that could neither be represented by convergent series or asymptotic expansions with explicitly known coefficients nor as definite integrals of simpler functions, and for which explicit differentiation and recursion formulae between "neighbouring" functions do not exist. From the analytical results Meixner and Schäfke developed approximation schemes which did allow numerical results to be derived with moderate efforts, at least by the standards of the electronic computing that had become available by that time. It gradually emerged that the above-mentioned "anti-properties" of Mathieu functions and spheroidal wavefunctions were shared by the entire class of functions that are solutions of what is now called the "Heun class of ordinary differential equations" (named after Karl Heun, 1859–1929, professor of theoretical mechanics at the Karlsruhe Institute of Technology from 1902 till 1922). This class of functions features prominently in *Special Functions: A Unified Theory Based on Singularities* by Serguei Yu. Slavyanov and Wolfgang Lay.

Although only one step in complexity beyond the "hypergeometric class" of special functions, the Heun class is much more diversified than the hypergeometric class. When I began to collect known results on the various subclasses I soon realized how disconnected the various approaches to the Heun class were. From the literature it was not clear to what extent the techniques developed by Meixner–Schäfke and their predecessors for Mathieu functions and spheroidal wave functions were applicable to other cases, where generalizations were possible, and where new techniques had to be conceived. Neither between the mathematicians working on the various subclasses nor between researchers interested in applications seemed to be much cross-fertilization. It occurred to me that the hundredth anniversary of the publication of Heun's seminal paper on what we now call Heun's equation [K. Heun, Zur Theorie der Riemannschen Funktionen zweiter Ordnung mit vier Verzweigungspunkten, *Math. Annalen* 33, 51 (1889)] might provide an opportunity to bring together the (quite limited) number of active researchers in the field. In pursuing this idea I was fortunate enough to win the collaboration of one of my research assistants, Dr. Wolfgang Lay, who on top of his research work on Heun-class equations was willing to shoulder the main administrative burdens connected with organizing such a meeting, and of Professors Peter Lesky of the Faculty of Mathematics of Stuttgart University and André Ronveaux of the Facultés universitaires de Nôtre-Dame de la Paix in Namur (Belgium), as well as the financial support of the Stiftung Volkswagenwerk (Hannover Germany) and of the Max-Planck-Gesellschaft zur Förderung der Wissenschaften e. V. The "Centennial Workshop on Heun's Equation – Theory and Applications" took place from September 3rd to 8th, 1989. The location, the splendidly situated Schloss Ringberg in Bavaria, a bequest of a member of the ducal family of Bavaria to the Max-Planck-Gesellschaft, was highly appropriate to the occasion, since Heun's 1889 paper was written when he was a lecturer in Munich, the capital of Bavaria. A limited number of copies of the workshop proceedings (including the list of participants) are still available and may be obtained by writing to the present writer (Fax Nr. ++49 711 689 1932). A further outcome of the workshop is the book *Heun's Differential Equations* (R. Ronveaux, ed., Oxford University Press, Oxford 1995) with chapters written by the leading experts in the various subfields and an account of Heun's life and scientific work by M. von Renteln.

Another streak of fortune was that quite some time ago I had been able to acquire in a Moscow bookshop a copy of *Spheroidal and Coulomb spheroidal Functions* by I. V. Komarov, L. I. Ponomarev, and S. Yu. Slavyanov (Nauka, Moscow 1976, in Russian), that the senior author of that book, L. I. Ponomarev, responded to my invitation by commending S. Yu. Slavyanov for participation in the intended workshop, and that from this a very fruitful collaboration developed. Numerous joint papers resulted from this, some of them introducing new concepts of lasting importance (so we think), e.g. the notion of the s-rank of a singularity and the distinction between strong and weak confluence of singularities of linear differential equations. The present book, also a fruit of this collaboration, testifies to the perfect match of the interests and the backgrounds of the groups in St Petersburg and Stuttgart.

As expressed in the title, *Special Functions: A Unified Theory Based on Singularities* emphasizes the key rôle of the singularities in linking physics to mathematics. In mapping a complex physical situation onto manageable mathematics, location and character of the singularities reflect the essentials of the situation, whereas the parameters not directly associated with the singularities usually carry incidental information, e.g. on the physical properties of the specific material under consideration. Recognizing this led to a new appreciation of the importance of asymptotic expansions and of the Stokes' phenomenon. On the mathematical side, it is the singularities of the differential equations resulting from the mapping that determine the character of the solutions. Although this has been known for a long time, a complete classification of the differential equations of the Heun class based on the number and the ranks of their singularities has been given only recently (cf. the Appendix, by S. Yu. Slavyanow, W. Lay, and A. Seeger, of Heun's Differential Equations). This classification is the starting point of the data base of the "special functions" developed by S. Yu. Slavyanov and his associates in St. Petersburg. *SFTools*, briefly described in the Addendum A3 of the present book, is being developed further and may well turn out to be the answer to the question of how electronic computing can most efficiently help the researcher who wants to use the special functions beyond the hypergeometric class (which, of course, is included in SFTools).

Recognition of the paramount importance of the singularities suggested another extension beyond the "Jahnke–Emde functions" referred to above. Consider the second-order ordinary differential equation

$$d^2 y(z)/dz^2 = F(dy/dz, y, z),$$ (1)

where F is a rational function of dy/dz and y, with coefficients that are analytic in z. Paul Painlevé (1863–1929) raised the question: Under what conditions do the solutions of 1 not have movable critical points? In this context, critical points mean branch points and essential singularities; "movable" means that their position depends on the choice of the constants of integration and is, therefore, not just a property of the differential equation. Differential equations satisfying the conditions are said to have the Painlevé property. The only "movable" singularities of the solutions of a differential equation having the Painlevé property are thus poles. The absence of movable critical points allows us to cut the complex z-plane in a manner depending only on the differential equation in such a way that 1 admits uniform solutions, clearly an important feature from the point of view of physics.

P. Painlevé and his research student B. Gambier showed that there are 50 distinct types of 1 possessing the Painlevé property and that the solutions of 44 of them may be expressed in terms of known functions, viz. elliptic functions and functions defined by solutions of linear differential equations (sometimes referred to as "classical transcendents"). The remaining six equations, the so-called Painlevé equations, define "non-classical transcendents", now known as Painlevé transcendents.

Painlevé equations and Painlevé transcendents were the subject of very active research during the first two decades of the 20th century. Recently, interest in them was re-awakened when their relationship to soliton theory, and hence to several branches of physics, was discovered. Thus, it was clearly appropriate to include the Painlevé transcendents in the SFTool data base. However, the field has recently become so diversified that its full treatment requires a monograph of its own (eagerly awaited by the scientific community). In Chapter 5 of their book the authors have concentrated on the connection between the Painlevé equations and the Heun class of differential equations: The Painlevé equations can be considered as Lagrange equations governing the classical motions of non-relativistic quantum systems whose Schrödinger equations fall into the Heun class.

Among my good wishes accompanying the publication of *Special Functions: A Unified Theory Based on Singularities* that the book will make applied mathematicians, physicists, theoretical chemists, and engineers familiar with new classes of "non-Jahnke–Emde functions", encourage them to use these functions, and stimulate the pursuance of the path in computing taken by SFTools.

Stuttgart, April 2000   Alfred Seeger

PREFACE

Books on special functions serve as basic tools for physicists and engineers. Most of them are written in the style of the 19th century and are about special functions as solutions of differential equations which belong to the hypergeometric class. Meanwhile, in various fields of modern physics, solutions of more complicated linear and nonlinear differential equations give rise to their inclusion in the list of special functions. For instance, the Painlevé transcendents—solutions of the nonlinear Painlevé equations—were recently acknowledged as special functions of nonlinear physics. The formulae for the more sophisticated basic equations and the corresponding special functions become rather complicated and can hardly be found in the literature without a good guide. On the other hand, modern information technologies propose to use for reference purposes, instead of books, program packages where the search, cross-referencing and other actions concerning the information are effected uniformly by electronic tools.

In this book, the authors present a theory of special functions as a field of knowledge lying at the intersection of mathematics, physics, and computer science. The book is accompanied by a software package (also available on the market) which the reader can use to learn the technical part of the subject with the help of a computer. While it was not intended to set great store by rigorous proofs, attention is paid to various links between equations and functions, thus providing good insight into the theory. Illustrations of the theory by examples, tables, schemes, and figures are widely spread in the text, enabling—hopefully—a better understanding of the material.

Special functions appear as solutions of differential equations, of difference equations, of integrals, in group representation theories, etc. Here, ordinary differential equations are dealt with—in particular, second-order equations.

Readers experienced in the field might wonder why they will not find a historical presentation of the equations, or standard notations and conventional derivations of formulas. The reason for this is that, in order to present the new theory, new structural elements had to be chosen as a basis. Primarily, these are the following:

- The notions of the s-rank of a singularity and of the s-multisymbol of a differential equation are efficient tools for the classification of the equations. They permit the division of linear second-order ordinary differential equations (ODE) into classes and into types. They also appear to be useful in studying the (nonlinear) Painlevé equations.

- The form of an equation, distinguishing the various elements of the set of equivalent equations, is another structural element that is important for practical needs. The standard forms: (i) canonical, (ii) normal, and (iii) self-adjoint—which frequently appear in physical theories—are emphasized. Formulas for the standard forms are more compact than those for more general forms.

- The third—so to say, dynamic—structural element consists of confluence and reduction processes of equations. The *confluence process* controls the coalescence of two singularities of an equation accompanied by limiting processes in the parameter space. It permits the extension of properties of solutions of the basic equation to confluent equations. The *reduction process* of an equation is a mechanism for changing the type of an equation by specializing parameters. If two singularities coalesce without limiting processes in the parameter space of the equation, then we speak of a *merging* process.

Several quite recent researches of the authors are included in the book. These are: (i) new asymptotics (S. Slavyanov, Chapter 3), (ii) generalization of Jaffé expansions (W. Lay: Chapter 1, Chapter 3, Addendum), (iii) new integral relations (S. Slavyanov: Chapter 3), (iv) generalizations of Riemann's scheme (W. Lay, S. Slavyanov: Chapter 1), (v) links between Heun equations and Painlevé equations (S. Slavyanov: Chapter 5), (vi) recurrent calculations of matrix elements (S. Slavyanov: Addendum), (vii) merging processes (W. Lay, S. Slavyanov: Chapter 3). Full references including the names of the co-authors of the relevent publications can be found in the bibliography. Yet the book should not be regarded as a personal study. Its main goal is to fit the demands of the reader using higher special functions as a tool in his or her own field.

Most of the book was written in Baden-Württemberg—the land where the great philosopher G.W.F. Hegel was born. Perhaps this was the reason why many structural elements appear as triads. Without writing their precise definitions (which can be found below), we exhibit the most important of them:

- Hypergeometric equations – Heun equations – Painlevé equations.
- Fuchsian equations – confluent equations – reduced confluent equations.
- Class of equations – type of equation – form of equation.
- Differential equation – difference equation – integral equation.
- Regular singularity – irregular ramified singularity – irregular unramified singularity.
- Local solution – eigensolution – polynomial solution.
- Local parameter – scaling parameter – accessory parameter.
- Confluence process – reduction process – merging process.

During the preparation of the book, two young scientists from Saint Petersburg State University, Alexey Akopyan and Alexey Pirozhnikov, elaborated the software package SFTools supporting the text. This package is described in the Appendix D. The conditions of ordering the corresponding CD ROM may also be found there.

In a book like this—with a large amount of explicit formulas and a rather few intermediate calculations—typing errors are inevitable. However, the authors hope that the reader will check calculations if a result is really needed, and will be tolerant on reading if he or she enjoys only the spirit of the field.

Here are some comments on the contents of this book.

- The first chapter presents the basic general theory dealing with linear homogeneous second-order differential equations with polynomial coefficients and their solutions. The reader is assumed to know basic facts in differential equations and complex variable theory.

- The next two chapters deal with equations and special functions belonging to the hypergeometric and Heun class respectively.

- In the fourth chapter, various examples of applications in physics are presented. The authors tried to choose these examples from different areas.

- The fifth chapter is an introductory text to Painlevé equations. No preliminary knowledge is required by the reader.

- Several items which do not fit the chapters mentioned, but still appear to be useful from the point of view of the authors, are collected in the Appendices A–C.

The termination of theorems, lemmas, proofs, and examples is indicated by the sign $\square$. By ϵ throughout this book is meant an arbitrary small positive constant. When it seemed to be convenient for the reader, derivatives of functions are indicated by primes, the number of which indicates the order of the derivative. In these cases, the independent variable is given explicitly.

A book like this is not exclusively the work of the authors but the result of contributions from various people and institutions. So, this book would not have been written without the support of many collaborators and colleagues. It was the late Professor Felix M. Arscott who brought our attention to Heun's differential equation and let us participate in his wide knowledge of this topic. We thank Professor Ronveaux, who brought up the idea of writing the first book on Heun's equation, published also by Oxford University Press.

We are pleased to express our gratitude to Karlheinz Bay who carried out most of the numerical calculations appearing in the concrete examples. Almost all of these calculations were performed for the first time, and thus required his deep knowledge of numerical analysis, which he always made readily available to us. He was assisted partly by Alexey Akopyan from Saint Petersburg State University. For valuable discussions we thank Professor Sir Michael Berry (Bristol), Professor Wolfgang Bühring (Heidelberg), Professor Alexander Kazakov (Saint Petersburg), Professor Peter Lesky sen. (Stuttgart), Professor André Voros (Saclay), Professor Bruno Salvy (INRIA), and Professor Dieter Schmidt (Essen). We would like to thank the Institut für Theoretische und Angewandte Physik (ITAP) of the Universität Stuttgart for technical support, and the Max-Planck Institut für Metallforschung (Stuttgart) for hospitality. For financial support, the authors are grateful to the Deutsche Forschungsgemeinschaft (DFG, Bonn), to the Max-Planck Gesellschaft (MPG), to the International Scientific Foundation, and to the Russian Foundation for Fundamental Research. All these institutions enabled the authors travel to have the interaction necessary for scientific discussions. One of us (S. Slavyanov) thanks Dr. Michael Zaiser for hospitality in Stuttgart.

In particular, thanks are due to Professor Alfred Seeger (Stuttgart). He was an organizer of the Centennial Workshop on Heun's equation (Schloss Ringberg, Germany) from which collaboration of the two authors arose. Moreover, he was a co-author of several of our publications, scrupulously improving texts of the articles on which the text of the book is partly based, and kept an interest in the project from the beginning. Last, but not least, he took the job of editing the text of the book. One of the authors (S. Slavyanov) is indebted to him for personal and financial support.

We would like to express our gratitude to Oxford University Press (OUP, Oxford) for publishing the book to its renowned quality standards.

Stuttgart   S. Yu. S.
April 2000   W. L.

CONTENTS

1

LINEAR SECOND-ORDER ODES WITH POLYNOMIAL COEFFICIENTS

1.1 Regular singularities and Fuchsian equations

1.1.1 Regular and Fuchsian singularities

The starting point and the major object of study of this book is the ordinary linear homogeneous second-order differential equation

$$L_z y(z) := P_0(z)y''(z) + P_1(z)y'(z) + P_2(z)y(z) = 0 \quad (z \in \mathbb{C}), \qquad (1.1.1)$$

with polynomial coefficients $P_0(z)$, $P_1(z)$, $P_2(z)$ considered in the complex z plane. The primes in (1.1.1) indicate differentiation with respect to z. It is supposed that the polynomials in (1.1.1) do not possess common factors depending on z. The degrees of the polynomials P_0, P_1, P_2 are denoted by k_0, k_1, k_2. Beyond the z plane as the range of definition for eqn (1.1.1), the Riemann sphere can be considered. This allows us to include the point $z = \infty$ in our studies. The complex z plane including the point $z = \infty$ is denoted by $\mathbb{C}$. Equation (1.1.1) has two linearly independent solutions. The general solution of eqn (1.1.1) is an arbitrary linear combination of these particular solutions.

All points of $\mathbb{C}$ in the neighbourhood of which the Cauchy problem (i.e. the initial value problem with arbitrary initial data: $y(z_0) = y_0$, $y'(z_0) = y'_0$) for eqn (1.1.1) can be solved are called *ordinary points* of the equation. The point $z = \infty$ is studied after the substitution $z = \frac{1}{\zeta}$, with a further study of $\zeta = 0$, as to whether the latter is an ordinary point or not. In the neighbourhood of an ordinary point, all particular solutions of eqn (1.1.1) are holomorphic functions (see [15]).

The Cauchy problem cannot be solved at *singularities* of eqn (1.1.1). Singularities (singular points) of (1.1.1) are the zeros of the polynomial $P_0(z)$—denoted z_j ($j = 1, \ldots, n$)—and possibly the point $z = \infty$.

Example: The Bessel equation

$$z^2 y''(z) + zy'(z) + (z^2 - v^2)y(z) = 0 \qquad (1.1.2)$$

has two singularities, namely, at $z = 0$ and at $z = \infty$. All the other points of $\mathbb{C}$ are ordinary points of eqn (1.1.2). $\square$

In general, the singular points are branching points of at least one particular solution of eqn (1.1.1). An example of exception is the so-called *apparent singularity*, which is discussed below.

A function $f(z)$ having a singularity at $z = z^*$ ($\neq \infty$) is called a function of *finite order* at z^* if there exists a real number ρ such that, on any ray in $\mathbb{C}$ with the endpoint z^*,

$$\lim_{z \to z^*} f(z)(z - z^*)^\rho = 0$$

holds. In the case $z^* = \infty$, the preceding condition is replaced by

$$\lim_{z \to \infty} f(z)(z)^{-\rho} = 0.$$

Examples: The functions $z^3 \log z$ and $z^{7/3} e^z$ are of finite order at $z = 0$. The functions $\sin \frac{1}{z}$ and $z^3 \exp \frac{1}{z^2}$ are not of finite order at $z = 0$, and the function $z^{7/3} e^z$ is not of finite order at $z = \infty$. □

If both linearly independent particular solutions of eqn (1.1.1) are functions of finite order at a singularity $z = z^*$, then it is called a *regular singularity*. Otherwise, it is called an *irregular singularity* of eqn (1.1.1).

Example: The point $z = 0$ is a regular singularity of (1.1.2), while the point $z = \infty$ is an irregular singularity. □

The above-given classification of singularities is based upon the properties of solutions of the underlying differential equation. Another classification of singularities is based on a study of the equation itself, namely of the polynomials $P_n(z)$ at $z = z^*$. If the function $P(z) = P_1(z)/P_0(z)$ has a pole of at most first order, and the function $Q(z) = P_2(z)/P_0(z)$ has a pole of at most second order at the singular point $z = z^*$, this point is called a *Fuchsian singularity*. The singular point at infinity is a Fuchsian singularity if

$$P(z) = O(z^{-1}), \qquad Q(z) = O(z^{-2}).$$

As already mentioned, the definition of the singularity at infinity is reduced to the definition at a finite point z^* on substituting $z^{-1} \mapsto z - z^*$.

Example: The singular point $z = 0$ is a Fuchsian singularity of (1.1.2); however, the point $z = \infty$ is not a Fuchsian singularity. □

Theorem 1.1 *A singularity of eqn (1.1.1) is regular if and only if it is Fuchsian.* □

The proof is given in many textbooks (see, for instance, [54]).

Theorem 1.1 states the equivalence of notions 'regular' and 'Fuchsian' for singularities of eqn (1.1.1).[1]

1.1.2 Fuchsian equations and their transformations

Equation (1.1.1) is *Fuchsian* if all its singularities are regular (Fuchsian). Equations with polynomial coefficients of the form (1.1.1) which are not Fuchsian are called *non-Fuchsian equations* in this book.

---

[1]This proposition is not valid for systems of equations (see [19]). For them the notions of a regular singularity and of a Fuchsian singularity are not equivalent.

Examples: The Euler equation

$$z^2 y''(z) - (a + b - 1)zy'(z) + aby(z) = 0 \qquad (1.1.3)$$

with two regular singularities $z = 0$ and $z = \infty$ is a Fuchsian equation, whereas eqn (1.1.2) is not. $\square$

The property of an equation being Fuchsian is not affected by isomorphic transformations of $\mathbb{C}$ (i.e. Möbius transformations)

$$z \mapsto \xi, \qquad \xi = \frac{az + b}{cz + d}, \qquad ab - cd \neq 0, \qquad (1.1.4)$$

which define an equivalence relation within the set of Fuchsian equations. Therefore, without loss of generality, it is assumed (unless declared otherwise) that, for Fuchsian equations to be studied below, the point $z = \infty$ is singular. The set of singularities of (1.1.1) is denoted as $\{z_j; \infty\}$ ($j = 1, \ldots, n$). In order to avoid further exceptions, it is also assumed that[2] $n \geq 1$.

For Fuchsian equations, specific linear transformations of the dependent variable called *s-homotopic transformations* can be defined. The notion of s-homotopic transformation will be extended below also to non-Fuchsian equations.

Suppose that

$$S : y \mapsto v, \qquad y(z) = G(z)v(z), \qquad (1.1.5)$$

where $G(z)$ is a solution of an auxiliary first-order equation with polynomial coefficients

$$Q_0(z)G'(z) + Q_1(z)G(z) = 0. \qquad (1.1.6)$$

Suppose, in addition, that the set of the zeros of the polynomial $Q_0(z)$ coincides with the set of the points $\{z_j\}$, and that all zeros are simple ones, so that $Q_0(z)$ may be expressed as

$$Q_0(z) = \prod_{j=1}^{n} (z - z_j),$$

and that $Q_1(z)$ is a polynomial of degree not higher than $n - 1$, with arbitrary coefficients. Under these conditions we say that eqn (1.1.6) is *associated* with the Fuchsian equation (1.1.1). Equation (1.1.6) has an explicit solution

$$G(z) = \prod_{j=1}^{n} (z - z_j)^{\mu_j} \qquad (1.1.7)$$

with arbitrary μ_j, some of which may be zero. The transformation (1.1.5–1.1.6)

---

[2] Equation (1.1.1) has at least one singularity.

converts eqn (1.1.1) into the new Fuchsian equation

$$P_0 v''(z) + \left(P_1 - 2P_0 \frac{Q_1}{Q_0} \right) v'(z)$$

$$+ \left(P_2 + \left[\frac{Q_1^2 + Q_0' Q_1}{Q_0^2} - \frac{Q_1}{Q_0} \right] P_0 - \frac{Q_1}{Q_0} P_1 \right) v(z) = 0. \tag{1.1.8}$$

We call a Fuchsian equation an *irreducible Fuchsian equation* if there is no transformation (1.1.5–1.1.6) for which the number of singularities in (1.1.8) is less than the number in (1.1.1). Otherwise, the equation is a *reducible Fuchsian equation*. Singularities which disappear under at least one transformation of the form (1.1.5–1.1.6) are called *removable Fuchsian singularities*.

The transformation (1.1.5–1.1.6) applied to an irreducible Fuchsian equation is called an *s-homotopic transformation* of the equation. The s-homotopic transformations determine the second equivalence relation in the set of irreducible Fuchsian equations.

The equation which is equivalent to a given irreducible Fuchsian equation is called a *form* of this equation. All forms of an original equation are generated either by Möbius transformations of the independent variable, or by s-homotopic transformations of the dependent variable, or by both.

The equation that is equivalent to a given irreducible Fuchsian equation and for which all zeros of the polynomial $P_0(z)$ are simple is called a *canonical form* of the original equation. A canonical form of a given equation is not unique. There is a discrete finite set of Möbius transformations and s-homotopic transformations preserving a canonical form. A transformation which converts an arbitrary irreducible Fuchsian equation into a canonical form is constructed in the next paragraph.

Examples: (1) For the Fuchsian equation

$$(1 - z^2) y''(z) - y'(z) + y(z) = 0, \tag{1.1.9}$$

the associated equation reads

$$(1 - z^2) G'(z) + (az + b) G(z) = 0 \tag{1.1.10}$$

with arbitrary a and b.

(2) The equation

$$z(z-1)^2 y''(z) - [a(z-1)^2 + 2bz(z-1)] y'(z)$$
$$+ [b(b-1)z + ab(z-1)] y(z) = 0$$

is reducible, since the singular point $z = 1$ can be removed by the transformation $y(z) = (z-1)^{-b} v(z)$.

(3) The equation

$$v''(z) + (a - b + 1) v'(z) = 0$$

is equivalent to eqn (1.1.3) and is its canonical form. The transformation $S : v(z) := z^{b-a} w(z)$ preserves the canonical form. □

An equation which is equivalent to a given irreducible Fuchsian equation and for which the point $z = \infty$ is a regular singularity we say is in its *natural form*. Any Fuchsian equation may be converted by Möbius transformation of z to its natural form (in a non-unique way).

1.1.3 Characteristic exponents

At any finite regular singularity z_j of eqn (1.1.1), the so-called *indicial equation*—a second-order algebraic equation—can be constructed:

$$\rho(\rho - 1) + p_j\rho + q_j = 0, \tag{1.1.11}$$

where

$$p_j = \text{Res}_{z=z_j} \frac{P_1(z)}{P_0(z)}, \qquad q_j = \text{Res}_{z=z_j}(z - z_j)\frac{P_2(z)}{P_0(z)}.$$

By $\text{Res}_{z=z_j} F(z)$ the residue of a function $F(z)$ at the point z_j is denoted. The roots $\rho_m(z_j)$ $(m = 1, 2)$ of the indicial equation (1.1.11) are called *characteristic exponents* or *Frobenius exponents* at the point z_j. At a regular singularity at infinity, the indicial equation reads

$$\rho(\rho + 1) + p_\infty\rho + q_\infty = 0, \tag{1.1.12}$$

where

$$p_\infty = -\text{Res}_{z=\infty} P_1(z)/P_0(z), \qquad q_\infty = \text{Res}_{z=\infty} z P_2(z)/P_0(z).$$

Characteristic exponents at infinity (Frobenius exponents at infinity) $\rho_m(\infty)(m = 1, 2)$ are the roots of eqn (1.1.12).

Lemma 1.1 *For each irreducible Fuchsian equation in natural form there exists a transformation (1.1.5) which converts it to a canonical form. This transformation is not unique.* □

Proof: The transformation is constructed explicitly by choosing the values μ_j in (1.1.7) to be any of the characteristic exponents at the point z_j. □

Example: Equation (1.1.3) is transformed into its canonical form (1.1.10) with the help of the transformation $y = z^{-b}w$. The transformation $y = z^{-a}w$ also transforms eqn (1.1.3) into a canonical form, but to one different from (1.1.10). □

Theorem 1.2 *The characteristic exponents at singular points of a Fuchsian equation obey Fuchs's condition [45]*

$$\sum_{j=1}^{n}\sum_{m=1}^{2} \rho_m(z_j) + \sum_{m=1}^{2} \rho_m(\infty) = n - 1. \tag{1.1.13}$$

□

Proof: Equation (1.1.13) is altered neither by Möbius transformations nor by s-homotopic transformations. Therefore a Fuchsian equation may be considered in a canonical natural form. According to (1.1.11–1.1.12)

$$\sum_{j=1}^{n}\sum_{m=1}^{2}\rho_m(z_j) = n - \sum_{j=1}^{n}p_j.$$

$$\rho_{1,2}(z_\infty) = -[1 - p_\infty]/2 \pm [(1 - p_\infty)^2/4 - q_\infty]^{1/2}.$$

The use of the residue theorem—which says that the sum of the residues of a meromorphic function over all singular points is zero—leads to eqn (1.1.13). □

A subset of *elementary* singularities [54] can be distinguished in the set of regular singularities of eqn (1.1.1). A regular singular point $z = z_k$ is called an *elementary singularity* of the underlying differential equation if the difference between the two characteristic exponents at this point is equal to $1/2$. The specific role of elementary singularities is based on the existence of an s-homotopic transformation (1.1.5–1.1.6) and a quadratic transformation $T : z \mapsto t$ $z - z_k = t^2$, of the independent variable, which together 'wipe out' the elementary singularity. An extensive use of elementary singularities can be found in the book [54]. The following examples are more precisely studied in the next chapter.

Examples: (1) The regular singular point $z = 1$ for eqn (1.1.9) is elementary.

(2) The Fuchsian equation with three singularities is the Riemann equation. The natural canonical form of this equation is the hypergeometric equation. The equation with one elementary singularity among other regular singularities may be transformed to the associated Legendre equation. The equation with two elementary singularities among other singularities may be transformed to the equation for the Tchebyshev polynomials. □

These examples are more precisely studied in the next chapter.

1.1.4 Frobenius solutions

In a neighbourhood of a finite regular singularity z_j, two linearly independent particular solutions called *Frobenius solutions* can be constructed. If the difference of the characteristic exponents at this singularity satisfies

$$\rho_1(z_j) - \rho_2(z_j) \neq l, \qquad l \in \mathbb{Z}, \tag{1.1.14}$$

then the Frobenius solutions have the form

$$y_m(z_j, z) = (z - z_j)^{\rho_m(z_j)} \sum_{k=0}^{\infty} c_k^m(z_j)(z - z_j)^k \quad \text{for } m = 1, 2. \tag{1.1.15}$$

If the singularity is located at the point $z = \infty$, expansions (1.1.15) are replaced by

$$y_m(z_j, z) = z^{-\rho_m(\infty)} \sum_{k=0}^{\infty} c_k^m(\infty)z^{-k} \quad \text{for } m = 1, 2. \tag{1.1.16}$$

As already mentioned, Frobenius solutions are linearly independent.

If the coefficients in eqn (1.1.1) induce the condition

$$\rho_1(z_j) - \rho_2(z_j) = l, \qquad l \in \mathbb{Z}, \tag{1.1.17}$$

then a logarithmic term can appear in one of the Frobenius solutions:

$$y_2(z_j, z) = (z - z_j)^{\rho_2(z_j)} \sum_{k=0}^{\infty} c_k^2(z_j)(z - z_j)^k$$

$$+ A_j y_1(z_j, z) \log(z - z_j). \tag{1.1.18}$$

A similar construction is valid for the modification of (1.1.16) at $z = \infty$:

$$y_2(\infty, z) = z^{-\rho_2(\infty)} \sum_{k=0}^{\infty} c_k^2(\infty)z^{-k} + A_\infty y_1(\infty, z) \log z. \tag{1.1.19}$$

Theorem 1.3 *The coefficients $c_k^m(z_j)$ and $c_k^m(\infty)$ on the right-hand sides of (1.1.15–1.1.16) are obtained recursively in a unique manner by inserting the series into the differential equation and nullifying the coefficients of subsequent powers of $z - z_j$. Exceptions are $c_0^m(z_j)$ and $c_0^m(\infty)$, which may be fixed by normalization. Coefficients $c_k^2(z_j)$ and the constant A_j on the right-hand side of (1.1.18, 1.1.19) are obtained by the same procedure with exception of the coefficients c_0^2 and c_l^2. The former is fixed by normalization and the latter may be taken as zero. The series in (1.1.15, 1.1.18) converge within a circle the centre of which is $z = z_j$ and the radius of which is equal to the distance between z_j and the nearest neighbouring singularity of the equation. The series in (1.1.16, 1.1.19) converge at $z > d$, where $d = \max_{z_j} |z_j|$. $\square$*

The proof of this theorem can be found in many textbooks (see, for instance, [54]).

Taking the value of the coefficient c_l^2 as nonzero, we simply add a second solution to the first one having an appropriate multiplier. If the condition (1.1.17) holds but nevertheless $A_j = 0$ for the second Frobenius solution, then the point $z = z_j$ is called an *apparent singularity*. With the help of the substitution $y(z) = (z - z_j)^{\rho_2(z_j)} w(z)$, the original equation is transformed to an equation for which the general solution $w(z)$ is a holomorphic function at $z = z_j$. The first derivative of $w(z)$ at an apparent singularity $z = z_j$ is zero, which is why an arbitrary Cauchy problem is unsolvable at this point.

Example: The regular singular point $z = 0$ is an apparent singularity of eqn (1.1.2) for *half-integer* (i.e. half-odd-integer) values of ν larger than $1/2$. $\square$

1.2 Irregular singularities and confluent equations

1.2.1 The s-rank of a singularity

Every singular point (regular and irregular) of eqn (1.1.1) may be characterized by its *s-rank* (*rank of the singularity*).

The *s-rank* $R(z_j)$ *of a finite irregular point* z_j is defined as

$$R(z_j) = \max(K_1(z_j), K_2(z_j)/2), \tag{1.2.1}$$

where $K_1(z_j)$ is the multiplicity of the zero of the function $P_0(z)/P_1(z)$ at the point z_j, and $K_2(z_j)$ is the multiplicity of the zero of the function $P_0(z)/P_2(z)$ at the same point.

At an infinite irregular point $z = \infty$, the s-rank $R(\infty)$ is defined as

$$R(\infty) = \max(K_1(\infty), K_2(\infty)/2), \tag{1.2.2}$$

where $K_1(\infty) = k_1 - k_0 + 2$ and $K_2(\infty) = k_2 - k_0 + 4$. It is easy to verify that the definition (1.2.2) correlates with (1.2.2) by substituting $z \mapsto (z - z_j)^{-1}$.

S-ranks of irregular singular points are integers or half-integers (i.e. half-odd-integers) larger than unity.

Irregular singular points with half-integer s-rank are called *ramified* irregular singular points. In the case of an integer s-rank, the corresponding point is called *unramified*.

The *s-rank of a regular singular point*, unless it is distinguished specially in some other way, is defined as being equal to unity, which corresponds to definitions (1.2.1)–(1.2.2).

This definition of the s-rank for a regular singularity is based on the assumption that the subset of elementary singularities is distinguished within the set of regular singularities. The choice of the definition of the s-rank of elementary singular points is by no means unique. There are some arguments for choosing it as unity, as for other regular singularities. According to historical traditions, since in many classical books the exceptional role of elementary points is stressed, and according to the previous publications of the authors, the s-rank of elementary singularities should be chosen as $1/2$. Other arguments for this choice may be added.

In this book, however, for the sake of clarity and simplicity, we do not subdivide the set of regular singularities into elementary and non-elementary ones. As a consequence the s-rank of regular singularities is unity, irrespective as to whether they are elementary or not. However, how the subdivision may be done is illustrated by several examples.

Example: The s-rank of the singularity at infinity for the Bessel equation (cf. (1.1.2)) is $R(\infty) = 2$. The singularity is unramified. The s-rank of the singularity at infinity of the equation

$$y''(z) - (z + \lambda)y(z) = 0 \tag{1.2.3}$$

is $R(\infty) = 5/2$. The singularity is ramified. □

Another definition of the s-rank is based on the behaviour of the particular solutions of eqn (1.1.1) at its singularities. The symbol $T(z, D)$—a polynomial in two variables z and D—can be associated with eqn (1.1.1) according to

$$T(z, D) := P_0(z)D^2 + P_1(z)D + P_2(z), \tag{1.2.4}$$

where D is the differentiation operator and z is the independent variable.

The practical classification of types of singularities may be performed on the basis of a treatment of the properties of the zeros of the *symbolic indicial equation*

$$T(z, D) = 0. \tag{1.2.5}$$

Since eqn (1.2.5) is of degree 2 in D, its solutions can be represented in the neighbourhood of singularities by Poisson series of the following types:

$$D_m(z_j) = (z - z_j)^{-\mu_{mj}} \sum_{k=0}^{\infty} h_{mk}(z_j)(z - z_j)^{k/2}, \qquad h_{m0}(z_j) \neq 0,$$

$$D_m(\infty) = z^{\mu_{m\infty}-2} \sum_{k=0}^{\infty} h_{mk}(\infty)z^{-k/2}, \qquad h_{m0}(\infty) \neq 0,$$

for $m = 1, 2$ and $j = 1, \ldots, n$, with integer or half-integer μ_{mj} satisfying

$$\frac{1}{2} \leq \mu_{mj}.$$

The first term of these expansions gives the leading behaviour of the logarithmic derivative of the particular solutions related to (1.1.1) in a vicinity of the corresponding singularity:

$$|\ln y_m(z_j, z)| \leq K|z - z_j|^{-\mu_{mj}-\epsilon} \quad \text{for } \mu_{mj} \geq 1, \qquad K > 0, \quad \epsilon > 0$$

with an appropriate constant K determined by the coefficients of polynomials in (1.1.1) and arbitrarily small ϵ. In a special case,

$$|\ln y_m(z_j, z)| \leq K|z - z_j|^{-1-\epsilon} \quad \text{for } \mu_{mj} = 1/2, 1.$$

If

$$\mu_{mj} \leq 1, \quad \text{for } m = 1, 2,$$

holds for the singularity $z = z_j$, it is called a *regular (Fuchsian) singularity*.

The s-rank of a finite singularity is defined as

$$R(z_j) = \max_{m=1,2} \mu_{mj}. \tag{1.2.6}$$

The s-rank of the singularity at infinity is defined as

$$R(\infty) = \max_{m=1,2} \mu_{m\infty}. \tag{1.2.7}$$

Examples: The symbolic indicial equation for eqn (1.1.2) reads

$$z^2 D^2 + zD + (z^2 - v^2) = 0.$$

The corresponding values $\mu_{m\infty}$ are $\mu_{m\infty} = 2$. $\quad \square$

1.2.2 Confluent and reduced confluent equations

The set of the s-*ranks* of singular points of eqn (1.1.1) constitutes its *s-multisymbol*.

Irreducible equations characterized by the same s-multisymbol belong to the same *type* of equations.

Equation (1.1.1) is called a *confluent case* if, besides regular singular points, it contains at least one irregular singular point.

Equation (1.1.1) is called a *reduced confluent case* if it contains at least one ramified irregular singularity.

Although reduced confluent cases belong to confluent ones, the second notion will be used in practical examples mainly for those equations which do not contain ramified singularities.

As in the case of Fuchsian equations, it is assumed below that one of the singularities is located at infinity, and moreover that there are n singularities located at finite points.

Examples: (1) Equation (1.1.2) is characterized by the s-multisymbol $\{1; 2\}$ and is a confluent case.

(2) Equation (1.2.3) is characterized by s-multisymbol $\{5/2\}$ and is a reduced confluent case. $\square$

1.2.3 The s-homotopic transformation

A generalization of the transformation (1.1.5–1.1.6) for Fuchsian equations may be introduced for confluent cases. It is also based on an associated first-order equation as (1.1.5–1.1.6), but with other constraints on the polynomials Q_0, Q_1.

The transformation of eqn (1.1.1) is defined by the formulae

$$S: \; y(z) \mapsto v(z), \; L_z \mapsto \tilde{L}_z,$$

$$y(z) = G(z)v(z), \qquad L_z = G(z)\tilde{L}_z G^{-1}, \qquad \tilde{L}_z v(z) = 0, \tag{1.2.8}$$

$$Q_0(z)G(z) + Q_1(z)G(z) = 0 \tag{1.2.9}$$

and is called an *s-homotopic transformation*, iff the following three conditions hold:

(1) the set of zeros of the polynomial $Q_0(z)$ is included in the set of singularities $\{z_j\}$ of eqn (1.1.1);

(2) the multiplicities $r(z_j)$ of zeros z_j ($j = 1, \ldots, n$) of the polynomial $Q_0(z)$ in (1.1.8) do not exceed the s-ranks of the singularities z_j ($j = 1, \ldots, n$) of eqn (1.1.1);

(3) the difference between degrees of the polynomials $Q_0(z)$ and $Q_1(z)$ (correspondingly l_0 and l_1) satisfies the condition

$$l_1 - l_0 \leq \max(K_1(\infty), K_2(\infty)/2) - 2.$$

The ratio Q_1/Q_0 may be expanded in partial fractions:

$$\frac{Q_1(z)}{Q_0(z)} = \sum_{j=1}^{n} \sum_{k=0}^{r(z_j)-1} h_{jk}(z - z_j)^{-k-1} + \sum_{k=1}^{l_1-l_0+1} f_k z^{k-1}. \tag{1.2.10}$$

From (1.2.9–1.2.10), the explicit expression of the function $G(z)$, determining the s-homotopic transformation (1.2.8), is found to be

$$G(z) = \exp\left(-\int \frac{Q_1}{Q_0}\,dz\right) = \exp\left(\sum_{k=1}^{l_1-l_0+1} f_k \frac{z^k}{k}\right)$$

$$\cdot \prod_{j=1}^{n}(z-z_j)^{h_{j0}} \cdot \exp\left(\sum_{k=1}^{r(z_j)-1} h_{jk}\frac{(z-z_j)^{-k}}{-k}\right). \qquad (1.2.11)$$

Example: The associated equation of eqn (1.1.2) is

$$zG'(z) - (az+b)G(z) = 0. \quad \square$$

In the course of the s-homotopic transformation of the initial equation, the s-rank of the singular point either is conserved or decreases.

If there exists an s-homotopic transformation that decreases the s-rank of a singular point, then this singularity is called *reducible*. Otherwise the point is called *irreducible*.

A confluent equation for which all singularities are irreducible is called *irreducible*. Otherwise it is *reducible*.

Example: (1) Equation (1.1.2) is irreducible.

(2) The equation

$$zy''(z) + (2z+ba)y'(z) + (z+a)y(z) = 0$$

is reducible. $\square$

S-homotopic transformations determine an equivalence relation between irreducible confluent equations. Another equivalence relation is determined by Möbius transformations of the independent variable.

Equivalent confluent equations are called *forms* of the initial equation.

If the opposite is not declared in the text explicitly, all singular points of confluent cases are assumed to be irreducible, as was the case above for Fuchsian equations.

1.2.4   Asymptotic solutions at irregular singularities

In a vicinity of an irregular singular point, local 'solutions' in the form of—in general divergent—formal asymptotic series (called Thomé solutions) can be constructed [115]. If the irregular singular point is unramified and finite, these series are of the form

$$y_m(z_j, z) = (z-z_j)^{\alpha_{m0}(z_j)} \exp\left(\sum_{k=1}^{R(z_j)-1} \frac{\alpha_{mk}(z_j)}{(-k)}(z-z_j)^{-k}\right)$$

$$\cdot \sum_{k=0}^{\infty} c_{mk}(z_j)(z-z_j)^k \quad \text{for } m = 1, 2. \qquad (1.2.12)$$

Here only the general case is considered. At special values of the coefficients, as in the case of Frobenius solutions, a logarithmic term can appear. Series (1.2.12) are called *normal solutions*. If the irregular singular point is ramified and finite, then the Thomé expansions are of the form

$$y_m(z_j, z) = (z - z_j)^{\alpha_{m0}(z_j)} \exp \left(\sum_{k=1/2}^{R(z_j)-1} \frac{\alpha_{mk}(z_j)}{(-k)} (z - z_j)^{-k} \right)$$

$$\cdot \sum_{k=0}^{\infty} c_{mk}(z_j)(z - z_j)^{k/2} \quad \text{for } m = 1, 2, \qquad (1.2.13)$$

where summation in the exponent is performed over both integer and half-integer values of the summation index k. Series (1.2.13) are called *subnormal solutions*. If the singularity is located at the point $z = \infty$, expansion (1.2.12) should be replaced by

$$y_m(\infty, z) = z^{-\alpha_{m0}(\infty)} \exp \left(\sum_{k=1}^{R(\infty)-1} \frac{\alpha_{mk}(\infty)}{k} z^k \right)$$

$$\cdot \sum_{k=0}^{\infty} c_{mk}(\infty)z^{-k} \quad \text{for } m = 1, 2, \qquad (1.2.14)$$

and expansion (1.2.13) should be replaced by

$$y_m(\infty, z) = z^{-\alpha_{m0}(\infty)} \exp \left(\sum_{k=1/2}^{R(\infty)-1} \frac{\alpha_{mk}(\infty)}{k} z^k \right)$$

$$\cdot \sum_{k=0}^{\infty} c_{mk}(\infty)z^{-k/2} \quad \text{for } m = 1, 2. \qquad (1.2.15)$$

A rigorous mathematical understanding of (1.2.12–1.2.15) will be discussed in Section 1.4. Here, only formal properties of the coefficients in (1.2.12–1.2.15) are discussed.

The multipliers in front of the series (1.2.12)–(1.2.15) are called *characteristic multipliers* for Thomé solutions.

The coefficients α_{mk} in (1.2.12) and (1.2.14) are called *characteristic exponents of the second kind* of order k at unramified irregular singular points. They constitute two *characteristic sequences*, which may be calculated explicitly in terms of the coefficients of the equation with the help of a recursion.

The coefficients α_{mk} in (1.2.13) and (1.2.15) are called *characteristic exponents of second kind* of order k at ramified irregular singular points. They constitute two *characteristic sequences* which may be calculated explicitly in terms of the coefficients of the equation with the help of a recursion.

Lemma 1.2 *The characteristic exponents of the second kind at an ramified irregular singular point satisfy the following conditions:*

(1) characteristic exponents of integer order:

$$\alpha_{1k} = \alpha_{2k},$$

(2) characteristic exponents of half-integer order:

$$\alpha_{1k} = -\alpha_{2k}. \quad \square$$

Proof: Consider a finite ramified irregular singular point z_j. Represent the characteristic multiplier at this point in the form $\exp[\int (\phi_1(z) + \phi_2(z))\,dz]$, where the function $\phi_1(z)$ includes only integer characteristic exponents and the function $\phi_2(z)$ includes only half-integer characteristic exponents. By substituting Thomé solutions into the equation, select terms which include leading singularities. Subsequent cancelling of these terms determines $\alpha_{mk}(z_j)$. Divide these terms into two parts: one part containing only integer powers of $z - z_j$ and the other part containing only half-integer powers. This leads to the equations

$$\phi_2^2 + \phi_1^2 + P_2/P_0 + \epsilon_2 = 0, \tag{1.2.16}$$

$$2\phi_2\phi_1 + \phi_2 P_1/(2P_0) + \phi_2' + \epsilon_1 = 0, \tag{1.2.17}$$

where $\epsilon_1(z)$ and $\epsilon_2(z)$ do not contain leading singularities at $z = z_j$. From (1.2.16–1.2.17), the results follows. $\quad \square$

Lemma 1.3 *Ramified singular points are irreducible.* $\quad \square$

Proof: Comparing characteristic multipliers for Thomé solutions which include half-integer powers in the exponent with the function $v(z)$ in the s-homotopic transformation which includes only integer powers in the exponent gives the required result. $\quad \square$

1.2.5 Canonical forms

On the basis of representations (1.2.12–1.2.15) it is possible to construct the definition of a canonical form of an irreducible confluent case. Equation (1.1.1) is in a *canonical form* at a finite unramified irregular singular point z_j if one of the sequences of characteristic exponents in the Thomé solutions at this point contains exclusively zeros.

Equation (1.1.1) is in a *canonical form* at an unramified irregular singular point $z = \infty$ if, for one of the Thomé solutions at this point, the characteristic exponents of the second kind of order larger than zero are zero.

Equation (1.1.1) is in a *canonical form* at a finite ramified irregular singular point z_j if, for both Thomé solutions at this point, the characteristic exponents of second kind of positive integer order are equal to zero and the characteristic exponents of zeroth order are equal to $R(z_j)/2$.

Equation (1.1.1) is in a *canonical form* at an ramified irregular singular point if, for both of the Thomé solutions at this point, the characteristic exponents of the second kind of order larger than zero are zero.

Equation (1.1.1) is in a *canonical form* at a finite regular singular point if one of the characteristic exponents of Frobenius solutions at this point is zero.

A confluent case of a Fuchsian equation is in its *natural form* if the point having the maximal s-rank is located at infinity. If there are more than one singularity having the maximal s-rank, then the natural form is not unique.

The *canonical natural form* of an irreducible confluent case of a Fuchsian equation is an equation which is canonical at each singularity and for which the point having maximal s-rank is located at infinity.

Lemma 1.4 *In order for a confluent case to be canonical at an unramified finite singular point z_j, it is necessary and sufficient that the multiplicity of the zero of the function $P_0(z)/P_1(z)$ coincides with the s-rank of this point and the multiplicity of the zero of the function $P_0(z)/P_2(z)$ does not exceed the s-rank at this point.* $\square$

Lemma 1.5 *In order for a confluent case of a Fuchsian equation to be canonical at a ramified finite irregular singular point z_j, it is necessary and sufficient that the function $P_0(z)/P_1(z)$ has no zero at this point.* $\square$

Proof: The proposition follows from the analysis of eqn (1.2.17) for characteristic exponents of integer order. They are equal to zero if P_1/P_0 has no singularities at the point z_j. If, conversely, they are equal to zero, then there can be no such singularities. $\square$

Lemma 1.6 *Any confluent irreducible Fuchsian equation by superposition of isomorphisms of the z plane and s-homotopic transformations can be converted to a natural canonical form.* $\square$

Proof: The transformation into a natural form is evident. A canonical form at any elementary or regular singular point is achieved by the method shown in Section 1.1. Consider a finite unramified point. At this point, a function $v(z)$ is chosen that gives an s-homotopic transformation which converts the equation into a canonical form at this point. $v(z)$ is chosen as the characteristic multiplier of both of the Thomé solutions at this point. No characteristic exponent at other singularities is changed by this transformation besides the characteristic exponent of zeroth order at infinity. Consider a finite ramified point z_j. For this case, the function $v(z)$ is chosen such that it yields the s-homotopic transformation that converts the equation into a canonical form at this point according to

$$v(z) = (z - z_j)^{\alpha_{10}(z_j)} \exp\left(\sum_{k=1}^{R(z_j)-3/2} \frac{\alpha_{mk}(z_j)}{(-k)} (z - z_j)^{-k} \right),$$

where k takes only integer values and $\alpha_{mk}(z_j)$ are characteristic exponents of integer order at this point. $\square$

In the case of a singular point at infinity it is necessary to exclude the multiplier $(z - z_j)^{\alpha_{10}(z_j)}$ from the function performing the s-homotopic transformation. Otherwise, it will disturb the behaviour at some finite point.

1.2.6 A generalization of Fuchs's theorem

A generalization of Fuchs's theorem for confluent cases will be formulated. It is proved using the definitions presented above.

Theorem 1.4 *Suppose that the number of regular points of a confluent case of an irreducible equation of type (1.1.1) equals n_r, the number of finite irregular points equals n_i, and the equation is written in its natural form. Then the following equality holds [74]:*

$$\sum_{j=1}^{n_r}\sum_{m=1}^{2}\rho_m(z_j) + \sum_{j=1}^{n_i}\sum_{m=1}^{2}\alpha_{m0}(z_j) + \sum_{m=1}^{2}\alpha_{m0}(\infty) = \sum_{j=1}^{n_i+n_r} R(z_j) + R(\infty) - 2.$$

$$(1.2.18)$$

On the left-hand side of this equality, all characteristic exponents of zeroth order for both Frobenius and Thomé solutions are summed; on the right-hand side, all s-ranks of singularities of the equation are added. □

Proof: The s-homotopic transformations can change neither on the left-hand side nor on right-hand side of the equality. Therefore, one can start with a canonical natural form of the equation. Suppose at first that the equation under consideration comprises no ramified singularities. For functions $f(z)$ having a pole at the point z_j, a decomposition of the Laurent series at this point into principal part G_j and regular part 'r.p.' may be introduced, namely

$$P_1/P_0 = G_j + \text{r.p.}, \qquad G_j = \sum_{k=1}^{R(z_j)} a_{-k}(z_j)(z - z_j)^{-k}. \qquad (1.2.19)$$

Let T_j be the logarithmic derivative of the characteristic multiplier at the point z_j

$$T_j = \alpha_{m0}(z_j)(z - z_j)^{-1} + \sum_{k=1}^{R(z_j)-1} \alpha_{mk}(z_j)(z - z_j)^{-k-1}. \qquad (1.2.20)$$

Substitute into eqn (1.1.1) the Thomé solutions, the characteristic multiplier of which should not be unity, and select terms containing leading singularities. The following equation for the recursive calculation of the characteristic exponents of second kind holds:

$$T_j^2 + G_j T_j + T_j' = 0. \qquad (1.2.21)$$

By choosing $\alpha_{1k}(z_j)$, one eliminates singularities of order $-2R(z_j)$ up to order $-R(z_j) - 1$ inclusively. The recurrence relation for α_{1k} reads

$$\alpha_{1,k-1}(z_j) = -a_{-k}(z_j) \quad \text{for } k = 2, \ldots, R(z_j);$$
$$\alpha_{10}(z_j) = R(z_j) - a_{-1}(z_j). \qquad (1.2.22)$$

All the $\alpha_{2k}(z_j) = 0$ take into account that the form is canonical. In the same way all other finite irregular singularities are treated. In the case of a regular point, the same result, namely, formula (1.2.22) is valid.

Consider the irregular singularity at infinity. In its vicinity, the representation (1.2.19) should be replaced by

$$P_1/P_0 = G_\infty + O(z^{-2}), \qquad G_\infty = \sum_{k=-1}^{R(\infty)-2} b_k z^k,$$

$$P_2/P_0 = c_\infty z^{-2} + O(z^{-3}),$$

and eqn (1.2.21) should be replaced by

$$T_j^2 + G_\infty T_j + T_j' + \frac{c_\infty}{z^2} = 0.$$

The characteristic exponents of zeroth order are given by

$$\alpha_{20}(\infty) = -\frac{c_\infty}{b_{R(\infty)-2}} + (R(\infty) - 2) + \sum_{j=1}^{n_i+n_r} a_{-1}(z_j),$$

$$\alpha_{10}(\infty) = \frac{c_\infty}{b_{R(\infty)-2}}. \tag{1.2.23}$$

By adding equalities (1.2.22) and (1.2.23), the final formula (1.2.18) is obtained.

Recomputing the left- and the right-hand side of (1.2.18) after an isomorphic transformation of $\mathbb{C}$, one may verify that the initial choice of the natural form of the equation is significant only for the *presentation* of the result but not for the result itself.

Assume now, that there are ramified irregular singularities of eqn (1.1.1). Consider a finite ramified singular point z_j. At this point, the function P_2/P_0 has no singularity and eqn (1.2.17) may be used once again to find the characteristic exponents $\alpha_{m0}(z_j)$. Their values are calculated as

$$\alpha_{m0}(z_j) = \frac{R(z_j)}{2} \quad \text{for } m = 1, 2. \tag{1.2.24}$$

It follows from eqn (1.2.24) that an addition of finite ramified singular points into eqn (1.1.1) does not influence eqn (1.2.18). Consider the change if the point $z = \infty$ is a ramified singular point. Equation (1.2.17) gives the following expressions for the coefficients $\alpha_{m0}(\infty)$

$$2\alpha_{m0}(\infty) = (R(\infty) - 2) + \sum_{j=1}^{n_r+n_i} a_{-1}^{(j)} \quad \text{for } m = 1, 2. \tag{1.2.25}$$

It follows from (1.2.25) that, in the case of ramified irregular singularities, formula (1.2.18) holds as well.   □

Formula (1.2.18) as well as Fuchs's theorem is a consequence of the residue theorem. It gives a simple test for checking the computations of characteristic exponents.

Example: For the Bessel equation (1.1.2), $\alpha_{m0}(\infty) = 1/2$ and $\rho_m(0) = \pm\nu$. On the other hand, $R(0) + R(\infty) = 3$. Formula (1.2.18) holds. $\quad\Box$

1.3 Confluence and reduction processes

1.3.1 Strong and weak confluence. A confluence theorem

In a broad sense, by *confluence process* is meant the coalescence of two singularities of eqn (1.1.1) accompanied by a parametric variation (with respect to the distance between the coalescing singularities) of the coefficients of the polynomials P_0, P_1, P_2 of eqn (1.1.1) (the free parameters of the equation) resulting in one final singularity the s-rank of which is larger than the s-ranks of initial coalescing singularities.

Some refining comments are needed:

- First, additional preliminary s-homotopic transformations may be used in order to achieve what is intended.

- Under confluence processes, reducible singularities may appear which can be eliminated by an appropriate succeeding s-homotopic transformation.

- In order to avoid the above-mentioned difficulties, a preliminary 'preparation' of the equation should be applied, namely, a transformation to the form where the confluence process is most simply traced.

- Furthermore, it is necessary to take care of the number of parameters. It is supposed that the original equation has the maximal number of free parameters (i.e. the coefficients of the polynomials P_0, P_1, P_2) that are compatible with the s-multisymbol and the form of the equation.

- If the number of free parameters of the equation under a confluence process decreases by one, it is called a *strong confluence process*.

- If the number of free parameters of the equation decreases more than by one, it is called a *weak confluence process*.

The following theorem about additivity of the s-rank under strong confluence processes (under weak confluence processes we have sub-additivity) can be proved:[3]

Theorem 1.5 *Consider a non-Fuchsian equation in its normal form for which one singular point located at $z_1 = 0$ is characterized by the s-rank R_1, and the other singular point located at $z_1 = \epsilon$ is characterized by the s-rank R_2. Due to the limiting process $\epsilon \to 0$, a strong confluence process is initiated. Then it leads to a resulting singularity $z_c = 0$ with the s-rank $R_c = R_1 + R_2$.* $\quad\Box$

---

[3]In order to avoid some particular cases, the initial equation is taken in its *normal form*—i.e. without the term of the first derivative.

Proof: We select the principal parts of the Laurent series of the function P_2/P_0 in the vicinities of two chosen singularities:

$$P_2(z)/P_0(z) = G_1(z) + G_2(z) + \text{r.p.},$$

$$G_1(z) = \sum_{k=1}^{2R_1} a_k z^{-k}, \qquad G_2(z) = \sum_{k=1}^{2R_2} b_k (z - \epsilon)^{-k}. \qquad (1.3.1)$$

By 'r.p.' we denote the regular parts. The sum $G_1(z) + G_2(z)$ may be converted into a fraction

$$G_1(z) + G_2(z) = \frac{\sum_{k=1}^{2R_1} a_k z^{2R_1 - k} (z - \epsilon)^{2R_2} + \sum_{k=1}^{2R_2} b_k (z - \epsilon)^{2R_2 - k} z^{2R_1}}{z^{2R_1} (z - \epsilon)^{2R_2}}.$$

$$(1.3.2)$$

If none of the coefficients a_k or b_k depend on ϵ, then, at the limit $\epsilon \to 0$, a singular point appears, the s-rank of which is equal to the maximum of the s-ranks of the points in the starting equation. For a confluence process to happen, it is necessary that these coefficients grow as certain powers of ϵ^{-1}. It follows from the expression in the denominator of (1.3.2) that it is impossible to make the s-rank of the resulting singularity larger than R_c. Therefore the proof of the theorem is reduced to proving possibility of choosing the functional dependence of the coefficients a_k and b_k on ϵ as of sufficient finite order to realize a strong confluence process. This can be achieved by explicit calculation: Suppose that the numerator in (1.3.1) becomes, at the limit $\epsilon \to 0$, a polynomial $G_c(z)$ in reciprocal powers of z:

$$G_c(z) = \sum_{k=1}^{2(R_1 + R_2)} f_{2(R_1 + R_2) - k} z^{-k}.$$

Then, by equating coefficients at separate powers of z to zero, the system of equations may be obtained. It is more convenient to write this system as two subsystems: one from which the coefficients a_k are recursively obtained as polynomials in ϵ^{-1}, i.e.

$$a_{2R_1}(-\epsilon)^{2R_2} = f_{2R_2 + 2R_1},$$

$$a_{2R_1 - 1}(-\epsilon)^{2R_2} + \Omega_1(a_{2R_1}, \epsilon^{-1}) = f_{2R_2 + 2R_1 - 1},$$

$$\cdots$$

$$a_1(-\epsilon)^{2R_2} + \Omega_{2R_1 - 1}(a_{2R_1}, \ldots, a_2, \epsilon^{-1}) = f_{2R_2 + 1}, \qquad (1.3.3)$$

and the other one from which the coefficients b_k are recursively obtained as polynomials in ϵ^{-1}, i.e.

$$b_1 + \Omega_{2R_1}(a_{2R_1}, \ldots, a_1, \epsilon^{-1}) = f_1,$$

$$b_2 + \Omega_{2R_1 + 1}(a_{2R_1}, \ldots, a_1, b_1, \epsilon^{-1}) = f_2,$$

$$\cdots$$

$$b_{2R_2} + \Omega_{2R_1 - 1}(a_{2R_1}, \ldots, a_1, b_1, b_{2R_2 - 1}, \epsilon^{-1}) = f_{2R_2}. \qquad (1.3.4)$$

Here, Ω_k are functions which are linear in the first argument and are polynomials in the other arguments. Choosing the coefficients a_k and b_k recursively—according to (1.3.3–1.3.4)—as linear functions of the free parameters f_k and as polynomials in ϵ^{-1}, in the limit $\epsilon \to 0$ we get a strong confluence process in eqn (1.1.1). $\square$

1.3.2 A confluence principle

Suppose that solutions of eqn (1.1.1) satisfy explicit identities (homogeneous and inhomogeneous) beyond eqn (1.1.1) which could be written as

$$T^{(j)} y(z) = 0. \tag{1.3.5}$$

Then the following informal 'confluence principle' can be formulated. The operator-function $T^{(j)}$ should be taken as the function depending on the same free parameters as $L^{(j)}$ and should be parametrized in the same way. This parametrization we denote by $T^{(j)}(\epsilon)$. Suppose that $L^{(j)}$ and $T^{(j)}$ are initial operators, and $L^{(k)}$ and $T^{(k)}$ are confluenced operators, which means that, in addition to (1.3.5), we have

$$L^{(j)}(\epsilon) \to \epsilon^n L^{(k)}(1 + o(1)), \qquad T^{(j)}(\epsilon) \to \epsilon^m T^{(k)}(1 + o(1)),$$

for appropriate n and m. Furthermore, $y(z)$ is a solution for equations with initial operators, and $\tilde{y}(z)$ is a solution of equation with operator $L^{(k)}$. Then, it is most likely that, under further assumptions, $\tilde{y}(z)$ is a solution of equation

$$T_z^{(k)} \tilde{y}(z) = 0. \tag{1.3.6}$$

This principle can be widely applied for various practical computations resulting in more order in the world of special functions.

1.3.3 Reduction of an equation

Ince [54] has shown that any singularity of eqn (1.1.1) may be regarded as a result of a confluence process (or a sequence of confluence processes) starting with elementary singularities. Since then, confluence of elementary singularities served as a basis for classification of second-order ordinary differential equations. From our point of view, the use of only this sort of confluence process gives no way to construct classes of differential equations and corresponding classes of special functions having similar properties. Moreover, elementary singularities do not seem to be sufficiently convenient to serve as basic singularities. Regular singularities are much better in this respect.

If regular singularities are taken as a basic set, then only the simultaneous use of two processes, namely confluence processes and reduction processes, enables one to obtain easily all possible non-Fuchsian equations, both confluent and reduced confluent. By a *reduction process* at a singular point with integer s-rank we mean a specialization of the coefficients of polynomials P_1 and P_2 which decreases the s-rank of this singular point by one half, thus converting an unramified irregular point into a ramified one.[4]

---

[4]It is also possible to consider a reduction of a regular singularity that is not elementary into an elementary one.

Not each form of an equation allows one to perform reduction processes. For instance, for a canonical form, it is not possible to perform a reduction process. A preliminary s-homotopic transformation applied to a normal form is needed. The reason is that, for a ramified singularity, Lemma 1.2 must hold.

On the basis of parallel use of confluence and reduction processes, in the next paragraph, we define the notion of a class of second-order linear homogeneous equations with polynomial coefficients.

1.3.4 Classes and types of equations

By a *generic equation* for the class M_n of second-order linear homogeneous equations with polynomial coefficients, we mean a Fuchsian equation with $n+1$ regular singular points having coefficients of the polynomials P_0, P_1, P_2 most generally compatible with the chosen form of the equation.

Equations equivalent to this generic equation and arising under specification of parameters, reduction, and confluence processes constitute a *class M_n* of equations with polynomial coefficients. A more precise definition is based on the notion of the s-multisymbol. For each multisymbol a quantifying characteristic feature can be defined:

$$\tau = \sum R(z_j) + R(\infty), \tag{1.3.7}$$

which takes integer and half-integer values. To the class M_n we attribute those equations for which

$$n - 1 \leq \tau \leq n. \tag{1.3.8}$$

In the case when $n - 1 = \tau$, it is additionally necessary that at least two s-ranks are half-integer.

Examples: (1) The equation

$$L_z^{\{;1\}} y(z) = D^2 y(z) = 0$$

is generic for the class M_0, and the Euler equation

$$L_z^{\{1;1\}}(\rho) y(z) = (zD^2 - \rho D) y(z) = 0$$

is generic for the class M_1. However, both classes have solutions in terms of elementary functions, and the corresponding equations should be excluded in a systematic presentation of special functions.

(2) The hypergeometric equation—the Fuchsian equation with three singularities—

$$L_z^{\{1,1;1\}}(a, b; c) y(z) = (z(1 - z)D^2 + (c - (a + b + 1)z)D - ab) y(z) = 0$$

is generic for the hypergeometric class M_2. Although the number of parameters is adequate, the way they enter the equation is somewhat surprising. The reason is the simplification of the formulae for its characteristic exponents.

(3) The Heun equation—Fuchsian equation with four singularities—

$$L_z^{\{1,1,1;1\}}(a, b; , d; t)y(z) - \lambda y(z) =$$
$$(z(z-1)(z-t)D^2 + (c(z-1)(z-t) + dz(z-t)$$
$$+ (a+b+1-c-d)z(z-1))D + (abz - \lambda))y(z) = 0$$

is generic for the Heun class M_3. The parameter λ is distinguished among other parameters, since it is an accessory that plays the role of a spectral parameter. □

The generic equation for the class M_n may be written as

$$L_z^{\{1,\dots;1\}}y(z) - \lambda y(z) = \prod_{j=1}^{n}(z-z_j)y''(z) + \sum_{j=1}^{n}c_j\prod_{i=1,i\neq j}^{n}(z-z_i)y'(z)$$
$$+ \left(\sum_{j=1}^{n-2}a_jz^j - \lambda\right)y(z) = 0, \quad z_1 = 0 \ z_2 = 1. \quad (1.3.9)$$

Once again the parameter λ is distinguished from the others. The number of parameters characterizing the generic equation for the class M_n, as can be easily counted from eqn (1.3.9), is equal to $3(n-1)$.

Inside the class of equations, the *types* of equations are distinguished.

The *type of an equation* is uniquely characterized by its *s-multisymbol*.

To count the number of types of equations is a purely combinatorial problem. The result depends on whether elementary singular points are distinguished concerning their s-ranks from the other regular singularities. In the latter case, the hypergeometric class includes 5 types of equations. Their s-multisymbols are given in Table 1.3.1. The Heun class includes 10 types of equations. Their s-multisymbols are given in Table 1.3.2. A combined classification scheme of both the hypergeometric and the Heun classes is presented in Table 1.3.3.

1.3.5 Standard forms of equations

As has been already stated, any irreducible equation can be transformed by means of a Möbius transformation and an s-homotopic transformation to various forms. Canonical, canonical natural, normal, and normal natural forms of an equation have been defined above. Here this list is prolonged, and also s-homotopic transformations converting one form to another are listed.

For a given type of an equation, a *general form* can be distinguished as an equation characterized by a maximal number of parameters compatible with this type. The generic equation of the class M_n in its general form is characterized by $4n$ parameters. Every strong confluence process of two singularities diminishes the number of parameters by two, and every reduction process diminishes the number of parameters by one. A general form of an equation is not always convenient for practical computations. In the previous sections, the canonical natural form of an equation was defined. It belongs to the *standard forms*. The other standard forms are *normal natural forms* and *self-adjoint natural forms*.

Table 1.3.1. s-rank multisymbols of the hypergeometric equations in natural form. The thick lines indicate specializations of parameters. The thin lines indicate confluence processes. The relations between the original and the resulting equations under specializations of parameters and confluence processes are indicated by arrows: they point into the direction of the resulting equations

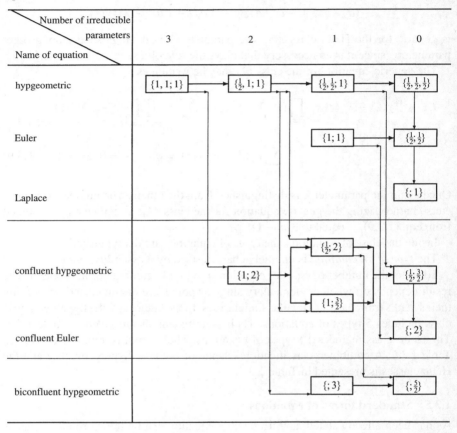

Number of irreducible parameters / Name of equation	3	2	1	0
hypgeometric	$\{1,1;1\}$	$\{\frac{1}{2},1;1\}$	$\{\frac{1}{2},\frac{1}{2};1\}$	$\{\frac{1}{2},\frac{1}{2};\frac{1}{2}\}$
Euler			$\{1;1\}$	$\{\frac{1}{2};\frac{1}{2}\}$
Laplace				$\{;1\}$
confluent hypgeometric		$\{1;2\}$	$\{\frac{1}{2};2\}$ $\{1;\frac{3}{2}\}$	$\{\frac{1}{2};\frac{3}{2}\}$
confluent Euler				$\{;2\}$
biconfluent hypgeometric			$\{;3\}$	$\{;\frac{5}{2}\}$

A *normal form* of an equation of a certain type is an equation equivalent to a general form of this type and containing no first derivative. If, in addition, the singularity with the highest s-rank lies at ∞, while other singularities, if existing, are located at $z = 0$ and $z = 1$, then it is called a *normal natural form*.

An equation with polynomial coefficients of a given type can also be transformed to a self-adjoint form.

Below, transformations from any form, including the canonical one, to the normal form and to the self-adjoint form are obtained. Suppose that we rewrite the starting equation as

$$L_z y(z) = [R(z)D^2 + P(z)D + Q(z)]y(z) = 0, \qquad D = \frac{\mathrm{d}}{\mathrm{d}z}. \qquad (1.3.10)$$

Table 1.3.2. s-rank multisymbols of the equations of Heun's class. The thick lines indicate specializations of parameters which reduce the number of the irreducible parameters. The thin lines indicate confluence processes. The relations between the original and the resulting equations under specializations of parameters and confluence processes are indicated by arrows: they point into the direction of the resulting equations

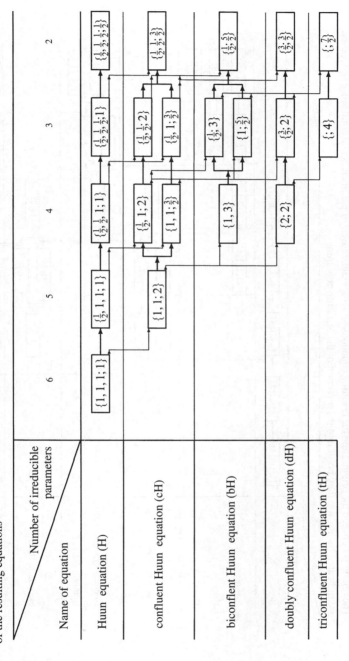

Table 1.3.3. s-rank multisymbols of the equations of Heun and hypergeometric classes. The thick lines indicate specializations of parameters which reduce the number of the irreducible parameters. The thin lines indicate confluence processes. The relations between the original and the resulting equations under specializations of parameters and confluence processes are indicated by arrows: they point into the direction of the resulting equations.

Number of irreducible parameters / Name of equation	6	5	4	3	2	1	0
Huun	$\{1,1,1;1\}$	$\{\tfrac12,1,1;1\}$	$\{\tfrac12,\tfrac12,1;1\}$	$\{\tfrac12,\tfrac12,\tfrac12;1\}$	$\{\tfrac12,\tfrac12,\tfrac12,\tfrac12\}$		
hypergeometric				$\{1,1,1\}$	$\{\tfrac12,1,1\}$	$\{\tfrac12,\tfrac12,1\}$	
Euler							$\{\tfrac12,\tfrac12,\tfrac12\}$
Euler						$\{1;1\}$	$\{;1\}$
confluent Heun		$\{1,1;2\}$	$\{\tfrac12,1;2\}$ / $\{1,1;\tfrac12\}$	$\{\tfrac12,\tfrac12;2\}$ / $\{\tfrac12,1;\tfrac12\}$	$\{\tfrac12,\tfrac12;\tfrac12\}$		
confluent hypergeometric					$\{1;2\}$	$\{\tfrac12;2\}$	$\{\tfrac12;\tfrac32\}$
confluent Euler						$\{1;\tfrac12\}$	$\{;2\}$
biconfluent Heun			$\{1;3\}$	$\{\tfrac12;3\}$ / $\{1;\tfrac52\}$	$\{\tfrac12;\tfrac52\}$		
biconfluent hypergeometric						$\{;3\}$	
doubly confluent Heun			$\{2;2\}$	$\{2;\tfrac23\}$	$\{3;\tfrac23\}$		
triconfluent Heun				$\{;4\}$	$\{;\tfrac72\}$		

Equation (1.3.10) with the help of the s-homotopic transformation

$$y(z) = \exp(-\int P/(2R)\,dz)\,y(z) = G(z)w(z) \qquad (1.3.11)$$

may be transformed to the normal form

$$N_z w(z) = (D^2 w(z) + \tilde{Q}(z))w(z) = 0, \qquad N_z = G(z)L_z G^{-1}(z), \qquad (1.3.12)$$

where

$$\tilde{Q}(z) = Q - P^2/(4R) - P'/2. \qquad (1.3.13)$$

By means of the slightly different s-homotopic transformation

$$y(z) = R^{1/2}\exp(-\int P/(2R)dz)\,y(z) = H(z)v(z), \qquad (1.3.14)$$

the same equation may be transformed to the self-adjoint form

$$M_z v(z) = (DR(z)Dw(z) + \hat{Q}(z))v(z) = 0, \qquad M_z = H(z)L_z H^{-1}(z), \qquad (1.3.15)$$

where

$$\hat{Q}(z) = (R'' - P')/2 - (P - R')^2/(4R) + Q. \qquad (1.3.16)$$

Example: Suppose that equation (1.3.10) is a Fuchsian equation in a canonical form. In this case, formulae (1.3.11–1.3.13) and (1.3.14–1.3.16) simplify to

$$R(z) = \prod_{j=1}^{n}(z - z_j), \qquad \frac{P(z)}{2R(z)} = \sum_{j=1}^{n}\frac{c_j}{2(z - z_j)},$$

$$c_j = \mathrm{Res}_{z=z_j}\frac{P(z)}{R(z)}. \quad \square$$

A canonical natural form, a normal natural form, and a self-adjoint natural form are considered as standard forms of an equation of a given type. A canonical form is usually used for convergent expansions of solutions, a normal form is used for asymptotics, and a self-adjoint form is used to study boundary problems.

1.3.6 Invariants of s-homotopic transformations

There is a function of the independent variable z of a differential equation which is preserved under s-homotopic transformations:

Lemma 1.7 *The polynomial*

$$F(z) = 4R(z)Q(z) - (P(z) - R'(z))^2 - 2(P'(z) - R''(z)) \qquad (1.3.17)$$

is an invariant of s-homotopic transformations. $\quad\square$

Proof: An arbitrary s-homotopic transformation converts a differential operator L_z in eqn (1.3.10) to the new operator $\hat{L}_z = S^{-1}(z)L_zS(z)$, where the function $S(z)$ determines the s-homotopic transformation. Recomputing the polynomials $R(z)$ and $P(z)$, $Q(z)$ to $\hat{R}(z)$, $\hat{P}(z)$, and $\hat{Q}(z)$, we obtain

$$\hat{R}(z) = R(z),$$

$$\hat{P}(z) = P(z) + \frac{2S'(z)R(z)}{S(z)},$$

$$\hat{Q}(z) = Q(z) + \frac{S''(z)R(z)}{S(z)} + \frac{S'(z)P(z)}{S(z)}.$$

These calculations result in $\hat{F}(z) = F(z)$. $\square$

Clearly, every coefficient of the polynomial $F(z)$ in (1.3.17) is also invariant under s-homotopic transformations. As a consequence, for instance, differences between Frobenius exponents at a particular regular singularity, and between zeroth-order exponents of Thomé solutions, are preserved.

1.4 Types of solutions

1.4.1 Eigenfunctions of singular Sturm–Liouville problems

In the preceding sections, local solutions of eqn (1.1.1) defined by convergent or divergent series in a vicinity of a singularity have been studied. This study will be continued, but also some global solutions will be defined in the following.

Suppose that two singularities z_1, and z_2 are located on the real axis, and that the interval $[z_1, z_2]$ contains no other singularities. Either of these singularities can be located at infinity. It may also be that the interval is $]-\infty, +\infty[$. The parameters of the equation are considered to be real. Suppose that the given equation is written in a self-adjoint form

$$(-Dr(z)D + q(z))v(z) = \lambda v(z). \tag{1.4.1}$$

The parameter λ plays the role of a spectral parameter. Then, under certain requirements on the parameters of the equation, additional conditions distinguishing an appropriate local behaviour of the solution $v(z)$ at the endpoints of the studied interval can define a singular Sturm–Liouville problem for eqn (1.4.1) on the interval $[z_1, z_2]$. If eqn (1.4.1) is Fuchsian with $n + 1$ singularities, then in general $n + 1$ Sturm–Liouville problems may be posed. For arbitrary values of parameters, these are equivalent since, by means of a Möbius transformation, any interval $[z_1, z_2]$ can be mapped onto $[0, 1]$.

In the case of the hypergeometric equation, eigenfunctions are polynomials (see Chapter 2), namely (under simple substitutions) Jacobi polynomials. In the hypergeometric class, two more sets of eigenfunctions for two types of confluent equations arise. The first set is expressed in terms of Laguerre polynomials, and the second set is expressed in terms of Hermite polynomials. In all mentioned cases, there are explicit expressions for eigenvalues. No well-posed singular Sturm–Liouville problems[5] exist for reduced confluent equations belonging hypergeometric class.

The eigenfunctions of singular Sturm–Liouville problems for equations of Heun class are, in general, not polynomials. They are studied in the third chapter.

The question arises whether there are polynomial solutions for equations of Heun class. The answer is positive if one of the characteristic exponents of zeroth order takes integer values, but the number of the polynomial solutions is finite (corresponding to the mentioned integer). Another possibility to pose the boundary problem with polynomial solutions is to turn to a multispectral problem [121].

1.4.2 Central and lateral connection problems

More general than the singular Sturm–Liouville problem is the central two-point connection problem. Suppose that two linearly independent local solutions $v_1(z_1, z)$ and $v_2(z_1, z)$ of eqn (1.4.1) are constructed in the vicinity of the singularity z_1, and that two other solutions mutually linearly independent $\tilde{v}_1(z_2, z)$ and $\tilde{v}_2(z_2, z)$ are constructed in the vicinity of the singularity z_2. These pairs of solutions are related to each other with the help of a transition matrix Ω_{12}:

$$\tilde{\mathbf{V}} = \Omega_{12}\mathbf{V}. \qquad (1.4.2)$$

Here $\mathbf{V}$ denotes the vector $(v_1(z_1, z), v_2(z_1, z))$, whereas $\tilde{\mathbf{V}}$ denotes the vector $(\tilde{v}_1(z_2, z), \tilde{v}_2(z_2, z))$. To find the matrix Ω_{12} means to solve the central two-point connection problem. If one of the nondiagonal matrix elements of Ω_{12} is zero, then it could relate to the solution of the corresponding singular Sturm–Liouville problem. The central two-point connection problem, in the case of equations of hypergeometric class, can be solved in terms of the gamma function. In the case of equations belonging to the Heun class, there are only asymptotic formulae at small or large values of parameters. Another possibility is to calculate the matrix Ω_{12} numerically. The numerical algorithm enabling us to obtain the eigenvalues as well as the eigenfunctions is studied in Sections 1.6 and 3.6.

Beyond the central connection problem, the lateral connection problem can be considered. Suppose that there is a simple loop encircling anticlockwise a singularity of eqn (1.4.1). Without loss of generality, this singularity is assumed to be at $z = z_j$. At the starting point z^* of this loop, the set of two solutions is locally defined: $\mathbf{U}(z) = (u_1(z^*, z), u_2(z^*, z))$. After winding around the singularity z_j, the analytic continuation of the initial set of solutions results in a new set of solutions $\hat{\mathbf{U}}(z) = (\hat{u}_1(z^*, z), \hat{u}_2(z^*, z))$. Both sets are related to each other by a matrix Υ_j:

$$\hat{\mathbf{U}} = \Upsilon_j\mathbf{U}. \qquad (1.4.3)$$

Clearly, the matrix Υ_j does not depend on the point z^*. If this point is a regular singularity the initial set of solutions can be chosen consisting of Frobenius solutions.

---

[5]In the sense described above.

Then, the matrix Υ_j is diagonal:

$$\Upsilon_j = \begin{pmatrix} e^{i2\pi\rho_1} & 0 \\ 0 & e^{i2\pi\rho_2} \end{pmatrix} \tag{1.4.4}$$

with ρ_1 and ρ_2 being Frobenius exponents at[6] z_j.

For a Fuchsian equation, the problem of analytic continuation of any set of solutions fixed at one singularity to any other set of solutions fixed at another other singularity is reduced to the knowledge of different matrices Ω_{ik} and Υ_j. Among these, only certain basic ones are needed. A more detailed study requires the methods of group theory, which is beyond the scope of this book. Here it is only important that these basic matrices represent the *monodromy data* for the equation under consideration.

These monodromy data in an explicit expression (in terms of the gamma function) are known only in cases of two and three regular singularities.

A more complicated situation arises if irregular singularities are taken into account. In a neighbourhood of an irregular singularity, unless the coefficients of the equation take specific values, two solutions of eqn (1.1.1) can be constructed:

$$y_m(z, z = z_j) = (z - z_j)^{\rho_m(z_j)} \sum_{-\infty}^{\infty} c_k^m (z - z_j)^k \qquad \text{for } m = 1, 2. \tag{1.4.5}$$

The coefficients $\rho_m(z_j)$ are called *path-multiplicative characteristic exponents* or also *Floquét exponents*, and the solutions themselves are called *path-multiplicative solutions* or *Floquét solutions*. The difference to the Frobenius solutions in the neighbourhood of a regular singularity is that the Taylor expansion is replaced by a Laurent expansion. However, this causes a more important difference. While Frobenius exponents are calculated as roots of the indicial equation, there is, in general, no method to calculate the path-multiplicative exponents in (1.4.5). It is also difficult to distinguish solutions (1.4.5) from one another from a theoretical as well as from a numerical point of view. Hence other solutions are more often used to expose the monodromy data, namely, those studied in the next paragraph.

1.4.3 Stokes lines at singularities. Stokes matrices

In the first section of this chapter, Thomé solutions at irregular singularities were defined. But these—in contrast to Frobenius solutions—are purely formal objects. The reason is that the series are in general divergent although asymptotic. But, even more important is that these asymptotic series do not represent any solution of the equation in a whole vicinity of the singularity. This is due to the so-called Stokes phenomenon. In this book, we treat this phenomenon in an abridged way, leaving a scrupulous presentation to other sources (see [43], [115]).

Suppose that an irregular singularity with the s-rank $R(z_j)$ is studied. Without loss of generality, the singularity is located at infinity, and the equation is taken in a

---

[6]We assume that the difference of characteristic exponents is not an integer. In this case the matrix Υ_j may have the structure of a Jordan block.

normal form:

$$(D^2 - q(z))\, w(z) = 0. \qquad (1.4.6)$$

The leading term in the function $q(z)$ can be taken as

$$q(z) = z^{2R-4}(1 + O(z^{-1})).$$

Then the leading term in the characteristic multiplier of the Thomé solution would be

$$f(z) = \exp(\pm\varphi(z)) = \exp\left(\pm\int\sqrt{z^{2R-4}}\,\mathrm{d}z\right) = \exp\left(\pm z^{R-1}/(R-1)\right).$$
$$(1.4.7)$$

On the $2R-2$ rays γ_l $(l = 0, \ldots, 2R-1)$, defined by

$$\arg z = \frac{\pi l}{R-1}, \qquad (1.4.8)$$

the function φ in (1.4.7) is real. Moreover

$$\varphi\left(z e^{i\pi l/(R-1)}\right) = e^{i\pi l}\varphi(z).$$

The defined rays (1.4.8) are called *Stokes rays*. They serve as borders of the *Stokes sectors*, defined as

$$S_l \ (l = 0, \ldots, 2R-1): \ \frac{\pi(l+1)}{R-1} > \arg z > \frac{\pi l}{R-1}. \qquad (1.4.9)$$

In each Stokes sector S_l, another ray can be considered dividing it into two halves:

$$\delta_l \ (l = 0, \ldots, 2R-1): \ \arg z = \frac{\pi(l+1/2)}{R-1}. \qquad (1.4.10)$$

On the ray (1.4.10), the function $\varphi(z)$ is purely imaginary. These rays are called *Anti-Stokes rays*.

On the Stokes rays, the function $f(z)$ exponentially increases when the plus sign is taken, and exponentially decreases when the minus sign is taken. Accordingly, the corresponding Thomé solution is called *dominant* at the Stokes ray in the first case and is called *recessive* at the Stokes ray in the second case.

On the Anti-Stokes rays, Thomé solutions are *oscillating*.[7]

Theorem 1.6 *For each Stokes ray, there exists one solution of eqn (1.4.6) asymptotically approximated in the sense of Poincaré on this ray by a recessive Thomé solution. This approximation is also valid in the sense of exponential asymptotics. For each Stokes sector, there exist two solutions of eqn (1.4.6) asymptotically approximated in the sense of Poincaré on this sector by corresponding Thomé solutions. This approximation is also valid in the sense of exponential asymptotics.* ☐

---

[7]That means that their real and imaginary parts take zero values at an unbounded set of points along each ray.

We do not give a proof of this basic theorem but refer to books on asymptotics written on the basis of Poincaré's definition of asymptotics (see e.g. [41], [43]). The proof in the sense of exponential asymptotics is based on resurgence theory [124]. □

Beyond the proof, some explanations are needed. Suppose that we have a Thomé solution in the form (1.2.14). Then the Poincaré definition of an asymptotic expansions may be defined as

$$
y_{ml}(z, z = \infty) \sim z^{-\alpha_0^m(\infty)} \exp\left(\sum_{k=1}^{R(\infty)-1} \frac{\alpha_k^m(\infty)}{k} z^k\right) \sum_{k=0}^{\infty} c_k^m(\infty) z^{-k},
$$

$$
\text{for} \quad m = 1, 2, \quad l = 0, \ldots, 2(R-1), \quad |z| \to \infty, \quad z \in S_l
$$

$$
\Longleftrightarrow \text{ for any } N: y_{ml}(z, z = \infty) z^{\alpha_0^m(\infty)} \exp\left(-\sum_{k=1}^{R(\infty)-1} \frac{\alpha_k^m(\infty)}{k} z^k\right)
$$

$$
-\sum_{k=0}^{N} c_k^m(\infty) z^{-k} = O(z^{-N-1}). \tag{1.4.11}
$$

In the case of a subnormal solution,

$$
y_{ml}(z, z = \infty) \sim z^{-\alpha_0^m(\infty)} \exp\left(\sum_{k=1/2}^{R(\infty)-1} \frac{\alpha_k^m(\infty)}{k} z^k\right) \sum_{k=0}^{\infty} c_k^m(\infty) z^{-k/2},
$$

$$
\text{for} \quad m = 1, 2, \quad l = 0, \ldots, 2(R-1), \quad |z| \to \infty, \quad z \in S_l
$$

$$
\Longleftrightarrow \text{ for any } N: y_{ml}(z, z = \infty) z^{\alpha_0^m(\infty)} \exp\left(-\sum_{k=1/2}^{R(\infty)-1} \frac{\alpha_k^m(\infty)}{k} z^k\right)
$$

$$
-\sum_{k=0}^{N} c_k^m(\infty) z^{-k/2} = O(z^{-(N+1)/2}). \tag{1.4.12}
$$

In the sense of exponential asymptotics, the following proposition can be formulated.

Theorem 1.7 *For each Stokes ray, there exists one solution of eqn (1.4.6) asymptotically approximated in the sense of exponential asymptotics on this ray by the dominant Thomé solution.* □

A rigorous proof of this theorem is based on artificial summation methods for divergent asymptotic series (see [124], [8], [14]). However, practically dealing with the actual solutions of an equation and the corresponding dominant Thomé solutions can always be arranged on a more heuristic level. Corresponding results can be found in the next chapter.

1.5 Generalized Riemann scheme

1.5.1 Introduction

In this section, a generalization of the Riemann scheme (Riemann P-symbol) is proposed. This scheme is valid for Fuchsian as well as for non-Fuchsian equations (cf. Section 1.1). The scheme describes

(i) s-ranks of the singularities,

(ii) the locations of the singularities,

(iii) Frobenius characteristic exponents,

(iv) Thomé characteristic exponents of normal and subnormal solutions.

The conventional Riemann scheme is used for Fuchsian equations only. It is a table exhibiting local characteristics of the equation—the locations of singularities and the characteristic exponents at these singularities. It clarifies the properties of an equation and simplifies transformations between its equivalent forms.

The Riemann equation with three regular singularities reads

$$y''(z) + \sum_{j=1}^{3} \frac{1 - \rho_1(z_j) - \rho_2(z_j)}{z - z_j} y'(z)$$

$$+ \prod_{j=1}^{3}(z - z_j)^{-1} \sum_{j=1}^{3} \rho_1(z_j)\rho_2(z_j) \frac{\prod_{i \neq j}(z_j - z_i)}{z - z_j} y(z) = 0. \qquad (1.5.1)$$

The corresponding Riemann scheme

$$\begin{pmatrix} z_1 & z_2 & z_3 \\ \rho_1(z_1) & \rho_1(z_2) & \rho_1(z_3) & ; z \\ \rho_2(z_2) & \rho_2(z_2) & \rho_2(z_3) \end{pmatrix} \qquad (1.5.2)$$

exhibits in the first row the location of its singularities z_j ($j = 1, 2, 3$), and in the two subsequent rows the Frobenius characteristic exponents $\rho_m(z_j)$ ($m = 1, 2$; $j = 1, 2, 3$). The scheme (1.5.2) determines eqn (1.5.1) completely in the sense that the equation may be recovered by the scheme. This, however, is not possible for equations with a larger number of singularities.

Since the Riemann equation belongs to the Fuchsian class of equations, the characteristic exponents obey Fuchs's condition (see Theorem 1.1)

$$\sum_{j=1}^{3} \sum_{m=1}^{2} \rho_m(z_j) = 1. \qquad (1.5.3)$$

This means that the number of independent parameters both in eqn (1.5.1) and in eqn (1.5.2) is 8.

The s-homotopic transformation of the dependent variable

$$S: y \mapsto w, \quad y = uw, \quad u = \prod_{j=1}^{3} (z - z_j)^{\mu_j},$$

including the additional condition that the point at infinity is to keep an ordinary point

$$\sum_{j=1}^{3} \mu_j = 0,$$

conserves the Riemann equation and alters the two last rows of the Riemann scheme according to

$$\begin{pmatrix} z_1 & z_2 & z_3 & \\ \rho_1(z_1) + \mu_1 & \rho_1(z_2) + \mu_2 & \rho_1(z_3) + \mu_3 & ; z \\ \rho_2(z_2) + \mu_1 & \rho_2(z_2) + \mu_2 & \rho_2(z_3) + \mu_3 & \end{pmatrix}.$$

The Möbius transformation of the independent variable

$$M: z \mapsto x, \quad x = \frac{az + b}{cz + d}, \quad ad - bc \neq 0$$

alters the first row of (1.5.2) according to

$$\begin{pmatrix} x_1 & x_2 & x_3 & \\ \rho_1(z_1) & \rho_1(z_2) & \rho_1(z_3) & ; x \\ \rho_2(z_2) & \rho_2(z_2) & \rho_2(z_3) & \end{pmatrix},$$

where

$$x_1 = (az_1 + b)/(cz_1 + d),$$
$$x_2 = (az_2 + b)/(cz_2 + d),$$
$$x_3 = (az_3 + b)/(cz_3 + d).$$

1.5.2 Generalized Riemann scheme

The generalization of the conventional Riemann scheme consists of adding Thomé exponents of irregular singular points to the scheme.

The *generalized Riemann scheme* (GRS) is a record in the form of a table consisting of columns of varying length. Each column except the last one corresponds to a singularity of the underlying equation. In the first row are written the s-ranks of the singularities. In the second row are exhibited the locations of singular points. All other rows contain characteristic exponents at singularities put in a sequence from lower to higher order. In the last column stands the independent variable, and below this is placed the accessory parameters (not more than one in this book).

Two more remarks should be made:

(1) In the case of an irregular singularity with half-integer s-rank, only one Thomé exponent $\alpha_{(k)}^{1}(z_j)$ $(k = 1/2, 1, \ldots)$ of a prescribed order is presented in the scheme. The other Thomé exponent related to the other solution is not written explicitly, since it may be recovered by Lemma 1.2.

(2) In the case of an irregular singularity with an integer s-rank, Thomé exponents create two characteristic sequences corresponding to two normal solutions. The terms of these sequences interchange each other in the columns of GRS. This means that it is not allowed to interchange two characteristic exponents with each other without interchanging the others for a given singularity.

According to the definition, a column in a GRS related to the singularity z_j with the s-rank $R(z_j)$ includes $2R(z_j) + 2$ numbers (algebraic symbols) of which $2R$ are characteristic exponents.

In the following, we give several examples for better understanding.

Example: The GRS for the Riemann equation (1.5.1) changes from (1.5.2) to

$$\begin{pmatrix} 1 & 1 & 1 \\ z_1 & z_2 & z_3 & ; z \\ \rho_1(z_1) & \rho_1(z_2) & \rho_1(z_3) \\ \rho_2(z_2) & \rho_2(z_2) & \rho_2(z_3) \end{pmatrix} . \quad \square$$

Example: Suppose that two singularities in eqn (1.5.1), located at z_2 and at z_3, coalesce, resulting in an equation with a regular singularity located at z_1 and an irregular singularity z_2 the s-rank of which is $R(z_j) = 2$. The corresponding GRS reads

$$\begin{pmatrix} 1 & 2 \\ z_1 & z_2 & ; z \\ \rho_1(z_1) & \alpha_1^{(0)}(z_2) \\ \rho_1(z_1) & \alpha_2^{(0)}(z_2) \\ & \alpha_1^{(1)}(z_2) \\ & \alpha_2^{(1)}(z_2) \end{pmatrix} . \quad \square$$

Example: Consider an equation

$$y''(z) + ay'(z) + (bz + c)y(z) = 0 \tag{1.5.4}$$

which possesses only one singularity, namely at the point $z = \infty$, the s-rank of which is $R(\infty) = 5/2$. Solutions of this equation may be expressed in terms of Airy functions. The Thomé exponents may be calculated straightforwardly as

$$\alpha_m^{(3/2)}(\infty) = \pm\sqrt{b}, \qquad \alpha_m^{(1)}(\infty) = -\frac{a}{2},$$

$$\alpha_m^{(1/2)}(\infty) = \pm\sqrt{b}\left(\frac{c}{2} - \frac{a^2}{8}\right), \qquad \alpha_m^{(0)}(\infty) = \frac{1}{4}.$$

It corresponds to the GRS

$$\begin{pmatrix} 5/2 \\ \infty \\ 1/4 \\ 1/4 \\ b^{1/2}(c/2 - a^2/8) \\ -a/2 \\ b^{1/2} \end{pmatrix} ; z . \qquad (1.5.5)$$

□

1.5.3 Applications

The next step of our study is to show the use of the GRS in describing simple transformations of the dependent and the independent variables. Möbius transformations, s-homotopic transformations, and quadratic transformations are considered.

The s-homotopic transformation is interpreted in terms of the GRS by means of substitution of the corresponding GRS instead of the functions y and w. In practice, in order to exhibit the local behaviour of the functions, we simply need to multiply the multipliers in front of the series. In the GRS this leads to the addition of the corresponding characteristic exponents.

Example: For the equation with the s-multisymbol $\{1; 2\}$, the s-homotopic transformation reads

$$(z - z_1)^\mu e^{\nu^{(1)}z} \begin{pmatrix} 1 & 2 \\ z_1 & \infty \\ \rho_1(z_1) & \alpha_1^{(0)}(\infty) \\ \rho_1(z_1) & \alpha_2^{(0)}(\infty) \\ & \alpha_1^{(1)}(\infty) \\ & \alpha_2^{(1)}(\infty) \end{pmatrix} ; z$$

$$= \begin{pmatrix} 1 & 2 \\ z_1 & \infty \\ \rho_1(z_1)+\mu & \alpha_1^{(0)}(\infty)-\mu \\ \rho_1(z_1)+\mu & \alpha_2^{(0)}(\infty)-\mu \\ & \alpha_1^{(1)}(\infty)+\nu^{(1)} \\ & \alpha_2^{(1)}(\infty)+\nu^{(1)} \end{pmatrix} ; z . \quad □$$

Example: In the case of eqn (1.5.4) and the corresponding GRS (1.5.5), the s-homotopic transformation reads

$$
e^{\nu^{(1)}z}
\begin{pmatrix}
& 5/2 & \\
& \infty & ; z \\
& 1/4 & \\
& 1/4 & \\
& \sqrt{b}(c/2 - a^2/8) & \\
& -a/2 & \\
& \sqrt{b} &
\end{pmatrix}
=
\begin{pmatrix}
& 5/2 & \\
& \infty & ; z \\
& 1/4 & \\
& 1/4 & \\
& \sqrt{b}(c/2 - a^2/8) & \\
& -a/2 + \nu & \\
& \sqrt{b} &
\end{pmatrix}. \quad \square
$$

Consider the linear transformation of the independent variable by

$$ L : z \mapsto az + b. $$

It changes the location of the finite singularities according to $z_j \mapsto az_j + b$ but does not change the Thomé exponents of zeroth order and the Frobenius exponents. Thomé exponents of higher orders transform according to

$$ \alpha_m^{(j)}(\infty) \mapsto a^j \alpha_m^{(j)}. $$

Example: Consider the equation having the s-multisymbol $\{1; 2\}$. Then

$$
\begin{pmatrix}
1 & 2 & \\
z_1 & \infty & ; az + b \\
\rho_1(z_1) & \alpha_1^{(0)}(\infty) & \\
\rho_1(z_1) & \alpha_2^{(0)}(\infty) & \\
& \alpha_1^{(1)}(\infty) & \\
& \alpha_2^{(1)}(\infty) &
\end{pmatrix}
$$

$$
=
\begin{pmatrix}
1 & 2 & \\
az_1 + b & \infty & ; z \\
\rho_1(z_1) & \alpha_1^{(0)}(\infty) & \\
\rho_1(z_1) & \alpha_2^{(0)}(\infty) & \\
& a\alpha_1^{(1)}(\infty) & \\
& a\alpha_2^{(1)}(\infty) &
\end{pmatrix}. \quad \square
$$

Suppose that z_0 and ∞ are singular points. The inverse transformation

$$ I : z \mapsto \frac{1}{z - z_0} $$

interchanges their locations. It results in an interchange of two corresponding columns in the GRS. Any other singularity z_j changes its location according to $z_j \mapsto (z_j - z_0)^{-1}$, preserving its characteristic exponents.

A Möbius transformation can be considered as a superposition of a linear transformation and an inverse transformation.

The quadratic transformation reads

$$Q : z \mapsto (z - z_0)^2. \tag{1.5.6}$$

Let z_0 be an elementary singularity. The transformation (1.5.6) converts it either to an ordinary point of the equation or to a removable singular point. The singularities beyond z_0 and ∞ are doubled. Their s-ranks are preserved and their locations alter according to (1.5.6). The s-rank of the singularity at infinity is changed from $R(\infty)$ to $2R(\infty) - 1$. It is necessary to stress that a quadratic transformation can bring the initial equation out of the considered class.

Here are two examples of quadratic transformations.

Example: Consider the GRS related to associated Legendre equation (cf. Chapter 2). The following equality holds:

$$
\begin{pmatrix}
1/2 & 1 & 1 & \\
0 & 1 & \infty & ; z^2 \\
0 & m/2 & (1/2 - v)/4 & \\
1/2 & -m/2 & (3/2 + v)/4 &
\end{pmatrix}
$$

$$
=
\begin{pmatrix}
1 & 1 & 1 & \\
-1 & 1 & \infty & ; z \\
m/2 & m/2 & (1/2 - v)/2 & \\
-m/2 & -m/2 & (3/2 + v)/2 &
\end{pmatrix}. \tag{1.5.7}
$$

$\square$

It follows from (1.5.7) that the associated Legendre equation is generated by an equation with the s-multisymbol $\{1/2, 1; 1\}$.

Example: Consider the equation

$$z^2 y''(z) + z y'(z) + \left(\frac{z^2}{4} - \frac{v^2}{4} \right) y(z) = 0.$$

Under a quadratic transformation it converts to Bessel's equation. This follows from the identity

$$
\begin{pmatrix}
1 & 3/2 & \\
0 & \infty & ; z^2 \\
v/2 & 1/4 & \\
-v/2 & 1/4 & \\
 & i &
\end{pmatrix}
=
\begin{pmatrix}
1 & 2 & \\
0 & \infty & ; z \\
v & 1/2 & \\
-v & 1/2 & \\
 & i & \\
 & -i &
\end{pmatrix}.
$$

It shows that Bessel's equation in essence may be regarded as an equation with the s-multisymbol $\{1; 3/2\}$. $\square$

The theorems describing the properties of the characteristic exponents may be simplified in terms of the GRS.

We propose a generalization of Fuchs's theorem (See Theorem 1.2).

Theorem 1.8 *Suppose that among the singularities of a given equation there are no reducible singular points. Then*

$$\sum_{m=1}^{2} \left(\sum_{j} \rho_m(z_j) + \sum_{j} \alpha_m^{(0)}(z_j) \right) = \sum_{j} R(z_j) - 2, \qquad (1.5.8)$$

where j denotes summation over all singular points. □

Proof: The s-homotopic transformation can alter neither the left nor the right part of the identity (1.5.8), since what is added to characteristic exponents at finite singular points is subtracted from the characteristic exponents of the singularity at infinity. Suppose that the equation is transformed to the normal form without the term of the first derivative. In this form, the characteristic exponents are calculated by means of the formula

$$\rho_m(z_j) = 1/2 \pm \theta_j. \qquad (1.5.9)$$

Suppose that there is an irregular singularity at infinity. Standard asymptotic calculations at this point give Thomé exponents as

$$\alpha_m^{(0)}(\infty) = (R(\infty) - 2)/2 \pm \theta_\infty. \qquad (1.5.10)$$

By summing (1.5.9) and (1.5.10), the needed result is obtained. □

Under the above mentioned assumptions, the sum of the elements in the third and fourth rows is equal to the sum of the elements in the first row.

The GRS can also clarify confluence processes.

Lemma 1.8 *At confluence processes of two singularities, the sum of the number of the characteristic exponents at the resulting point does not exceed the sum of the number of the characteristic exponents of the initial points.* □

Proof: First, the theorem of subadditivity of the s-rank under confluence processes is applied (Theorem 1.4). Then the definition of the GRS is taken into account. □

The given examples show that the GRS

(1) explicitly exhibits all characteristic properties of local solutions,
(2) is particularly appropriate for the search of those solutions which have polynomial or other specialized characteristic properties.

1.6   Central two-point connection problems (CTCPs)

1.6.1   Introduction

Boundary (eigenvalue) problems for linear ordinary second-order differential equations undoubtedly belong to the most important problems in the field with respect to applications. Generally this means the search for certain values of a parameter (or a set of parameters) of the differential equation such that the corresponding solution behaves in a prescribed manner at two points. Normally, these are located at the endpoints of an interval on the real axis. This interval may be finite, half-finite, or infinite. If the coefficients of the studied equation are arbitrary functions, then the problem is usually studied by methods typical for functional analysis. However, in the case of differential equations with polynomial coefficients studied in this book, specific formulations of the problem and methods of its solution based on complex analysis can be used.

If singularities of the underlying differential equation play the role of the relevant endpoints, and a particular asymptotic behaviour of the solutions is required while approaching them, we speak of *the central two-point connection problem* (CTCP).[8]

Suppose that we have a second-order differential equation with polynomial coefficients, and thus an instance of eqn (1.1.1),

$$ P_0(z) \frac{d^2 y(z)}{dz^2} + P_1(z) \frac{dy(z)}{dz} + P_2(z) \, y(z) = 0 \quad \text{for } z \in \mathbb{C}, \qquad (1.6.1) $$

where the coefficient $P_2(z) = P_2(z; \lambda)$ is a linear function of a parameter λ while $P_0(z)$ and $P_1(z)$ are independent of λ. Figure 1.1 demonstrates a typical situation of a CTCP where the two relevant endpoints are finite singularities located at $z = 0$ and at $z = 1$. The boundary problem is to be solved thus on $[0, 1]$. It is important to mention that this seemingly restricted situation is not affected by a loss of generality (see below). The problem is to find the values λ_i for the parameter λ for which the corresponding solutions of (1.6.1) behave in a prescribed manner (e.g. decreasing) while approaching $z = 0$ and $z = 1$.

In the following presentation, we exclusively deal with non-ramified singularities.

There may occur a variety of specific situations according to the location and the type of the endpoints. All of them may be reduced to the following three cases that we distinguish as generic:

(i) Both relevant endpoints in the CTCP are regular singularities.

(ii) One relevant endpoint in the CTCP is a regular singularity, while the other is an irregular singularity.

(iii) Both relevant endpoints in the CTCP are irregular singularities.

In fact, our analysis stays valid if one of the two relevant singularities in case (i), and the relevant regular singularity in case (ii), simplifies to an ordinary point of the differential equation.

The two endpoints of the relevant interval on which the CTCP is to be solved are denoted as z_0 and z_1.

---

[8] In the literature, usually a CTCP means the search for the connection matrix binding the local solutions at the two relevant endpoints. But, as far as we know, this more general problem has no solution yet except for equations belonging to the hypergeometric class. Therefore we restrict ourselves to the problem of seeking only the eigenvalues.

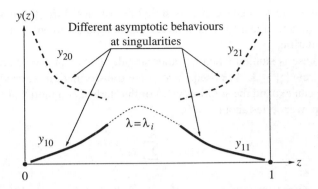

FIG. 1.1. CTCP on a finite interval with two singularities.

The main tools of the proposed approach are:

(i) a Möbius transformation of the independent variable placing the relevant singularities to $z_0 = 0$ and to $z_1 = 1$ as supposed above and an s-homotopic transformation of the dependent variable, making one particular solution of the underlying differential equation at each of the relevant endpoints being bounded and the other unbounded,[9]

(ii) a mapping of the differential equation to a difference equation,

(iii) an asymptotic investigation of all the particular solutions of this difference equation.

1.6.2 Two regular singularities as relevant endpoints

As already mentioned above without loss of generality it is assumed that, by means of a Möbius transformation, the endpoints of the relevant interval are located at

$$z_0 = 0, \qquad z_1 = 1,$$

and at infinity there is either a regular singularity or an ordinary point of the underlying differential equation. The first pair of characteristic exponents of the Frobenius solutions at the two relevant singularities are supposed to be $\rho_1(0) = 0$ and $\rho_1(1) = 0$. This can always be achieved by appropriate s-homotopic transformations. Moreover, we may assume that

$$\operatorname{Re} \rho_2(0) \leq 0, \qquad \operatorname{Re} \rho_2(1) \leq 0. \qquad (1.6.2)$$

In most practical problems, the characteristic exponents are real, and (1.6.2) simplifies to $\rho_2(0) \leq 0$, $\rho_2(1) \leq 0$.

A consequence of these conditions is that, at each relevant singularity, there is one bounded particular solution of the differential equation and one is unbounded,

---

[9]We call the combination of these two transformations a *Jaffé transformation* (see below).

as was required above. The CTCP is posed as follows: Find those values of the parameter λ (eigenvalues) for which there exist solutions (eigenfunctions) bounded at both singularities.

First, the case is studied when no other singularity except $z = 0$ and $z = 1$ lies in the unit circle $|z| \le 1$. A consequence of the transformations carried out so far is that we now can expand the solution $y_1(z)$ of the studied equation that is bounded at $z_0 = 0$ in a power series about $z = 0$:

$$y_1(z) = \sum_{n=0}^{\infty} a_n z^n. \tag{1.6.3}$$

The series in (1.6.3) is convergent within the unit circle $|z| < 1$. At $z = 1$, it is in general divergent, since the other relevant singularity is located there.

The variable-asymptotic behaviour of the function $y_1(z)$ for $z \to 1$ along the real axis is determined by the index-asymptotic behaviour of the coefficient a_n in (1.6.3) for $n \to \infty$. Thus, to fulfill the second boundary condition at $z_1 = 1$, it is necessary to study this index-asymptotic behaviour.

The coefficients a_n in (1.6.3) obey a difference equation of the form

$$\sum_{k=0}^{J} \Omega_k(n)\, a_{n+k} = 0, \tag{1.6.4}$$

where

$$\Omega_k(n) = \Omega_k + \frac{\alpha_k}{n} + \frac{\beta_k}{n^2}, \quad \text{for } k = 0, 1, \ldots, J, \tag{1.6.5}$$

holds as well as

$$\Omega_J \ne 0 \quad \text{and} \quad \Omega_0 \ne 0. \tag{1.6.6}$$

Equation (1.6.5) means that we have

$$\lim_{n \to \infty} \Omega_k(n) = \Omega_k$$

for all k, i.e. the defining property for (1.6.4) to be a difference equation of *Poincaré–Perron type* (see [132]).

The order J of eqn (1.6.4) is determined by the number of singularities of (1.6.1) and their s-ranks. It is necessary to mention that, in addition to the difference equation, there are $J - 1$ initial conditions such that (1.6.4) is solvable recursively, after having fixed a_0.

Difference equations of Poincaré–Perron type are accompanied by *characteristic equations* (see [102] p. 9). Their meaning consists in the fact that, under certain conditions in the limit $n \to \infty$, they represent the ratio of two consecutive terms of all their particular solutions of a fundamental system of the underlying difference equation. If the characteristic equation has exclusively simple roots, then the related difference equation of Poincaré–Perron type is called *regular*. If there are multiple roots it is called *irregular* (see [6]). The elaboration of the above-mentioned conditions is the central concern of Oskar Perron's early work [101]–[103].

If all the singularities of the underlying differential equation are regular, then the related difference equation of Poincaré–Perron type is regular. This is assumed in the following. Then, according to the above-mentioned publications of Perron, the index-asymptotic behaviour $n \to \infty$ of the solutions of eqn (1.6.4) is determined by the roots of the characteristic equation

$$\sum_{k=0}^{J} \Omega_k \, t^k = 0. \tag{1.6.7}$$

Suppose that these roots are ordered as $|t_1| \geq |t_2| \cdots \geq |t_J|$. Under our supposition that $z_0 = 0$ and $z_1 = 1$, and that no other singularity lays within the unit circle, we have

$$t_1 = 1, \qquad |t_m| < 1 \quad \text{for } m = 2, \ldots, J.$$

The solutions of the characteristic equations (1.6.7) give a first overview of the asymptotic behaviour of the particular solutions of the difference equations (1.6.4). However, in order to get the full asymptotic behaviour, one has to resort to an asymptotic representation of a fundamental system of the difference equation (1.6.4) for $n \to \infty$, called the *Birkhoff set* (see [132] p. 274). Generally, it has a rather complicated form. However, for a Fuchsian equation having a characteristic equation with distinct roots, the form of the set's elements simplifies to

$$s^{(m)}(n) = n^{r_m} \sum_{i=0}^{\infty} C_{mi} \, n^{-i} \quad \text{for } m = 1, \ldots, J. \tag{1.6.8}$$

The various solutions constituting the Birkhoff set $s^{(m)}(n)$ are called *Birkhoff solutions*. They are asymptotic particular solutions of the underlying difference equation for $n \to \infty$. This item will be discussed in more detail in the following subsection. As a consequence, this means

$$a_n^{(m)} \sim s^{(m)}(n) \quad \text{as } n \to \infty, \quad \text{for } m = 1, \ldots, J.$$

The particular solution $a_n^{(1)}$ of the difference equation (1.6.4) related to the root t_1 we call *maximum solution* $a_{n,max}$.

The eigenvalue condition in the underlying case means that the solution $y(z)$, according to (1.6.3, 1.6.8), has to be holomorphic at $z_1 = 1$. As a consequence, this means that the series (1.6.3) has to be convergent in a larger domain than the unit circle. The least step of enlargement (i.e. an increase of the radius of convergence from unity to reach the nearest singularity) means that precisely *one* particular solution of the difference equation (1.6.4), namely the maximum solution, vanishes. Writing the general solution of (1.6.4) as

$$a_n = \sum_{m=1}^{J} L_m a_n^{(m)} \quad \text{for } m = 1, \ldots, J, \tag{1.6.9}$$

and supposing that $a_n^{(1)}$ is the maximum solution, means that

$$L_1(\lambda = \lambda_i) = 0 \tag{1.6.10}$$

is the exact eigenvalue condition for the CTCP; i.e. λ_i are the eigenvalues of the CTCP.

Thus we have the result:

> The exact eigenvalue condition for the central two-point connection problem is that the coefficient L_1 in front of the maximum solution $a_n^{(1)} = a_{n,max}$ of the difference equation (1.6.4) in (1.6.9) must vanish.

The reader may ask what is the modification of the proposed method in the case when some non-relevant singularities of the differential equation are found within the unit circle? The eigenvalue condition (1.6.10) is no longer valid, since the radius of convergence for the series (1.6.3) is smaller than needed. The recipe for solving this problem is to use the following Möbius transformation of the dependent variable z, which keeps the points $z = 0$ and $z = 1$ at their places but puts the non-relevant singularities (lying in the unit disk) outside the unit circle:

$$\zeta = \frac{z(1 - z_*)}{z - z_*}, \tag{1.6.11}$$

where z_* is the singularity nearest to the origin.[10] The further considerations remain the same as in the previous part of this section.

1.6.3 One regular singularity and one irregular singularity as the endpoints

We consider a differential equation, of the form (1.6.1), having a regular singularity at zero and an irregular singularity at infinity, the s-rank of which is denoted $R(\infty)$. Between these singularities, i.e. on the positive real axis, is posed a CTCP. As shown in Chapter 4, such a problem arises quite often in physical applications. From singularity analysis outlined in the preceding sections, we know that there are two different possible asymptotic behaviours, while approaching each of the two singularities, that

---

[10]The case $|z_*| = 1$ with $z_* \neq 1$ needs special treatment.

may be admitted by the solutions of eqn (1.6.1). At $z = 0$ we have[11]

$$y^{(m)}(z) = z^{\pm\rho} [1 + O(z)] \quad \text{for } m = 1, 2,$$

where $\pm\rho$ are the characteristic exponents of the Frobenius solutions at $z = 0$; here $m = 1$ relates to $+\rho$ and $m = 2$ to $-\rho$, and ρ is supposed to be real positive.

At infinity we have

$$y^{(m)}(z) = \exp\left(\pm \alpha \, z^{R(\infty)-1} + \sum_{j=1}^{R(\infty)-2} \alpha_{j,m} z^j \right) \cdot$$

$$z^{\alpha_{0,m}} \left[1 + O\left(\frac{1}{z}\right) \right] \quad \text{for } m = 1, 2,$$

where α is real positive and where $m = 1$ relates to $+\alpha$ and $m = 2$ to $-\alpha$. The quantities α, $\alpha_{j,m}$, $\alpha_{0,m}$ are the characteristic exponents of the normal solutions at $z = \infty$.

With the help of these formulae, the CTCP may be formulated more precisely: For arbitrary values of the eigenvalue parameter λ, find a solution of (1.6.1) that behaves at the origin $z = 0$ like

$$y(z) = z^{+\varrho} [1 + O(z)] \quad \text{as } z \to 0$$

(i.e. decreasing) the asymptotic behaviour of which at infinity is given by

$$y(z) = A(\lambda) \exp\left(+ \alpha \, z^{R(\infty)-1} + \sum_{j=1}^{R(\infty)-2} \alpha_{j,1} z^j \right) \cdot z^{\alpha_{0,1}} \left[1 + O\left(\frac{1}{z}\right) \right]$$

$$+ B(\lambda) \exp\left(- \alpha \, z^{R(\infty)-1} + \sum_{j=1}^{R(\infty)-2} \alpha_{j,2} z^j \right) \cdot z^{\alpha_{0,2}} \left[1 + O\left(\frac{1}{z}\right) \right]$$

for $z \to \infty$, with $A(\lambda)$ and $B(\lambda)$ being arbitrary constants in z but depending on λ. The CTCP consists in the question of how to find those values λ_i of λ for which $A(\lambda_i) = 0$. Here, we have tacitly assumed that at least one such value exists.

The first step in tackling this problem is to split off the full asymptotic behaviour of the requested solution at *both* relevant singularities. In applications, normally, this is an asymptotic decreasing behaviour, and we will follow this route.

As one can easily see, to split the full asymptotic behaviour is by no means trivial, since generally the behaviour is power-like at both singularities. However, the multiplication of two power terms (yielding one power term in the result) cannot describe

---

[11] For the sake of simplicity, we assume that the absolute values of the two exponents for plus and minus are the same. This is no loss of generality, since it can always be achieved by an appropriate s-homotopic transformation.

the two power-like behaviours at the two relevant singularities! This problem may be solved by carrying out the following transformation

$$y(z) = \exp\left(- \alpha\, z^{R-1} + \sum_{j=1}^{R-2} \alpha_{j,2} z^j \right) \cdot z^{+\varrho} \cdot (z - z_*)^\beta \cdot w(z) \qquad (1.6.12)$$

with

$$\varrho + \beta = \alpha_{0,2}. \qquad (1.6.13)$$

Here z_* may be a regular singularity of the studied differential equation or an ordinary point. In the latter case, the differential equation for $w(z)$ has a singularity at $z = z_*$. Thus, in this case, (1.6.12) is a *singularity-generating transformation*. Note that z_* must not lie on the relevant interval, i.e. on the positive real axis.

After this transformation of the dependent variable, a Möbius transformation of the independent variable is carried out:

$$z \mapsto x = \frac{z}{z - z_*}. \qquad (1.6.14)$$

The resulting differential equation for $w(x)$ is characterized by the following features:

- The relevant regular singularity is kept at the origin: $z = 0 \to x = 0$.
- The irregular relevant singularity is put to unity: $z = \infty \to x = 1$.
- The relevant interval on which the boundary problem is to be solved changes from $\{z|[0, \infty]\}$ to $\{x|[0, 1]\}$.
- The resulting differential equation for $w(x)$ has to have a singularity at infinity. If the original equation has singularities beyond the two relevant ones, a non-relevant one is to be taken for this aim. If the original equation only has two singularities, both of which are relevant, one has to generate a regular singularity according to (1.6.12, 1.6.13).
- All the other non-relevant singularities (if there are any) of the resulting differential equation are to be located outside the unit circle (cf. (1.6.11)).

A transformation (1.6.12–1.6.14) having the formulated properties is called a *Jaffé transformation*.

The reader will find the above-discussed situation sketched in Figure 1.2. The filled (black) circles indicate the locations of the singularities, while the white ones indicate that there may be an ordinary point or a singularity. The figure shows how the singularities are moved under the M obius transformation $z \to x$ going from the complex z plane to the complex x plane. The relevant interval is marked as a hatched region.

A consequence of the transformations carried out so far is that the CTCP is reduced now to find the function $w(x)$ that is holomorphic at the endpoint $x = 0$ of the

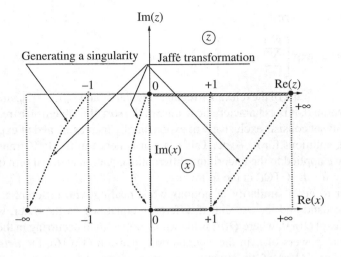

FIG. 1.2. Jaffé and singularity-generating transformation.

relevant interval $[0, 1]$. Therefore we may expand the function $w(x)$ in a power series about $x = 0$:

$$w(x) = \sum_{n=0}^{\infty} a_n x^n. \tag{1.6.15}$$

The series in (1.6.15) is called a *Jaffé expansion* for the original equation and is always convergent within the unit circle. At $x = 1$, it is in general divergent, since the irregular relevant singularity is located there. As will turn out below, the question about the convergence of (1.6.15) at $x = 1$ is a crucial one for the CTCP in this case.

The coefficients a_n in (1.6.15) obey again a difference equation of Poincaré–Perron type (see (1.6.4–1.6.6)). If there are no other singular points beyond $x = 0$, $x = 1$, and the regular singularity at infinity, then the order of the difference equation is R, where R is the s-rank of the irregular singularity.

We should mention that, in addition to the difference equation, there are $R - 1$ initial conditions such that the difference equation is solvable recursively, after having fixed a_0.

The main difference to the former case treated in the foregoing subsection is that the difference equations are of *irregular* type here. This means that their characteristic equations have multiple roots.

It was G.D. Birkhoff [17] who has shown that there are formal solutions for difference equations of irregular Poincaré–Perron type in the form of fractional powers of the index n. The number of solutions of this sort coincides with the order of the associated difference equation. They are also called *Birkhoff solutions*. The totality of all Birkhoff solutions of such a difference equation is called *Birkhoff set* (see [132] p. 274).

They have the form[12]

$$s^{(m)}(n) = \exp\left[\sum_{j=1}^{R-1} \gamma_{mj} n^{\frac{j}{R}}\right] n^r \sum_{i=0}^{\infty} C_{mi} n^{-i/R} \quad \text{for } m = 1, \ldots, R, \qquad (1.6.16)$$

where R is the s-rank of the relevant irregular singularity and γ_{mj}, r, C_{mi} are constants to be determined. It is a characteristic feature of the sort of problems we treat here that the Birkhoff set consists exclusively of exponentially increasing and of exponentially decreasing solutions for $n \to \infty$. This is a consequence of the Jaffé transformation that we have applied to the underlying differential equation and will turn out to be of importance for the CTCP in the following.

In order to get a similarity of notions while dealing with differential equations on the one hand and the associated difference equations on the other, we call the quantities $\exp[Q(n)]$, where $Q(n)$ is the whole polynomial occurring in the exponent in fractional powers of n in the asymptotic expansion (1.6.16), the *determinating factors* and $\exp[Q(n)]n^r$ the *asymptotic factors* of the related series.

$$Q(n) = \sum_{j=1}^{R-1} \gamma_{mj} n^{\frac{j}{R}}.$$

As one can see in Section 3.6 and in Appendix B, the asymptotic factors depend only upon the α_k in (1.6.5) and not on the parameters β_k.

The power series of the Birkhoff solutions are in general divergent (see [6] p. 507). Hence there arises the question of whether (analogous to Thomé's normal solutions as formal solutions of differential equations at irregular singularities), for each particular solution $\{a_n^{(m)}\}$ of the related difference equation, the asymptotic behaviour of it, for $n \to \infty$, is given by the one of the Birkhoff solutions $s^{(m)}(n)$.

A substantial paper on this problem was published in 1932 by G.D. Birkhoff and W.J. Trjitzinsky [18], based upon two preceding papers by G.D. Birkhoff [16], [17] and one further one by C.R. Adams [6]. On the one hand, this paper is valid for even more general difference equations than those we are dealing with here. On the other hand, it is not easy to understand, even for specialists in the field. Wong and Li [133], for second-order equations, have given a new proof that the Birkhoff solutions represent asymptotic solutions for irregular difference equations. The ideas are based on similar questions concerning normal series of Thomé type as formal solutions of linear ordinary differential equations discussed by Olver [96].

Birkhoff solutions (1.6.16), as asymptotic representations of the fundamental system of the difference equation (1.6.4) for $n \to \infty$, moreover are the analogues to the normal and subnormal solutions of differential equations about irregular singularities.

Remark: The actual calculation of the coefficients γ_{mj}, r, C_{mi} of the Birkhoff set is achieved by the method of indeterminate coefficients. These calculations should be carried out by computer-algebraic methods because they are rather lengthy. □

---

[12]It is important, here, to assume that the differential equation has no other singularities beyond those at $x = 0$, $x = 1$, and $x = \infty$. Otherwise, other solutions may occur in the Birkhoff set.

The Birkhoff set gives the full asymptotic behaviour of all the particular solutions of the underlying differential equation. However, as just remarked, it is laborious to compute. As was already discussed in the foregoing subsection, there is a simpler way to get at least a part of the information about the asymptotic behaviour of the particular solutions for difference equations of Poincaré–Perron type by studying the ratio of two consecutive terms, i.e.

$$\frac{a_{n+1}^{(m)}}{a_n^{(m)}} \quad (m = 1, \ldots, J), \tag{1.6.17}$$

for each of the m particular solutions of the difference equations, and calculating the limit for $n \to \infty$ in (1.6.17) in the complex n plane. This is done by solving the characteristic equation. However, it is no longer sufficient to study characteristic equations of zeroth order in n, as was possible in the case of eqn (1.6.7); in the case of irregular difference equations, they have to be of first order in n^{-1}. Otherwise the result would be trivial. Thus we have to write

$$\sum_{k=0}^{R} \left(\Omega_k + \frac{\alpha_k}{n} \right) t^k = 0; \tag{1.6.18}$$

the index-asymptotic solutions of this equation is given by

$$t^{(m)}(n) = 1 + \frac{\gamma_m}{n^{\frac{1}{R}}} + O\left(\frac{1}{n^{\frac{2}{R}}} \right) \quad \text{for } n \to \infty \quad (m = 1, \ldots, R), \tag{1.6.19}$$

where the distinct values of γ_m are solutions of the Rth-order algebraic equation

$$\gamma^R = -\sum_{k=0}^{R} \alpha_k$$

and R is the s-rank of the relevant irregular singularity. The most important and most characteristic feature of the CTCP as it is formulated here is that all solutions (1.6.19) tend to $t = 1$ for $n \to \infty$. Moreover, if the coefficients of (1.6.18), and thus of (1.6.1), are real, there is always one solution that tends to $t = 1$ from outside the unit circle $|t| = 1$ *along the real axis*, which naturally is the maximum solution $t_{max}(n)$.

As far as we know, equations of the form (1.6.18), with the help of which one can calculate the ratio of two consecutive terms of irregular difference equations, were first introduced not in the mathematical literature but in a paper on physics. After G. Jaffé [56] (following an idea of F. Hund and W. Pauli) made use of eqn (1.6.18), D.R. Bates *et al.* [11] took this as a basis to calculate the spectra of the hydrogen-molecule ion, and E.A. Solov'ev [122] has shown for the first time that one can admit here complex parameters in the differential equations as well.

Eigenvalue condition
As in the former subsection, we also may define the *maximum solution* $a_{n,max}$ of the difference equation (1.6.4) as being that one which causes the solution $t(n) = t_{max}(n)$

in (1.6.18) to have the largest real part. However, this maximum solution does not play the same role as in the former case. Here, we have to introduce the following distinction. The Birkhoff sets consist exclusively of exponentially increasing and of exponentially decreasing solutions. The former we call *dominant*, and the latter we call *recessive*.

To solve a CTCP means to give a condition that yields the eigenvalues $\lambda = \lambda_i$. This condition, here, is derived in two steps:

The first step comes from singularity analysis as done earlier, in this chapter. From there we know that the general solution of the differential equation, after having split the asymptotic factors at $z = \infty$ or $x = 1$, consists of one exponentially increasing particular solution and of one particular solution for which all its characteristic exponents are zero, which therefore tends to a finite value for $x \to 1$. Just to get such a particular solution was the aim of splitting off the asymptotic factors. This latter solution is the eigensolution of the CTCP.

Thus, we have the following situation: When λ is not an eigenvalue, the function $w(x)$ represented by (1.6.15) behaves like

$$w(x) \sim \exp\left(+2\frac{\alpha}{(1-x)^{R-1}}\right) \tag{1.6.20}$$

for $x \to 1$. (The factor 2 comes from the splitting procedure.) As a consequence, (1.6.15) is *divergent* at $x = 1$. When λ becomes an eigenvalue, this behaviour changes to

$$w(x) \sim O(1) \tag{1.6.21}$$

for $x \to 1$ from inside the unit circle. This means that (1.6.15) becomes *convergent* at $x = 1$ or, more precisely, that

$$\lim_{N \to \infty} \sum_{n=0}^{N} a_n \tag{1.6.22}$$

is convergent, or that

$$\sum_{n=0}^{\infty} a_n \tag{1.6.23}$$

exists and thus has a finite value. This value then is $w(1)$.

The second step is based upon the asymptotic behaviour of the difference equations (1.6.4) for the a_n in (1.6.15) considered above: As one may learn from the Birkhoff sets, and as was already mentioned above, the particular solutions of the difference equations (1.6.4) for the a_n of (1.6.15) exclusively consist of exponentially increasing and of exponentially decreasing solutions. Taking into account the above-derived convergence condition for (1.6.15) at $x = 1$, this means that an eigenvalue causes the corresponding solution of the difference equation (1.6.4) to consist exclusively of recessive particular solutions. Writing the general solution of (1.6.4) as

$$a_n = \sum_{m=1}^{R} L_m \, a_n^{(m)},$$ (1.6.24)

this means that all the L_m in front of the dominant solutions must vanish.

As a result we have:

> The exact eigenvalue condition for the central two-point connection problem is that the coefficients in front of the dominant solutions of the difference equation (1.6.4) in (1.6.24) must vanish.

How this eigenvalue condition can be implemented into a numerical program in order to carry out concrete calculations will be covered in Section 3.6 when dealing with Heun equations.

Also, we should mention the following special case. When Birkhoff solutions exist having purely oscillating behaviour to first order, the decision as to whether these are exponentially increasing or exponentially decreasing can only be made by considering the second asymptotic order of the asymptotic factor. A concrete example is the triconfluent case, which is dealt with in Section 3.6: Among the four Birkhoff solutions, two oscillate to first (leading) asymptotic order. This is seen in Figure 1.3, which shows the particular solutions of the difference equation (1.6.4–1.6.6) in the t-plane. The next-higher order, however, shows that both of these oscillating solutions are exponentially increasing.

1.6.4 A proof

Above, we have used the relation between the variable-asymptotic behaviour $x \to 1$ of a function represented by a power series like (1.6.15) being convergent on the open

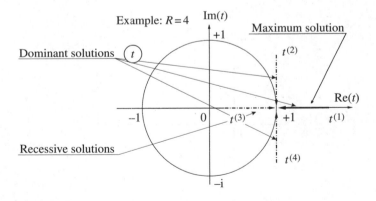

FIG. 1.3. Particular solutions in the complex t plane.

disk $|x| < 1$ and the index-asymptotic behaviour of its coefficient a_n for $n \to \infty$. This was a purely formal consideration that will be made more rigorous in the following. Readers not interested in mathematical proofs may omit the next section.

Theorem 1.9 *The index-asymptotic behaviour of the asymptotic factors of the Birkhoff sets (1.6.16) is determined by the variable-asymptotic behaviour of the asymptotic factors of the underlying differential equation.* □

To prove this theorem, we need a lemma.

Lemma 1.9 *Consider the function*

$$h(x) := e^{\frac{\alpha}{(1-x)^s}} (1 - x)^{-\beta} \quad (x \in \mathbb{C}) \tag{1.6.25}$$

with α and β being complex constants and s a positive integer. The function $h(x)$ may be expanded in a converging power series about $x = 0$:

$$h(x) = \sum_{n=0}^{\infty} c_n x^n, \tag{1.6.26}$$

the radius of convergence being unity. Then there is the following integral representation of the coefficients of the series in (1.6.26) (cf. [10] eqn (12)):

$$c_n = \frac{1}{2\pi i} \int_{\sigma-i\infty}^{\sigma+i\infty} \tau^{-\beta} (1 - \tau)^{-(n+1)} e^{\frac{\alpha}{\tau^s}} \, d\tau, \qquad 1 > \sigma > 0, \tag{1.6.27}$$

for sufficiently large n. The path of integration in the complex τ-plane is shown in Figure (1.4). □

Proof of the lemma: The following integral representation is valid for coefficients c_n:

$$c_n = \frac{1}{2\pi i} \oint_C \zeta^{-n-1} (1 - \zeta)^{-\beta} e^{\frac{\alpha}{(1-x)^s}} \, d\zeta. \tag{1.6.28}$$

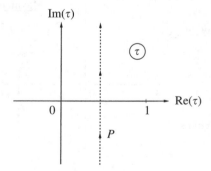

FIG. 1.4. Integration path P.

The contour C encircling the point $\zeta = 0$ anticlockwise may be deformed as the sum of the segment $[\sigma - iR, \sigma + iR]$ and the corresponding part of the circle $\omega_R : |\zeta| = \sqrt{\sigma^2 + R^2}$. For sufficiently large R and n,

$$\int_{\omega_R} h(\zeta)\zeta^{-n-1}\,d\zeta \to 0,$$

and the integral over the segment $[\sigma - iR, \sigma + iR]$, after the substitution $\zeta \mapsto 1 - \zeta$, tends to the integral in the right-hand side of (1.6.28). $\square$

Proof of the theorem: The coefficients of the expansion (1.6.15) may be presented in the same manner as in eqn (1.6.28):

$$a_n = \frac{1}{2\pi i}\oint_C \zeta^{-n-1} w(\zeta)\,d\zeta.$$

According to eqn (1.6.20) the leading behaviour of the function $w(x)$ in (1.6.15) at the singularity $x = 1$ is determined by the function $h(x)$. Similar to the proof of the lemma, we obtain

$$a_n = \frac{1}{2\pi i}\int_{\sigma-i\infty}^{\sigma+i\infty} \tau^{-\beta}(1-\tau)^{-(n+1)}e^{\frac{\alpha}{\tau^s}}u(\tau)\,d\tau + \sum_{k=1}^{K} I_k, \qquad 1 > \sigma > 0.$$

(1.6.29)

Here, I_k denote integrals over contours generated by singularities of the function $w(x)$ lying in the left half plane. The integrals in (1.6.27) may be evaluated by the saddle-point method (see e.g. [25] pp. 77). This means that, if we have an integral of the form

$$I = \int_{\mathcal{L}} \exp(\nu\psi(\tau))\,d\tau,$$

and if τ_0 is a saddle-point, and thus is a solution of

$$\frac{d\psi(\tau)}{d\tau} = 0,$$ (1.6.30)

then the contribution of this saddle point to the integral can be evaluated by[13]

$$I = \sqrt{\frac{\pi}{\nu}}\left|\frac{1}{2}\frac{d^2\psi(\tau)}{d\tau^2}\right|_{\tau=\tau_0}^{-\frac{1}{2}} e^{\nu\psi(\tau_0)-i/2\arg(\psi''(\tau_0))}.$$ (1.6.31)

From (1.6.25) and (1.6.30), we see that the locations of the saddle points τ_0 in the leading approximation are given by the solutions of

$$\frac{d\psi(\tau)}{d\tau} - \frac{s\alpha}{\tau^{s+1}} + \frac{n}{1-\tau} = 0,$$

[13]Primes indicate differentiation with respect to τ.

or in leading-order asymptotics:

$$\tau_0^{s+1} \sim \frac{s\alpha}{n} \quad \text{for } n \to \infty. \tag{1.6.32}$$

Equation (1.6.32) shows that, if α in eq. (1.6.29) is real positive, then there is always a saddle point on the interval $[0, 1]$ near $\tau = 0$. If α is real negative, then there is no saddle point on the positive real axis, while there are always (at least) two saddle points complex-conjugate to one another.

Thus only the saddle point of (1.6.32) which has the largest real part gives the leading-order behaviour of a_n, i.e. one term of the sort (1.6.31). Elaborating all the saddle points yields the Birkhoff set (cf. [78], [79], [83]) and thus the complete asymptotic factors of the large-n behaviour of a_n. The asymptotic impact of the integrals I_k is of lower order.

If the leading-order behaviour of a_n is vanishing in the general solution (see eqn (1.6.24)), this means that the solution of the differential equation $w(x)$ does not have growing asymptotics (1.6.20) and thus is the eigenfunction. $\quad\square$

A remark

In the standard situation for which we have elaborated the general procedure, the CTCP consisted in a differential equation (1.6.3) or (1.6.4), the relevant singularities of which were located at the origin and at infinity. The former was regular, and the latter irregular with s-rank R. There is a special case that can be dealt with on the basis of the method above without any crucial changes. This case is given when the regular relevant singularity is an ordinary point. Then the CTCP can be solved in complete analogy except for the following items:

- According to Painlevé's theorem (see [15] p. 11) all particular solutions of (1.6.4) at $z = x = 0$ are holomorphic. Hence, there, the Frobenius solution is replaced by a Taylor expansion.

- As a result, the number of initial conditions of the difference equation (1.6.4–1.6.6) is no longer $R - 1$ but $R - 2$. Here, for calculating the coefficients a_n of (1.6.15) recursively, one has to fix not only a_0 but also a_1. Only then is the spectrum of eigenvalues determined. If a_0 is fixed (for instance by means of normalizing the solution), then the spectrum changes while changing a_1.

1.6.5 Two irregular singularities

When both relevant singularities of (1.6.4) at $z = 0$ and at $z = \infty$ are irregular (having the s-ranks R_2 and R_1, respectively) the general procedure to solve the CTCP has to be changed with respect to one item, namely to the Jaffé transformation (1.6.9). The relevant singularity at $z = 0$, which now is irregular, can no longer be kept at the origin but has to be put to $x = -1$. Thus (1.6.14) changes to

$$z \mapsto x = \frac{z + z_*}{z - z_*}.$$

It is assumed below that there are no further singularities of the underlying differential equation. However this restriction is important only for formulating the results but not for the validity of the approach. The resulting differential equation, after having carried out the Jaffé and the singularity-generating transformations, thus has two irregular singularities (that are the relevant ones), located at $x = -1$ and at $x = +1$, and a non-relevant regular singularity placed at $x = -\infty$.

The Möbius and the singularity-generating transformation that result in an equation whose two relevant singularities are located on the unit circle we call a *Jaffé–Lay transformation*, and in this case (1.6.15) becomes a *Jaffé–Lay expansion*. The consequences for the general procedure are the following:

- The difference equation (1.6.4–1.6.6) still is an irregular one of Poincaré–Perron type. However, the order of which here is $R_1 + R_2$.

- The number of initial conditions of the corresponding difference equation (1.6.4–1.6.6) is $R_1 + R_2 - 2$. Here, for calculating the coefficients a_n of (1.6.15) recursively, one has to fix not only a_0 but also a_1. Only then, the spectrum of eigenvalues is determined.

  Although this situation seems to be similar to that in the above-mentioned triconfluent case discussed in the foregoing remark, where one relevant point was an irregular singularity while the other was an ordinary point, it is important to understand the difference: Suppose that a_0 is fixed, e.g. by means of a normalization condition. While, in the former case, for any a_1 there exists a spectrum, this is no longer the case here. Since we have put both relevant singularities on the unit circle, while $w(x)$ in (1.6.15) is still expanded about the origin $z = x = 0$, we can no longer meet one boundary condition a priori. As a consequence, a second boundary parameter must appear besides λ, with the help of which one may fulfill both boundary conditions. This second parameter may be chosen as a_1.

- The general solution of the difference equation (1.6.4) has two groups of solutions. One of them contains solutions which tend to $+1$ in the complex t-plane (cf. (1.6.17)), while the others tend to -1. As a consequence, the Birkhoff set has the form

$$s^{(m)}(n) = \exp\left[\sum_{j=1}^{R_1-1} \gamma_{mj} n^{\frac{j}{R_1}}\right] n^r \sum_{i=0}^{\infty} \frac{C_{mi}}{n^{\frac{i}{R_1}}} \quad \text{for } m = 1, \ldots, R_1.$$

$$s^{(m)}(n) = (-1)^n \exp\left[\sum_{j=1}^{R_2-1} \gamma_{mj} n^{\frac{j}{R_2}}\right] n^r \sum_{i=0}^{\infty} \frac{C_{mi}}{n^{\frac{i}{R_2}}},$$

for $m = R_1 + 1, \ldots, R_1 + R_2$.

- Each of these two groups has dominant and recessive solutions. Among those tending to $+1$, the dominant ones lie outside the unit circle in the complex t plane tending to unity. The recessive ones lie inside the unit circle in the complex t plane tending to unity. Among those tending to -1, the dominant ones also lie outside the unit circle in the complex t plane tending to unity, while the recessive ones once again lie inside the unit circle in the complex t plane tending to unity.

- The eigenvalue condition, here, consists of two parts: both subsets of dominant solutions in the two groups must vanish, or, more precisely, the coefficients L_m of (1.6.24) in front of the dominant solutions must vanish.

- This is a compound eigenvalue condition that is not appropriate for a numerical algorithm to compute the eigenvalues. It is possible to reduce this compound condition to a single one on the numerical level by applying backward recursion processes. This method will be outlined for the concrete example of the doubly confluent case of Heun's equation dealt with in Chapter 3. There the subsets of dominant and recessive solutions each contain one member related to $x = +1$ and one member related to $x = -1$. Thus, this equation is the most simple one among those having two relevant irregular singularities.

2

THE HYPERGEOMETRIC CLASS OF EQUATIONS

2.1 Classification scheme

2.1.1 General presentation

The *hypergeometric class* of second-order linear ODEs is generated by the *Riemann equation*—the Fuchsian equation with three singularities. It includes five types of equations which are considered mainly in their standard forms:

(i) The generating *hypergeometric equation* (hE).

(ii) Two types of confluent equations which we call the *confluent hypergeometric equation* (ChE) and *biconfluent hypergeometric equation* (BhE).

(iii) Two types of reduced confluent equations which we call *reduced confluent hypergeometric equation* (RChE) and *reduced biconfluent hypergeometric equation* (RBhE).

Among these equations, only the hE and ChE are 'independent'. All other equations may be obtained from the ChE by transformations of the independent variable and specialization of parameters.

Different forms of the ChE are known in the literature as e.g. the *Whittaker equation* or the *Slater equation*. Different forms of the BhE are known as the *parabolic cylinder equation*, the *Hermite equation*, and the *harmonic oscillator equation*. The RChE has no special name, but can be transformed by a quadratic transformation into a *Bessel equation*. The RBhE is known as the *Airy equation*.

Equations belonging to the hypergeometric class are characterized by local parameters only (there are no scaling parameters and no accessory parameters) starting with three for the hE. There are two parameters for the ChE and one parameter for the BhE and RChE.

The first three equations have polynomial solutions which also serve as eigen-solutions for appropriate boundary problems. These solutions generate three infinite sets of classical orthogonal polynomials: *Jacobi polynomials, Laguerre polynomials*, and *Hermite polynomials*. Other known sets of orthogonal polynomials—*Legendre polynomials, Gegenbauer polynomials, Chebyshev polynomials*—may be regarded as specializations of Jacobi polynomials.

Expansions of Frobenius type as particular solutions of the hE, ChE, and RChE generate two-term recurrence relations for the coefficients. They lead to explicit forms of the series and thus to a way of practical computation. In the case of the BhE and the RBhE, expansions about the ordinary point zero have the same significance.

Solutions of the above-mentioned equations except the RBhE satisfy, as functions of their parameters, three-term (second-order) difference equations. For those values of the parameters that correspond to polynomial solutions, these equations may be truncated from one side. They give recurrence relations generating the classical orthogonal polynomials.

By integral transformations, the kernels of which are elementary functions, the solutions of the equations belonging to the hypergeometric class may be represented in terms of elementary functions. This is a powerful tool for practical computation and also yields explicit formulae (up to the gamma function) for the lateral as well as central two-point connection problems (see below).

Here, we give the equations of the hypergeometric class in the main standard forms, summarizing more detailed information in several tables.

2.1.2 Hypergeometric equation

By means of the s-homotopic transformation of the dependent variable and the Möbius transformation of the independent variable, the *Riemann equation* may be transformed to its *canonical natural form*—the *hypergeometric equation* (also called *Gauss equation*)

$$L_z^{\{1,1;1\}}(a, b; c)y(z) = [z(1 - z)D^2$$
$$+ \{c - (a + b + 1)z\} D - ab]y(z) = 0, \qquad (2.1.1)$$

$$D = \mathrm{d}/\mathrm{d}z.$$

This equation is represented by the generalized Riemann scheme

$$\begin{pmatrix} 1 & 1 & 1 & \\ 0 & 1 & \infty & ; z \\ 0 & 0 & a & \\ 1 - c & c - a - b & b & \end{pmatrix}. \qquad (2.1.2)$$

The differential operator adjoint to (2.1.1) reads

$$\left(L_z^{\{1,1;1\}}(a, b; c)\right)^* := L_z^{\{1,1;1\}}(1 - a, 1 - b; 2 - c). \qquad (2.1.3)$$

With the help of the specialized s-homotopic transformation

$$S_{c \to n} : y = Gw, \ G^{-1}LG = N \quad \Leftrightarrow \quad y = z^{-c/2}(1 - z)^{-(c-a-b-1)/2}w,$$

eqn (2.1.1) may be transformed to the normal form

$$N_z^{\{1,1;1\}}(a, b; c)w(z) := \left(D^2 + \frac{1}{4}\left(\frac{1 - (1 - c)^2}{z^2} + \frac{1 - (c - a - b)^2}{(z - 1)^2}\right)\right.$$
$$\left. + \frac{2ab + c(c - a - b - 1)}{2z(z - 1)}\right) w(z) = 0, \qquad (2.1.4)$$

and with the transformation

$$S_{n \to s-a} : w = Hv, \ H^{-1}NH = M \quad \Leftrightarrow \quad w = z^{1/2}(1 - z)^{1/2}v,$$

it transforms to its self-adjoint form

$$M_z^{\{1,1;1\}}(a, b; c)v(z) := \left(Dz(z-1)D - \frac{z(z-1)}{4}\left(\frac{(1-c)^2}{z^2} + \frac{(c-a-b)^2}{(z-1)^2}\right) \right.$$
$$\left. + \frac{2ab + c(c-a-b)}{2z(z-1)} \right) v(z) = 0. \tag{2.1.5}$$

A complete list of standard forms of the hypergeometric equation is given in Table 2.1.1. Beyond these forms, a modified form of the hypergeometric equation can be studied which is more balanced in relation to the finite singularities. Suppose that the independent variable z is replaced by $(z+1)/2$, transforming the singularities $z = 0$ and $z = 1$ into $z = -1$ and $z = 1$. Additionally, instead of b, the parameter d is taken in eqn (2.1.1) satisfying the equality

$$a + b + 1 = c + d.$$

The modified canonical equation reads

$$[(1 - z^2)D^2 - \{c(z-1) + d(z+1)\}D - a(c+d-a-1)]y(z) = 0. \tag{2.1.6}$$

The modified normal and modified self-adjoint forms also may easily be written.

The standard solution of eqn (2.1.1), valid for $c \neq 0, -1, -2, \ldots$, is the so-called *hypergeometric function* $y_1^{\{1,1;1\}}(a, b; c; z = 0, z)$ with $y_1(0) = 1$, which in conventional notation is expressed as

$$y_1^{\{1,1;1\}}(a, b; c; z = 0, z) = F(a, b; c; z) = {}_2F_1(a, b; c; z). \tag{2.1.7}$$

Table 2.1.1. Standard forms of the hypergeometric equation

c	$L_z^{\{1,1;1\}}y(z) = 0$	$L_z^{\{1,1;1\}}(a, b; c, d) = z(1-z)D^2 +$ $(c(1-z) - dz)D - ab$	$a+b+1 =$ $c+d$
c	$\tilde{L}_z^{\{1,1;1\}}y(z) = 0$	$\tilde{L}_z^{\{1,1;1\}}(a, b; c, d) = D^2 + \left(\dfrac{c}{z} + \dfrac{d}{z-1}\right) +$ $\dfrac{ab}{z(z-1)}$	$\tilde{L} = \dfrac{L}{z(z-1)}$
n	$N_z^{\{1,1;1\}}w(z) = 0$	$N_z^{\{1,1;1\}}(a, b; c, d) = D^2 + \dfrac{1}{4}\left(\dfrac{1-(1-c)^2}{z^2} + \right.$ $\left.\dfrac{1-(1-d)^2}{(z-1)^2}\right) + \dfrac{ab - cd/2}{z(z-1)}$	$y = z^{-c/2}.$ $(1-z)^{-d/2}w$
s-a	$M_z^{\{1,1;1\}}v(z) = 0$	$M_z^{\{1,1;1\}}(a, b; c, d) = Dz(z-1)D -$ $\dfrac{z(z-1)}{4}\left(\dfrac{(1-c)^2}{z^2} + \dfrac{(1-d)^2}{(z-1)^2}\right) +$ $ab - (cd-1)/2$	$w = \sqrt{z(1-z)}v$

It is defined by the following power series constructed in the vicinity of zero (one of the Frobenius solutions):

$$y_1^{\{1,1;1\}}(a, b; c; z = 0, z) = \sum_0^\infty g_k^{\{1,1;1\}} z^k, \tag{2.1.8}$$

which leads to the following recurrence relation for the coefficients $g_k^{\{1,1;\}}$:

$$(k+1)(k+c)g_{k+1}^{\{1,1;1\}} = (k+a)(k+b)g_k^{\{1,1;1\}}, \quad g_0^{\{1,1;1\}} = 1$$

$$\implies g_k^{\{1,1;1\}} = \frac{\Gamma(k+a)\Gamma(k+b)\Gamma(c)}{\Gamma(k+1)\Gamma(k+c)\Gamma(a)\Gamma(b)}. \tag{2.1.9}$$

The series on the right-hand side of (2.1.8) converges within the unit circle. Outside this circle, the function $y_1^{\{1,1;1\}}(a, b; c; z = 0, z)$ may be defined by analytic continuation. In the complex z plane, it is holomorphic on any closed domain that does not include the points of the cut $[1, \infty[$ between the singularities of eqn (2.1.1).

The s-homotopic transformation $S : y = z^{1-c}\breve{y}$ transforms eqn (2.1.1) to another canonical natural form of the hypergeometric equation

$$\left[z(1-z)D^2 + \{2 - c - (a+b+3-2c)z\}D \right.$$
$$\left. - (a+1-c)(b+1-c)\right]\breve{y}(z) = 0. \tag{2.1.10}$$

For eqn (2.1.10), the solution (2.1.8) may also be constructed. Hence the second Frobenius solution of eqn (2.1.1) takes the form

$$y_2^{\{1,1;1\}}(a, b; c; z = 0, z) = z^{1-c} F(a+1-c, b+1-c; 2-c) := G(a, b; c). \tag{2.1.11}$$

The Wronskian of these two solutions may be obtained from (2.1.8)–(2.1.9) as

$$W(F, G) = (1-c)z^{-c}(1-z)^{c-a-b-1}. \tag{2.1.12}$$

Some troubles arise at integer values of the parameter c in the formulae (2.1.9) and (2.1.11). They can be avoided by the more convenient normalization $F \mapsto F/\Gamma(c)$, $G \mapsto /\Gamma(2-c)$. However, the case $c = 0$ changes the equation to a simpler one having two regular singularities.

Further major properties of the solutions of (2.1.1) are: integral representations, integral relations, difference equations, connection matrices, and polynomial solutions. These are studied in the following sections of this chapter.

2.1.3 Confluent equations

The *singly confluent hypergeometric equation*

$$L_z^{\{1;2\}}(a; c; 1)y(z) := (zD^2 + (c-z)D - a)y(z) = 0 \tag{2.1.13}$$

is generated by means of the confluence process $z \mapsto z/b$, with $b \to \infty$, of two regular singularities $z = 1$ and $z = \infty$, resulting in an irregular singularity at infinity, the s-rank of which is $R(\infty) = 2$. Equation (2.1.13) is represented by the GRS

$$
\begin{pmatrix}
1 & 2 & \\
0 & \infty & ; z \\
0 & a & \\
1 - c & a - c & \\
& 0 & \\
& 1 &
\end{pmatrix}.
\tag{2.1.14}
$$

It justifies the inclusion of unity in the list of parameters in (2.1.13), since this unity corresponds to one of the first-order characteristic exponents for one of the solutions. The differential operator L^* adjoint to (2.1.13) reads

$$
\left(L_z^{\{1;2\}}(c; a, 1)\right)^* := L_z^{\{1;2\}}(2 - c; 1 - a, -1).
\tag{2.1.15}
$$

Under the s-homotopic transformation

$$
S_{c \to n} : y = Gw, \quad G^{-1}LG = N \quad \Leftrightarrow \quad y = z^{-c/2} e^{z/2} w,
$$

eqn (2.1.13) transforms to its normal form

$$
N_z^{\{1;2\}}(a; c) w(y) = \left(D^2 + \left(-\frac{1}{4} - \frac{a - c/2}{z} + \frac{1 - (1 - c)^2}{4z^2} \right) \right) w(z) = 0,
\tag{2.1.16}
$$

which is also known as the Whittaker equation. A complete list of standard forms of the confluent hypergeometric equations is given in Table 2.1.2.

There are two standard solutions of the ChE. One solution (Frobenius solution)

$$
y_1^{\{1;2\}}(a; c; z = 0, z)
$$

is defined by its behaviour in a vicinity of zero

$$
y_1^{\{1;2\}}(a; c; z = 0, z) = \sum_0^\infty g_k^{\{1;2\}} z^k.
\tag{2.1.17}
$$

It is related to the so-called *confluent hypergeometric function*

$$
y_1^{\{1;2\}}(a; c; z = 0, z) = \Phi(a; c; z) = {}_1F_1(a; c; z).
\tag{2.1.18}
$$

The recurrence relation of the coefficients g_k may be obtained either by the confluence principle

$$
(b^k g_k^{\{1,1;1\}})_{b \to \infty} \to g_k^{\{1;2\}}
$$

Table 2.1.2. Standard forms of the confluent hypergeometric equation

	hE $\mapsto$ ChE		
	$z := \epsilon z,$	$\epsilon L^{\{1,1;1\}} \to L^{\{1;2\}},$	
	$b := \epsilon^{-1},$	$\epsilon^2 \tilde{L}^{\{1,1;1\}} \to \tilde{L}^{\{1;2\}},$	
	$\epsilon \to 0$	$\epsilon^2 N^{\{1,1;1\}} \to N^{\{1;2\}}$	
		$-\epsilon M^{\{1,1;1\}} \to M^{\{1;2\}}$	
c	$L_z^{\{1;2\}} y(z) = 0$	$L_z^{\{1;2\}}(a; c) = zD^2 + (c - z)D - a$	
c	$\tilde{L}_z^{\{1;2\}} y(z) = 0$	$\tilde{L}_z^{\{1;2\}}(a; c) = D^2 + \left(-1 + \dfrac{c}{z}\right)D - \dfrac{a}{z}$	$\tilde{L} = L/z$
n	$N_z^{\{1;2\}} w(z) = 0$	$N_z^{\{1;2\}}(a; c) = D^2 +$ $\left(-\dfrac{1}{4} - \dfrac{a - c/2}{z} + \dfrac{1 - (1-c)^2}{4z^2}\right)$	$y = z^{-c/2} \cdot$ $e^{z/2} w$
s-a	$M_z^{\{1;2\}} v(z) = 0$	$M_z^{\{1;2\}}(a; c) = DzD +$ $\left(-\dfrac{z}{4} - a + c/2 - \dfrac{(1-c)^2}{4z}\right)$	$w = z^{1/2} v$

or by direct substitution of (2.1.17) into eqn (2.1.13):

$$k(k + c)g_{k+1}^{\{1;2\}} = (k + a)g_k^{\{1;2\}}, \quad g_0^{\{1;2\}} = 1$$

$$\implies g_k^{\{1;2\}} = \frac{\Gamma(k + a)\Gamma(c)}{\Gamma(k)\Gamma(k + c)\Gamma(a)}. \tag{2.1.19}$$

The other solution $y_1^{\{1;2\}}(a; c; z = +\infty, z)$ (related to the Thomé solution at the Stokes line $\arg z = 0$) is defined by its behaviour at infinity. In the sector $|\arg z| \leq \pi$ it is expanded in the asymptotic series

$$y_1^{\{1;2\}}(a; c; z = +\infty, z) = z^{-a} \sum_0^\infty h_k^{\{1;2\}} z^{-k}.$$

The coefficients of this series $h_k^{\{1;2\}}$ satisfy the recurrence relation

$$(k + 1)h_{k+1}^{\{1;2\}} = -(k + a + 1 - c)(k + a)h_k^{\{1;2\}}, \quad h_0^{\{1;2\}} = 1 \tag{2.1.20}$$

$$\implies h_k^{\{1;2\}} = \frac{(-1)^k \Gamma(k + a + 1 - c)\Gamma(k + a)}{\Gamma(k + 1)\Gamma(a + 1 - c)\Gamma(a)}.$$

The recurrence relation (2.1.20) is similar to (2.1.19) but is solved in 'the other direction'. It leads—for the general case of coefficients—to the divergence of the series. Beyond this solution, the second Thomé solution

$$y_2^{\{1;2\}}(a;c;z=\infty,z) = e^z z^{a-c} \sum_0^\infty f_k^{\{1;2\}} z^{-k} \quad \text{for } 0 < \arg z < 2\pi \qquad (2.1.21)$$

can be constructed. This solution is dominant at the Stokes ray $\arg z = 0$ and is recessive at the Stokes rays $\arg z = \pi$ and $\arg z = -\pi$, respectively. The coefficients $f_k^{\{1;2\}}$ satisfy the recurrence relation

$$(k+1)f_{k+1}^{\{1;2\}} = (k-a+c)(k-a+1)f_k^{\{1;2\}}. \qquad (2.1.22)$$

Both series (2.1.20) and (2.1.21) are summable by Borel summation [124, 8].

Next comes the biconfluent hypergeometric equation. It is generated by eqn (2.1.13) under a confluence process of the regular singularity at zero and of the irregular singularity at infinity. The resulting irregular singularity has the s-rank $R(\infty) = 3$

$$L_z^{\{:3\}}(a)y(z) := (D^2 - zD - a)y(z) = 0. \qquad (2.1.23)$$

Equation (2.1.23) is known as the Hermite equation. The corresponding GRS reads

$$\begin{pmatrix} 3 \\ \infty \\ a \\ 1-a \\ 0 \\ 0 \\ 0 \\ 1 \end{pmatrix} ; z \qquad (2.1.24)$$

Under the s-homotopic transformation

$$S_{c\to n} : y = Gw, \quad G^{-1}LG = N \quad \Leftrightarrow \quad y = e^{z^2/4}w,$$

Table 2.1.3. Standard forms of the biconfluent hypergeometric equation

	ChE $\mapsto$ BhE	
	$z := \epsilon^{-1} + \epsilon^{-1/2}z,$	$L^{\{1;2\}} \to L^{\{:3\}},$
	$c := \epsilon^{-1},\ \epsilon \to 0$	$\epsilon M^{\{1;2\}} \to M^{\{:3\}},$
c	$L_z^{\{:3\}}y(z) = 0$	$L_z^{\{:3\}}(a) = D^2 - zD - a$
n, s-a	$N_z^{\{:3\}}w(z) = 0$	$N_z^{\{:3\}}(a) = D^2 + \left(\dfrac{1}{2} - a - \dfrac{z^2}{4}\right) \qquad y = e^{z^2/4}w$

eqn (2.1.23) transforms to its normal form

$$N_z^{\{;3\}}(a)w(z) = \left(D^2 + \left(\frac{1}{2} - a - \frac{z^2}{4}\right)\right)w(z) = 0, \qquad (2.1.25)$$

which is known (under the substitution $a \to -a$) as parabolic cylinder equation and, after the scaling $z \mapsto \sqrt{2}z$, as the harmonic oscillator equation.

The standard solution of eqn (2.1.23)—the function $y_1^{\{;3\}}(a; z = +\infty, z)$ (Thomé solution)—is defined by its behaviour at infinity. Within the sector $|\arg z| \le \pi/2$, it is determined by the asymptotic series

$$y_1^{\{;3\}}(a; z = +\infty, z) = z^{-a} \sum_0^\infty h_k^{\{;3\}} z^{-k}. \qquad (2.1.26)$$

The coefficients of this series h_k satisfy the recurrence relation

$$(k + 2)h_{k+2}^{\{;3\}} + (k + a)(k + a + 1)h_k^{\{;3\}} = 0,$$
$$h_0^{\{;3\}} = 1, \quad h_1^{\{;3\}} = 0$$
$$\implies \quad h_{2k}^{\{;3\}} = \frac{(-1)^k 2^k \Gamma(k + a/2)\Gamma(k + a/2 + 1/2)}{\Gamma(k + 1)\Gamma(a/2)\Gamma(a/2 + 1/2)}. \qquad (2.1.27)$$

The conventional subdominant solution of the parabolic cylinder equation

$$\left(D^2 + \left(\frac{1}{2} + v - \frac{z^2}{4}\right)\right)w(z) = 0 \qquad (2.1.28)$$

—the *parabolic cylinder function* $D_v(z)$—is expressed in terms of $y_1^{\{;3\}}(a; z = +\infty, z)$ as

$$D_v(z) = \exp(-z^2/4)y_1^{\{;3\}}(-v; z = +\infty, z). \qquad (2.1.29)$$

The dominant solution of the biconfluent equation is the function $y_2^{\{;3\}}(a; z = +\infty, z)$, which is expressed in terms of the second linear independent particular solution of the parabolic cylinder equation

$$V_v(z) := \frac{\Gamma(-v)}{\sqrt{2\pi}}(D_v(-z) - \cos\pi v \cdot D_v(z)), \qquad (2.1.30)$$

$$V_v(z) = \exp(-z^2/4)y_2^{\{;3\}}(-v; z = +\infty, z). \qquad (2.1.31)$$

Under the transformation of the independent variable $z^2 \mapsto 2z$, eqn (2.1.23) transforms to a special form of the ChE, namely to

$$\left(zD^2 + \left(\frac{1}{2} - z\right)D - \frac{a}{2}\right)y(z) = 0. \qquad (2.1.32)$$

Table 2.1.4. Correspondence between notations for hypergeometric and confluent hypergeometric functions

hE	$y_1^{\{1,1;1\}}(a, b; c; z = 0; z)$	$(y_1)_{z=0} = 1$	$F(a, b; c; z)$
	$y_2^{\{1,1;1\}}(a, b; c; z = 0; z)$	$\lim_{z \to 0} y_2(z)z^{c-1} = 1$	$G(a, b; c; z)$
ChE	$y_1^{\{1;2\}}(a; c; z = 0, z)$	$(y_1)_{z=0} = 1$	$\Phi(a; c; z)$
	$y_2^{\{1;2\}}(a; c; z = +\infty, z)$	$\lim_{z \to +\infty} y_2(z)z^a = 1$	$\Psi(a; c; z)$
BhE	$y_1^{\{;3\}}(a; z = +\infty, z)$	$\lim_{z \to +\infty} y_1(z)z^a = 1$	$e^{z^2/4}D_\nu(z)$
	$y_2^{\{;3\}}(a; z = +\infty, z)$	$\lim_{z \to +\infty} y_2(z)e^{-z^2/2} \cdot$ $z^{1-a} = 1$	$e^{z^2/4}V_\nu(z)$

This implies that there are relations between solutions of eqns (2.1.23) and (2.1.32). One of the relations may be found by comparing the series (2.1.39) and (2.1.26). Thus

$$y_1^{\{;3\}}(a; z = +\infty, z) = y_1^{\{1;2\}}(a/2; 1/2; z = +\infty, z^2/2). \qquad (2.1.33)$$

This fact simplifies the study of the solutions of the BhE.

2.1.4 Reduced confluent equations

First comes the reduced confluent hypergeometric equation

$$L_z^{\{1;3/2\}}(c)y(z) := (zD^2 + cD - 1)y(z) = 0. \qquad (2.1.34)$$

As a result of a weak confluence process from the hE,

$$z \mapsto z\epsilon^2, b \mapsto \epsilon^{-1}, a \mapsto \epsilon^{-1}, \epsilon \to 0 \implies \epsilon^2 L_z^{\{1,1;1\}}(a, b, c) \to L_z^{\{1;3/2\}}(c). \qquad (2.1.35)$$

The corresponding GRS reads

$$\begin{pmatrix} 1 & 3/2 & \\ 0 & \infty & ; z \\ 0 & c/2 - 1/4 & \\ 1 - c & c/2 - 1/4 & \\ & 1 & \end{pmatrix}. \qquad (2.1.36)$$

One of its Frobenius solutions $y_1^{\{1;3/2\}}(c; z = 0, z)$ is defined by the Taylor series at zero:

$$y_1^{\{1;3/2\}}(c; z = 0, z) = \sum_0^\infty g_k^{\{1;3/2\}} z^k. \qquad (2.1.37)$$

The recurrence relation for the coefficients $g_k^{\{1;3/2\}}$ are obtained by means of the confluence principle

$$(b^k a^k g_k^{\{1,1;1\}})_{b,a\to\infty} \to g_k^{\{1;3/2\}},$$

$$k(k+c)g_{k+1}^{\{1;3/2\}} = g_k^{\{1;3/2\}}, \qquad g_0^{\{1;3/2\}} = 1$$

$$\implies \quad g_k^{\{1;3/2\}} = \frac{\Gamma(c)}{\Gamma(k)\Gamma(k+c)}. \tag{2.1.38}$$

The second linearly independent particular solution $y_1^{\{1;3/2\}}(c; z = +\infty, z)$ (Thomé subnormal solution) is defined by its behaviour at infinity. In the sector $|\arg z| \le 2\pi$ it is expanded in the asymptotic series

$$y_2^{\{1;3/2\}}(c; z = +\infty, z) = z^{-c/2+1/4} \exp(-2z^{1/2}) \sum_0^\infty h_k^{\{1;3/2\}} z^{-k/2}. \tag{2.1.39}$$

The coefficients of this series $h_k^{\{1;3/2\}}$ satisfy the recurrence relation

$$4(k+1)h_{k+1}^{\{1;3/2\}} + (k+3/2-c)(k+c-1/2)h_k^{\{1;3/2\}} = 0,$$

$$h_0^{\{1;3/2\}} = 1 \quad \implies \quad h_k^{\{1;3/2\}} = \frac{(-1)^k}{2^{2k}} \frac{\Gamma(k+3/2-c)\Gamma(k+c-1/2)}{\Gamma(k+1)\Gamma(3/2-c)\Gamma(c-1/2)}. \tag{2.1.40}$$

Equation (2.1.34) is not 'independent', in the sense that, by means of the quadratic transformation $z \mapsto z^2$, the s-rank at infinity changes from $R(\infty) = 3/2$ to $R(\infty) = 2$, and therefore eqn (2.1.34) changes from the RChE to a form of the ChE. More precise computations show that

$$y_1^{\{1;3/2\}}(c; z = 0, z) = e^{-2\sqrt{z}} y_1^{\{1;2\}}(c - 1/2; 2c - 1; z = 0, 4\sqrt{z}) \tag{2.1.41}$$

and

$$y_1^{\{1;3/2\}}(c; z = +\infty, z) = e^{-2\sqrt{z}} y_1^{\{1;2\}}(c - 1/2; 2c - 1; z = +\infty, 4\sqrt{z}). \tag{2.1.42}$$

In applications, eqn (2.1.34) rarely appears. More often the solution of the Bessel equation

$$\{zD^2 + D + (z - v^2/z)\}v(z) = 0 \tag{2.1.43}$$

and solutions of the modified Bessel equation

$$\{zD^2 + D - (z + v^2/z)\}v(z) = 0 \tag{2.1.44}$$

Table 2.1.5. Standard forms of the reduced confluent hypergeometric equation

hE $\mapsto$ RChE			
$z := \epsilon^2 z,$	$\epsilon^2 L^{\{1,1;1\}} \to L^{\{;3/2\}},$		
$b = a := \epsilon^{-1},$	$\epsilon^4 \tilde{L}^{\{1,1;1\}} \to \tilde{L}^{\{1;3/2\}},$		
$\epsilon \to 0$	$\epsilon^4 N^{\{1,1;1\}} \to N^{\{;3/2\}}$		
	$-\epsilon^2 M^{\{1,1;1\}} \to M^{\{;3/2\}},$		
c	$L_z^{\{1;3/2\}} y(z) = 0$	$L_z^{\{1;3/2\}}(c) = zD^2 + cD - 1$	
c	$\tilde{L}_z^{\{1;3/2\}} y(z) = 0$	$\tilde{L}_z^{\{1;3/2\}}(c) = D^2 + \dfrac{c}{z}D - 1$	$\tilde{L} = L/z$
n	$N_z^{\{1;3/2\}} w(z) = 0$	$N_z^{\{1;3/2\}}(c) = D^2 - \dfrac{1}{z} + \dfrac{1 - (1-c)^2}{4z^2}$	$y = z^{-c/2} w$
s-a	$M_z^{\{1;3/2\}} v(z) = 0$	$M_z^{\{1;3/2\}}(c) = DzD - \dfrac{(1-c)^2}{4z} - 1$	$w = z^{1/2} v$

are studied. Both equations are written in the self-adjoint form. Equation (2.1.43) is obtained from eqn (2.1.44) at substitution $z \mapsto iz$. The standard Frobenius solution of eqn (2.1.44) is the Bessel function

$$J_\nu(z) = \left(\frac{z}{2}\right)^\nu \sum_0^\infty \frac{(-1)^k z^{2k}}{2^{2k}\Gamma(k+1)\Gamma(k+\nu+1)}. \qquad (2.1.45)$$

Other solutions defined by their behaviour at infinity are the two Hankel functions

$$H_\nu^1(z) = \sqrt{\frac{2}{\pi z}} \exp\left[i\left(z - \frac{\nu\pi}{2} - \frac{\pi}{4}\right)\right]\left[1 + \sum_0^\infty (i)^k h_k z^{-k}\right]$$

$$\text{with } h_k = \frac{\Gamma(k+\nu+1/2)\Gamma(k-\nu+1/2)}{2^k\Gamma(k)}$$

$$\text{for } -\pi/2 < \arg(z) < 3\pi/2, \qquad (2.1.46)$$

$$H_\nu^2(z) = \sqrt{\frac{2}{\pi z}} \exp\left[-i\left(z - \frac{\nu\pi}{2} - \frac{\pi}{4}\right)\right]$$

$$\cdot\left[1 + \sum_0^\infty \frac{(-i)^k\Gamma(k+\nu+1/2)\Gamma(k-\nu+1/2)}{2^k\Gamma(k)} z^{-k}\right]$$

$$\text{for } -3\pi < \arg(z) < \pi/2. \qquad (2.1.47)$$

The solutions of the modified Bessel equation are the *modified Bessel function*

$$I_\nu(z) = \left(\frac{z}{2}\right)^\nu \sum_0^\infty \frac{z^{2k}}{2^{2k}\Gamma(k+1)\Gamma(k+\nu+1)} \qquad (2.1.48)$$

and the *Macdonald function*

$$K_\nu(z) = \sqrt{\frac{2}{\pi z}} e^{-z}\left[1 + \sum_0^\infty h_k z^{-k}\right]$$

$$\text{for } -\pi < \arg(z) < \pi. \qquad (2.1.49)$$

On the other hand, under the transformation $z \mapsto 2\sqrt{z}$, eqn (2.1.44) transforms to

$$\left(zD^2 + D - \left(z + \frac{\nu^2}{z}\right)\right)v(z) = 0. \qquad (2.1.50)$$

According to Table 2.4, eqn (2.1.50) is the self-adjoint form of the RChE with $\nu = 1 - c$ or $-\nu = 1-c$. Hence the following relation between the functions (2.1.37) and (2.1.48) is valid:

$$I_\nu(z) = \frac{1}{\Gamma(\nu+1)}\left(\frac{z}{2}\right)^\nu y_1^{\{1;3/2\}}(1+\nu; z = 0, z^2/4). \qquad (2.1.51)$$

A similar relation holds between the functions (2.1.39) and (2.1.49):

$$K_\nu(z) = \frac{1}{\sqrt{\pi}}\left(\frac{z}{2}\right)^\nu y_1^{\{1;3/2\}}(1+\nu; z = +\infty, z^2/4). \qquad (2.1.52)$$

The final equation belonging to the hypergeometric class is the reduced biconfluent hypergeometric equation (RBhE) known in the literature as the Airy equation:

$$L_z^{(;5/2)}y(z) = (D^2 - z)y(z) = 0. \qquad (2.1.53)$$

It is represented by the GRS

$$\begin{pmatrix} 5/2 \\ \infty \\ 1/4 \\ 1/4 \\ 0 \\ 0 \\ 1 \end{pmatrix} ; z \qquad (2.1.54)$$

The standard solution of eqn (2.1.53)—the function $y_1^{\{;5/2\}}(z = +\infty, z)$ (Thomé solution)—is defined by its behaviour at infinity. In the sector $|\arg z| \le 2\pi/3$, it is

Table 2.1.6. Standard forms of the reduced biconfluent hypergeometric equation

RChE $\mapsto$ RBhE	
$z := \epsilon^{-2}z - \epsilon^{-3},$ $c := 2\epsilon^{-3}, \ \epsilon \to 0$	$\epsilon^4 M^{\{1;3/2\}} \to M^{\{5/2\}}$
c, n, s-a $\quad N_z^{\{;5/2\}}w(z) = 0$	$N_z^{\{;5/2\}} = D^2 - zD$

expanded in the asymptotic series

$$y_1^{\{;5/2\}}(a; z = +\infty, z) = z^{-1/4}\exp\left(-\frac{2}{3}z^{3/2}\right)\sum_0^\infty h_k^{\{;5/2\}}z^{-k}. \qquad (2.1.55)$$

The coefficients of this series $h_k^{\{;5/2\}}$ satisfy the recurrence relation

$$(k+3)h_{k+3}^{\{;5/2\}} + \frac{1}{16}(2k+1)(2k+5)h_k^{\{;5/2\}} = 0,$$

$$h_0^{\{;5/2\}} = 1, \ h_1^{\{;5/2\}} = 0, \ h_2^{\{;5/2\}} = 0$$

$$\implies \quad h_k^{\{;5/2\}} = \frac{\Gamma(k+a)}{\Gamma(k+1)\Gamma(a)}. \qquad (2.1.56)$$

The foregoing list of equations belonging to the hypergeometric class comprises equations with elementary singular points. Setting $c = 1/2$ in eqn (2.1.1), we reduce the point $z = 0$ to an elementary singularity. Under the substitution $z \mapsto z^2$, eqn (2.1.1) in this case transforms to the canonical form of the associate Legendre equation:

$$\{(1-z)^2 D^2 - 2(a+b+1)zD - 4ab\}y(z) = 0. \qquad (2.1.57)$$

Under the condition $a+b = 0$, the second singular point $z = 1$ becomes elementary, resulting in the equation

$$\{\sqrt{z(1-z)}D\sqrt{z(1-z)}D + a^2\}y(z) = 0, \qquad (2.1.58)$$

the solutions of which may be given in terms of elementary functions. For $a = 1/4$, the singular point at infinity becomes elementary. For $c = 1/2$, the ChE reduces to the equation which, under the substitution $z \mapsto z^2$, transforms to the canonical form of the biconfluent hypergeometric equation. At the same value of the parameter a, and under the same substitution, the RChE transforms to the equation for the modified Bessel functions with the index $1/2$.

Table 2.1.7. Correspondence between notations for reduced confluent hypergeometric functions

RChE	$y_1^{\{1;3/2\}}(c; z = 0, z)$	$(y_1)_{z=0} = 1$	$I_\nu(z) = \dfrac{1}{\Gamma(\nu + 1)}\left(\dfrac{z}{2}\right)^\nu \cdot$ $y_1(1 + \nu; z^2/4)$
	$y_2^{\{1;3/2\}}(c; z = \infty, z)$	$\lim\limits_{z \to +\infty} y_2(z)\sqrt{z} = 1$	$K_\nu(z) = \dfrac{1}{\sqrt{\pi}}\left(\dfrac{z}{2}\right)^\nu \cdot$ $y_2(1 + \nu; z^2/4)$
RBhE	$y_1^{\{1;5/2\}}(z = \infty, z)$	$\lim\limits_{z \to +\infty} y_1(z)z^{1/4} \cdot$ $\exp\left(\dfrac{2}{3}z^{3/2}\right) = 1$	$\mathrm{Ai}(z) = \dfrac{1}{2\sqrt{\pi}}y_1(z)$
	$y_2^{\{1;5/2\}}(z = \infty, z)$	$\lim\limits_{z \to +\infty} y_2(z)z^{1/4} \cdot$ $\exp\left(-\dfrac{2}{3}z^{3/2}\right) = 1$	$\mathrm{Bi}(z) = \dfrac{1}{\sqrt{\pi}}y_2(z)$

2.2 Difference equations

2.2.1 General consideration

Special functions belonging to the hypergeometric class may be linked not only to differential equations but also to several difference equations and differential–difference equations. The difference operator can act on a grid of a discrete variable which is dual to the independent complex variable in the differential equation. These types of difference equations, for instance, arise for coefficients of series expansions for solutions of the studied equations. They have been presented in the previous section. On the other hand, the independent discrete variable may be taken as any of the singular parameters a, b, c in eqn (2.1.1). In the latter case, the corresponding difference equations are usually called *contiguous relations*.

2.2.2 Difference equations for hypergeometric functions

We introduce three formal operators T, Z, and D, where Z is multiplication by z, D is differentiation over z, and T is defined by $T = Z(1 - Z)D$. In terms of these operators, the hypergeometric equation reads

$$(TD + (c - (a + b + 1)Z)D)\, y = a\, b\, y. \qquad (2.2.1)$$

The operator in the left-hand side is denoted by

$$L = TD + (c - (a + b + 1)Z)D.$$

Simple calculations lead to the following commutation relations:

$$LZ = ZL + 2T - (a + b + 1)Z + c, \qquad (2.2.2)$$
$$LT = TL - 2ZL + (a + b - 1)T + L.$$

Beyond the standard solution of the hypergeometric equation—the functions[1] $F(a, b; c; z)$—the functions with shifted parameter a are considered. It is important to stress that the characteristic exponents at the singular points $z = 0$ and $z = 1$ are fixed. Hence the shift in a causes an inverse shift in b, and the considered functions are $F(a + j, b - j; c; z)$ ($j = \pm1, \pm2, \ldots$). All the functions $F(a + j, b - j, c; z)$ are normalized by the condition

$$F(a + j, b - j; c; z)_{z=0} = 1. \tag{2.2.3}$$

For the sake of brevity, the following notation is introduced:

$$F_j := F(a + j, b - j; c; z), \qquad \lambda_j := (a + j)(b - j).$$

Equation (2.2.1), in the case of the set of the presented functions, changes to

$$LF_j = \lambda_j F_j. \tag{2.2.4}$$

Our goal is to prove the following.

Theorem 2.1 *Hypergeometric functions $F(a, b, c; z)$ satisfy the following difference and differential–difference equations:*

$$zF(a, b, c; z) = \frac{a(b - c)}{(a - b)(a - b + 1)} F(a + 1, b - 1, c; z)$$
$$+ \frac{c(a + b - 1) - 2ab}{(a - b + 1)(a - b - 1)} F(a, b, c; z)$$
$$+ \frac{b(a - c)}{(a - b)(a - b - 1)} F(a - 1, b + 1, c; z), \tag{2.2.5}$$

$$z(1 - z)F'(a, b, c; z) = \frac{ab(b - c)}{(a - b)(a - b + 1)} F(a + 1, b - 1, c; z)$$
$$+ \frac{ab(2c - a - b - 1)}{(a - b + 1)(a - b - 1)} F(a, b, c; z)$$
$$+ \frac{ab(a - c)}{(a - b)(a - b - 1)} F(a - 1, b + 1, c; z). \tag{2.2.6}$$

□

Proof: Suppose that the functions TF_0 and ZF_0 are expanded in the the series

$$TF_0 = \sum_{j=-\infty}^{\infty} g_j F_j, \qquad ZF_0 = \sum_{j=-\infty}^{\infty} h_j F_j. \tag{2.2.7}$$

[1] We use here the standard notation for the hypergeometric function.

By action of the operator L on both sides of equalities (2.2.7), we obtain

$$LZF_0 = \lambda_0 Z F_0 + 2T F_0 - (a+b+1)Z F_0 + c F_0 = \sum_{j=-\infty}^{\infty} \lambda_j h_j F_j, \quad (2.2.8)$$

$$LT F_0 = \lambda_0 T F_0 - 2\lambda_0 Z F_0 + (a+b-1)T F_0 + L F_0 = \sum_{j=-\infty}^{\infty} \lambda_j g_j F_j,$$

or

$$\sum_{j=-\infty}^{\infty} (\lambda_0 - \lambda_j - (a+b+1))h_j F_j + 2 \sum_{j=-\infty}^{\infty} g_j F_j + c F_0 = 0, \quad (2.2.9)$$

$$\sum_{j=-\infty}^{\infty} (\lambda_0 - \lambda_j + a + b - 1)g_j F_j - 2\lambda_0 \sum_{j=-\infty}^{\infty} h_j F_j + \lambda_0 F_0 = 0.$$

After equating to zero the coefficients in front of the functions F_j, the following inhomogeneous system for the coefficients g_0 and h_0 is found

$$(a+b-1)g_0 - 2abh_0 + ab = 0, \quad (2.2.10)$$
$$2g_0 - (a+b+1)h_0 + c = 0.$$

Solutions of the system (2.2.10) are

$$h_0 = \frac{c(a+b-1) - 2ab}{(a-b+1)(a-b-1)}, \qquad g_0 = \frac{ab(2c-a-b-1)}{(a-b+1)(a-b-1)}. \quad (2.2.11)$$

The coefficients g_j and h_j according to (2.2.9) satisfy a homogeneous system of equations

$$(\lambda_0 - \lambda_j + a + b - 1)g_j - 2\lambda_0 h_0 = 0, \quad (2.2.12)$$
$$2g_0 + (\lambda_0 - \lambda_j - (a+b+1))h_0 = 0.$$

For nontrivial solutions of these equations, the determinants Δ_j should be equal to zero:

$$(\lambda_j - \lambda_0)^2 + 2(\lambda_j - \lambda_0) - (a-b)^2 + 1 = 0, \quad (2.2.13)$$

which yields

$$(j^2 + j(a-b) + a - b - 1)(j^2 + j(a-b) - a + b - 1) = 0. \quad (2.2.14)$$

For arbitrary values of a and b, the determinant is equal to zero only at $j = 1$ and at $j = -1$. In order to calculate the coefficients g_1, g_{-1}, h_1, h_{-1}, we need to add the conditions

$$g_{-1} + g_0 + g_1 = 0, \qquad h_{-1} + h_0 + h_1 = 0, \quad (2.2.15)$$

resulting in the normalization condition (2.2.3) and in eqn (2.2.7). Taken together, eqns (2.2.12), (2.2.14), and (2.2.15) lead to the following values of g_1, g_{-1}, h_1, h_{-1}:

$$g_1 = \frac{ab(b-c)}{(a-b)(a-b+1)}, \qquad g_{-1} = \frac{ab(a-c)}{(a-b)(a-b-1)},$$

$$h_1 = \frac{a(b-c)}{(a-b)(a-b+1)}, \qquad h_{-1} = \frac{b(a-c)}{(a-b)(a-b-1)}. \tag{2.2.16}$$

The coefficients g_j and h_j obey the symmetry conditions

$$g_1(a,b) = g_{-1}(b,a), \qquad h_1(a,b) = h_{-1}(b,a).$$

While using infinite series, our considerations were formal. But with finite number of terms and known coefficients it is possible to act with the operator L on $T F_0$ and on $Z F_0$ and to verify that eqns (2.2.5) and (2.2.6) are satisfied. $\square$

If $a = -n$, $c = \alpha + 1$ and $b = \alpha + \beta + n + 1$, then the recurrence relations (2.2.4) and (2.2.5) convert to recurrence relations generating Jacobi polynomials $P_n^{(\alpha,\beta)}((1-z)/2)$ normalized by the above-mentioned condition.

2.2.3   Confluent hypergeometric functions

Applying a confluence process makes it easy to obtain from (2.2.5–2.2.6) the recurrence relations for the confluent hypergeometric function $\Phi(a, c; z)$ normalized as $\Phi(a; c; z)|_{z=1} = 1$:

$$z\Phi(a, c; z) = a\Phi(a+1, c; z) + (c - 2a)\Phi(a, c; z)$$

$$+(a - c)\Phi(a - 1, c; z), \tag{2.2.17}$$

$$z\Phi'(a, c; z) = a(\Phi(a+1, c; z) - z\Phi(a, c; z)). \tag{2.2.18}$$

In the case of the RChE, difference equations directly modified for Bessel functions are discussed. Consider two operators: S divides by z and D differentiates. If the operator $L = z^2 D^2 + zD - (z^2 + v^2)$ from eqn (2.1.44), is taken then the following commutation relations hold:

$$LD = DL - 2SL + D - 2v^2 S, \qquad LS = SL + S - 2D.$$

Suppose that

$$DI_v(z) = \sum_{k=-\infty}^{\infty} a_k I_{v+k}(z), \qquad SI_v(z) = \sum_{k=-\infty}^{\infty} b_k I_{v+k}(z). \tag{2.2.19}$$

Acting on terms on both sides of eqn (2.2.19) by the operator L, we get the equalities

$$\sum_{k=-\infty}^{\infty} ((k(2v + k) - 1)a_k + 2v^2 b_k)I_{v+k}(z) = 0,$$

$$\sum_{k=-\infty}^{\infty} (k((2v + k) - 1)b_k + 2a_k)I_{v+k}(z) = 0.$$

Clearly, functions $I_\nu(z)$ are linearly independent, which leads to the fact that, for any k, the following linear homogeneous system for a_k and b_k holds:

$$k(2\nu + k - 1)a_k + 2\nu^2 b_k = 0,$$
$$k(2\nu + k - 1)b_k + 2a_k = 0. \tag{2.2.20}$$

Only for $k = \pm 1$ is the determinant of this system equal to zero, enabling a nonzero solution of (2.2.20):

$$a_1 = -\nu b_1, \qquad a_{-1} = \nu b_{-1}. \tag{2.2.21}$$

Explicit values of a_1, b_1, a_{-1}, b_{-1} are obtained by direct comparison of the first terms of series expansions about zero. As a result, we get the equations

$$DI_\nu(z) = \frac{1}{2}\left(I_{\nu-1}(z) + I_{\nu+1}(z)\right), \qquad SI_\nu(z) = \frac{1}{2\nu}\left(I_{\nu-1}(z) - I_{\nu+1}(z)\right). \tag{2.2.22}$$

Similar equations are valid for Bessel functions:

$$DJ_\nu(z) = \frac{1}{2}\left(J_{\nu-1}(z) - J_{\nu+1}(z)\right), \qquad SJ_\nu(z) = \frac{1}{2\nu}\left(J_{\nu-1}(z) + J_{\nu+1}(z)\right). \tag{2.2.23}$$

In the case of the BhE, more well-known functions $D_\nu(z)$ are studied. They are solutions of eqn (2.1.28). Commutation relations can be obtained by denoting as L the differential operator related to this equation and by introducing operators D and Z as above:

$$LD = DL + Z/2, \qquad LZ = ZL + 2D.$$

Suppose that

$$DD_\nu(z) = \sum_{k=-\infty}^{\infty} a_k D_{\nu+k}(z), \qquad ZD_\nu(z) = \sum_{k=-\infty}^{\infty} b_k D_{\nu+k}(z). \tag{2.2.24}$$

Then, acting on both sides of the identities (2.2.24) by operator L, we arrive, as in previous cases, at homogeneous equations for the coefficients $a_{\nu+k}$ and $b_{\nu+k}$:

$$2a_k + kb_k = 0, \qquad \frac{1}{2}b_k + ka_k = 0. \tag{2.2.25}$$

Only for $k = \pm 1$ does the determinant of the corresponding system equal zero, enabling a nonzero solution:

$$b_1 = -2a_1, \qquad b_{-1} = 2a_{-1}. \tag{2.2.26}$$

Actual values of a_1 and a_{-1} are computed by comparison of the first terms of the corresponding asymptotic expansions at infinity. It leads to the relations

$$DD_\nu(z) = -\frac{1}{2}D_{\nu+1}(z) + \frac{\nu}{2}D_{\nu-1}(z), \qquad ZD_\nu(z) = D_{\nu+1}(z) + \nu D_{\nu-1}(z). \tag{2.2.27}$$

2.3 Integral representations and integral relations

2.3.1 Preliminary lemmas

We can derive several formulae involving integration and certain solutions of an equation belonging to the hypergeometric class. Here, we concentrate on the most general ones, which are valid for all equations and all local solutions, and can be traced by the confluence principle.

The following integral relation is considered as a basic one:

$$y(z) = \int_C K\{\phi(z, \xi)\} v(\xi)\, d\xi. \tag{2.3.1}$$

The kernel $K(\phi)$ is supposed to be expressed in terms of elementary functions, and the function $\phi(z, \xi)$ is one of the two simple monomials

$$\phi_1(z, \xi) = z\xi, \tag{2.3.2}$$

$$\phi_2(z, \xi) = z - \xi. \tag{2.3.3}$$

C is an appropriate contour in the complex ξ plane. The function $y(z)$ belongs to the hypergeometric class. If the function $v(\xi)$ may be expressed in terms of elementary functions, then eqn (2.3.1) is called an *integral representation*. If this function is a solution of an equation of the hypergeometric class, then eqn (2.3.1) is called an *integral relation*.

Possible applications of formulae (2.3.1) include:

 (i) derivation of monodromy matrices for the hypergeometric class of functions,

(ii) computation of special functions.

For convenience, we rewrite the differential equations belonging to the hypergeometric class:

$$L_z^{\{1,1;1\}} y(z) = [z(1-z)D^2 + \{c - (a+b+1)z\} D - ab]y(z) = 0, \tag{2.3.4}$$

$$L_z^{\{1;2\}} y(z) = (zD^2 + (c-z)D - a)y(z) = 0, \tag{2.3.5}$$

$$L_z^{\{1;3/2\}} y(z) = (zD^2 + cD - 1)y(z) = 0, \tag{2.3.6}$$

$$L_z^{\{;3\}} y(z) = (D^2 - zD - a)y(z) = 0, \tag{2.3.7}$$

$$L_z^{\{;5/2\}} y(z) = (D^2 - z)y(z) = 0. \tag{2.3.8}$$

Below, integral and differential operators are studied at a formal level. This means that all integrals involved are assumed to converge, and all constant terms arising under integration by parts vanish. This can be achieved by appropriate conditions on the parameters of the equation and an appropriate choice of the integration contour.

Lemma 2.1 *Let L_z and $\hat{L}_\xi$ be second-order linear differential operators with polynomial coefficients. Assume that the functions $y(z)$ and $v(\xi)$ satisfy the equations*

$$L_z y(z) = 0, \qquad (2.3.9)$$

$$\hat{L}_\xi v(\xi) = 0, \qquad (2.3.10)$$

and are related to each other by (2.3.1). Then the kernel $K(z, \xi)$ may be chosen as a solution of the partial differential equation

$$(L_z - f(z)\hat{L}_\xi^*)K(z, \xi) = 0 \qquad (2.3.11)$$

with an appropriate function $f(z)$. □

Proof: The proof follows from the equations

$$L_z y(z) = \int_C v(\xi)(L_z - f(z)\hat{L}_\xi^*)K(z, \xi)\,\mathrm{d}\xi + \int_C f(z)K(z, \xi)\hat{L}_\xi v(\xi)\,\mathrm{d}\xi$$

$$= \int_C v(\xi)(L_z - f(z)\hat{L}_\xi^*)K(z, \xi)\,\mathrm{d}\xi = 0. \quad \square$$

Lemma 2.2 *Suppose that $\tilde{L}_\xi$ is a first-order linear differential operator with polynomial coefficients, and that the function $v(\xi)$ is a solution of the equation*

$$\tilde{L}_\xi v(\xi) = 0. \qquad (2.3.12)$$

If the integral relation (2.3.1) holds, then the kernel $K(z, \xi)$ may be chosen as a solution of the partial differential equation

$$(L_z - f_1(z)\tilde{L}_\xi^* D_\xi - f_2(z)\tilde{L}_\xi^*)K(z, \xi) = 0, \qquad D_\xi = \mathrm{d}/\mathrm{d}\xi \qquad (2.3.13)$$

with appropriate functions $f_1(z)$, $f_2(z)$. □

Proof: The proof follows from the equations

$$L_z y(z) = \int_C v(\xi)(L_z - f_1(z)\tilde{L}_\xi^* D_\xi - f_2(z)\tilde{L}_\xi^*)K(z, \xi)\,\mathrm{d}\xi$$

$$+ \int_C K(z, \xi)(f_1(z)D_\xi \tilde{L}_\xi + f_2(z)N_\xi)v(\xi)\,\mathrm{d}\xi$$

$$= \int_C v(\xi)(L_z + f_1(z)\tilde{L}_\xi^* D_\xi - f_2(z)\tilde{L}_\xi^*)K(z, \xi)\,\mathrm{d}\xi = 0. \quad \square$$

Our next goal is the factorization of the equations (2.3.11) and (2.3.12) into ordinary differential equations in the variables (2.3.2) and (2.3.3). The crucial part of such a factorization is the factorization of the term with the second derivative, since it does not include any free parameters at our disposal.

2.3.2 Integral representations

The first kernel is taken in the form

$$K(z, \xi) = K(\eta), \qquad \eta = z\xi.$$

In this case, the construction (2.3.12)–(2.3.13) is used. Suppose that

$$\tilde{L}_\xi = \xi(\xi - 1)D_\xi + (\sigma - (\rho + \sigma)\xi), \qquad (2.3.14)$$

which corresponds to the solution $v(z) = z^\sigma (1 - z)^\rho$. Then, with $f_1(z) = z^{-1}$ and $f_2(z) = 0$, we get from (2.3.13) the representation

$$-L_z + \frac{1}{z}\tilde{L}_\xi^* D_\xi = \eta(\eta - 1)D_\eta^2 + \{(a + b + 1)\eta - (\sigma + 1)\} D_\eta$$

$$+ ab + \xi(\rho + \sigma + 2 - c)D_\eta. \qquad (2.3.15)$$

The last term in (2.3.15) is nullified under the condition $\rho = c - \sigma - 2$, which gives the value of the parameter ρ. By choosing specified values of the parameter σ, several modifications of the formula (2.3.14) may be obtained. The most useful and well-known result is as follows.

Theorem 2.2 *Suppose that* $\sigma = b - 1$ *in* (2.3.15). *Then equation* (2.3.12) *has a solution* $(1 - z\xi)^{-a}$. *Hence, under the conditions* $\operatorname{Re} c > \operatorname{Re} b > 0$, *the regular solution of equation* (2.3.4) *at* $z = 0$—*the hypergeometric function* $F(a, b, c; z)$—*has the integral representation*

$$y^{\{1,1;1\}}(a, b; c; z = 0, z) = F(a, b, c; z) =$$

$$B^{\{1,1;1\}} \int_0^1 (1 - z\xi)^{-a}(1 - \xi)^{c-b-1}\xi^{b-1}\,d\xi \qquad (2.3.16)$$

with $B^{\{1,1;1\}}$ *determined by the normalization*

$$(B^{\{1,1;1\}})^{-1} = \int_0^1 (1 - \xi)^{c-b-1}\xi^{b-1}\,d\xi = \frac{\Gamma(c - b)\Gamma(b)}{\Gamma(c)}. \qquad (2.3.17)$$

The integral on the right-hand side converges uniformly in any compact domain that does not contain points of the interval $[1, \infty[$. $\square$

Formulae arising from (2.3.16)–(2.3.17) under the substitution $a \leftrightarrow b$, with corresponding changes of the conditions on the parameters, are also valid. If $z = 1$ in (2.3.16), then the limiting value of the hypergeometric function $F(a, b, c; 1)$ at another singularity is obtained:

$$F(a, b, c; 1) = B^{\{1,1;1\}} \int_0^1 (1 - \xi)^{c-b-a-1}\xi^{b-1}\,d\xi = \frac{\Gamma(c)\Gamma(c - a - b)}{\Gamma(c - b)\Gamma(c - a)}. \qquad (2.3.18)$$

A staircase of integral relations for solutions of confluent cases of equations contained in the hypergeometric class (2.3.5–2.3.8) may be derived with the help of the confluence principle.

The integral representation of the confluent hypergeometric function reads

$$y^{\{1;2\}}(a;c;z=0,z)=\Phi(a,c;z)=B^{\{1;2\}}\int_0^1 e^{z\xi}\xi^{a-1}(1-\xi)^{c-a-1}\,d\xi$$

$$(2.3.19)$$

with

$$(B^{\{1;2\}})^{-1}=\int_0^1 (1-\xi)^{c-a-1}\xi^{a-1}\,d\xi=\frac{\Gamma(c-a)\Gamma(a)}{\Gamma(c)}.\qquad(2.3.20)$$

From (2.3.16), by substituting $\xi \mapsto -\sqrt{ab}\xi$, and letting $a,b \to \infty$, we also obtain

$$y^{\{1;3/2\}}(c;z=+\infty,z)=B^{\{;3/2\}}\int_0^\infty e^{-z\xi}\xi^{c-2}e^{-1/\xi}\,d\xi\qquad(2.3.21)$$

for an appropriately chosen solution of the equation related to (2.3.6).

From (2.3.19) and by substituting $\xi \mapsto -\xi(c)^{-1/2}$, we get

$$y^{\{;3\}}(a;z=+\infty,z)=B^{\{;3\}}\int_0^\infty e^{z\xi-\xi^2/2}\xi^{a-2}\,d\xi\qquad(2.3.22)$$

for an appropriately chosen solution of the equation related to (2.3.5).

In order to obtain an integral representation for the solution of the Airy equation (2.3.8), we use the transformation (2.3.21) along with the following substitution of the integration variable:

$$\sqrt{z}\xi+\frac{\xi}{\sqrt{z}}=-2(2c)^{-1/3}s.\qquad(2.3.23)$$

The result reads

$$y^{\{;5/2\}}(z=+\infty,z)=B^{\{;5/2\}}\int_{-\infty}^\infty \exp[i(zs+s^3/3)]\,ds.\qquad(2.3.24)$$

Limits of integration in (2.3.24) are chosen for the sake of convergence.

2.3.3 Integral relations

We start with the kernel $K(z,\xi)$ in (2.3.1) of the form

$$K(z,\xi)=K(\zeta),\qquad \zeta=z-t.\qquad(2.3.25)$$

Suppose that $L_z^{\{1,1;1\}}$ and $\hat{L}_\xi^{\{1,1;1\}*}$ are differential operators related to eqn (2.3.4) but with different sets of values of parameters: a, b, c and a', b', c', respectively. Then

$L_z^{\{1,1;1\}} - \hat{L}_\xi^{\{1,1;1\}*}$ is represented in the form

$$
\begin{aligned}
L_z^{\{1,1;1\}} - \hat{L}_\xi^{\{1,1;1\}*} = {} & [1 - (z + \xi)] \zeta D_\zeta^2 \\
& + \big[c + c' - (a + a' + b + b')(z + \xi)/2 \\
& + (a - a' + b - b')\xi/2\big] D_\zeta - (ab - a'b').
\end{aligned}
$$

$$(2.3.26)$$

Expression (2.3.26) simplifies to

$$
\begin{aligned}
L_z^{\{1,1;1\}} - \hat{L}_\xi^{\{1,1;1\}*} = {} & (1 - (z + \xi))(\zeta D_\zeta^2 - (v - 1)D_\zeta) \\
& + (a - a' + b - b')(\zeta D_\zeta - v)/2
\end{aligned}
$$

$$(2.3.27)$$

if the following relations between the coefficients a, b, c, and a', b', c' hold:

$$
a + a' + b + b' + 2v = 0, \qquad c + c' + v - 1 = 0,
$$

$$
2(ab - a'b') = v(a - a' + b - b').
$$

$$(2.3.28)$$

In this case the equation

$$
(L_z - \hat{L}_\xi^*)K(z, \xi) = 0
$$

$$(2.3.29)$$

has a solution

$$
K(z, \xi) = (z - \xi)^v
$$

$$(2.3.30)$$

with the arbitrary parameter v.

The system (2.3.28) has the following two sets of solutions:

$$
\begin{aligned}
a_1' &= -a - v, & b_1' &= -b - v, & c_1' &= 1 - c - v; \\
a_2' &= -b - v, & b_2' &= -a - v, & c_2' &= 1 - c - v.
\end{aligned}
$$

$$(2.3.31)$$

Consider the operators as operator-functions with parameters of the operator defined as complex arguments of the operator-function. First we note that, since $L(a, b, c)$ is a symmetrical function of a and b, it is sufficient to examine only one set of the roots (2.3.31). If the first set is chosen, then

$$
\hat{L}^{\{1,1;1\}*}(a, b, c) = L^{\{1,1;1\}}(-a - v, -b - v, 1 - c - v).
$$

$$(2.3.32)$$

The operator $\hat{L}^{\{1,1;1\}}$ is also related to equation (2.3.4) but with a new set of parameters a'', b'', c''. It may be expressed with the help of $\hat{L}^{\{1,1;1\}*}$ as follows:

$$
\hat{L}^{\{1,1;1\}}(a, b, c) = \hat{L}^{\{1,1;1\}*}(1 - a, 1 - b, 2 - c).
$$

$$(2.3.33)$$

Comparing (2.3.32) and (2.3.33), we obtain

$$
\hat{L}^{\{1,1;1\}}(a, b, c) = L^{\{1,1;1\}}(1 + a + v, 1 + b + v, 1 + c + v).
$$

$$(2.3.34)$$

Formulae (2.3.30) and (2.3.34) added by analysis of convergence and by a special choice of parameters generate several integral relations for the hypergeometric functions.

Suppose that C is a double loop encircling the points $\xi = 1$ and $\xi = z$ in both directions, and that the function $G(a, b, c; z)$ is a second particular solution of the hypergeometric equation branching at zero. Then the hypergeometric function $F(a, b; c; z)$ is related to $G(a, b, c; z)$ by

$$F(a, b, c; z) = D \int_C (\xi - z)^v G(1 + a + v, 1 + b + v,$$

$$2 + a + b + v - c; 1 - \xi)\, d\xi \qquad (2.3.35)$$

with an appropriate multiplier D determined by normalization. □

More useful and well-known results are obtained under a special choice of the parameter v.

From Fuchs's theorem, it follows that the hypergeometric equation (2.3.4) has a solution of the form

$$v(z) = z^\sigma (1 - z)^\rho, \qquad (2.3.36)$$

with nonzero σ and ρ, if and only if either $a = 1$ or $b = 1$. In addition, if, for instance, $b = 1$, then $\sigma = 1 - c$ and $\rho = c - b - 1$.

Therefore, if the parameter v in (2.3.31) takes the values $v = -a$ or $v = -b$, then the function $v(z)$ may be chosen as elementary having the form (2.3.36).

Suppose that parameters of the hypergeometric equation satisfy $\operatorname{Re} c > \operatorname{Re} b > 0$. Then the regular solution of equation (2.3.4) at $z = 0$—the hypergeometric function $F(a, b; c; z)$—has the integral representation

$$F(a, b; c; z) = D^{\{1,1;1\}} \int_1^\infty (\xi - z)^{-a}(\xi - 1)^{c-b-1}\xi^{a-c}\, d\xi \qquad (2.3.37)$$

with the multiplier $D^{\{1,1;1\}}$ determined by normalization.

Restrictions on parameters may be obviated by another choice of the integration contour.

The third possible type of statement arises by choosing $v = -a - n - 1$ with non-negative integer n. In this case the hypergeometric equation has a polynomial solution which leads to the following conjecture.

Suppose that $v = -a - n - 1$. Then the hypergeometric function $F(a, b, c; z)$ may be represented as

$$F(a, b, c; z) = \delta \int_1^\infty (\xi - z)^v F(-n, b - a - n, 1 + b - c - n; 1 - \xi)\, d\xi$$

$$(2.3.38)$$

with the multiplier δ determined by normalization

$$\delta = \frac{\Gamma(c)}{\Gamma(b)\Gamma(c - b)}. \qquad (2.3.39)$$

The equivalence of (2.3.34) and (2.3.38) easily follows by means of the substitution $\xi \mapsto \xi^{-1}$ under the integral sign.

2.4 Central two-point connection problems

The central two-point connection problem is studied in the context of this chapter in a way somewhat different to that of Chapters 1 and 3. The explanation of it lies in the particular properties of special functions belonging to the hypergeometric class, namely in the existence of explicit formulae in terms of the gamma function connecting local solutions at different singularities. Such formulae are missing in cases of higher classes, and are replaced by some infinite processes which can be implemented only numerically and are discussed generally in Section 1.6 and for the Heun class in Section 3.6.

2.4.1 Standard sets of solutions for the hypergeometric equation

At each regular singularity characterizing the hypergeometric equation, standard fundamental sets of Frobenius solutions can be constructed. In Section 2.1, two such solutions at $z = 0$ have been presented, namely $F(a, b; c; z)$ and $z^{1-c}F(a + 1 - c, b+1-c; 2-c; z)$. At each singularity $z = 1$ and $z = \infty$, a similar couple of local solutions can be defined. Each pair of these three sets can be linked with a matrix which is called the connection matrix. These connection matrices can be expressed in terms of gamma functions. The reason lies in the existence of integral representations for the hypergeometric function. However, these six solutions do not exhaust all possible local solutions. For instance, s-homotopic transformations preserving the canonical form of the equation can be performed at $z = 1$ and at $z = \infty$, thus giving rise to new local solutions at zero. However, they are the same functions although having different notation. For instance,

$$F(a, b; c; z) = (1 - z)^{c-a-b} F(c - a, c - b; c; z). \qquad (2.4.1)$$

In order to prove eqn (2.4.1), we need only to compare the Riemann schemes for the corresponding functions and their behaviour at zero. Moreover, the list of local solutions can be extended if, beyond s-homotopic transformations of the dependent variable, Möbius transformations of the independent variable are considered. In this way it is possible to obtain

$$F(a, b; c; z) = (1 - z)^{-a} F\left(a, c - b; c; \frac{z}{z - 1}\right)$$
$$= (1 - z)^{-b} F\left(c - a, b; c; \frac{z}{z - 1}\right). \qquad (2.4.2)$$

As a whole, 24 so-called Kummer solutions can be constructed. Among these, only six are basic, which were mentioned above. The latter are the subject of our further study.

In the neighbourhood of the origin, the fundamental set of solutions of eqn (2.1.1) is taken as

$$\vec{U}_0(z) = \begin{pmatrix} u_1(z) \\ u_2(z) \end{pmatrix} = \begin{pmatrix} F(a, b; c; z) \\ z^{1-c}F(a + 1 - c, b + 1 - c; 2 - c; z) \end{pmatrix}. \qquad (2.4.3)$$

It is assumed that the parameter c is not an integer.

In the neighbourhood of the other singularity $z = 1$, the corresponding set of solutions is

$$\vec{U}_1(z) = \begin{pmatrix} u_3(z) \\ u_4(z) \end{pmatrix} = \begin{pmatrix} F(a, b; a + b + 1 - c; 1 - z) \\ (1 - z)^{c-a-b} F(c - a, c - b; 1 + c - a - b; 1 - z) \end{pmatrix}.$$

$$(2.4.4)$$

On a formal level, this second set of solutions is obtained from the first one by substitutions

$$z \mapsto 1 - z, \qquad c \mapsto a + b + 1 - c.$$

The third set of solutions at infinity is

$$\vec{U}_\infty(z) = \begin{pmatrix} u_5(z) \\ u_6(z) \end{pmatrix} = \begin{pmatrix} F(a, b; a + b + 1 - c; 1/z) \\ z^{c-a-b} F(c - a, c - b; 1 - c - a - b; 1/z) \end{pmatrix}. \qquad (2.4.5)$$

Sets of solutions (2.4.3–2.4.4) are connected with each other by a connection matrix S^{10}:

$$\vec{U}_0(z) = S^{(10)} \vec{U}_1(z). \qquad (2.4.6)$$

The matrix elements $S_{jk}^{(10)}$ ($j = 1, 2$; $k = 1, 2$) are computed with the help of the formula

$$F(a, b; c; 1) = \frac{\Gamma(c)}{\Gamma(b)\Gamma(c - b)} \int_0^1 t^{b-1}(1 - t)^{c-a-b-1} \, dt$$

$$= \frac{\Gamma(c)\Gamma(c - a - b)}{\Gamma(c - b)\Gamma(c - a)}, \qquad (2.4.7)$$

which results from the integral representation of the hypergeometric function (see eqn (2.3.16)). Although the integral converges only when b and $c - a - b$ are positive, the final equality is valid unless the gamma functions take infinite values.

Suppose that the values of parameters satisfy the conditions

$$0 < c < 1, \qquad 0 < c - a - b < 1.$$

In this case, the solutions $u_2(z, z = 0)$ and $u_4(z, z = 1)$ tend to zero when z tends to zero and unity respectively. The connection relation (2.4.6) at $z = 0$ then yields

$$\begin{pmatrix} 1 \\ 0 \end{pmatrix} = \begin{pmatrix} S_{11}^{(10)} u_3(1) + S_{12}^{(10)} u_4(1) \\ S_{21}^{(10)} u_3(1) + S_{22}^{(10)} u_4(1) \end{pmatrix}.$$

At the point $z = 1$, we have

$$\begin{pmatrix} u_1(1) \\ u_2(1) \end{pmatrix} = \begin{pmatrix} S_{11}^{(10)} \\ S_{21}^{(10)} \end{pmatrix}.$$

These two relations allow us to obtain the matrix elements of the connection matrix:

$$S_{11}^{(10)} = u_1(1), \qquad S_{21}^{(10)} = u_2(1),$$

$$S_{22}^{(10)} = -\frac{u_2(1)u_3(1)}{u_4(1)}, \qquad S_{12}^{(10)} = \frac{1 - u_1(1)u_3(1)}{u_4(1)}. \qquad (2.4.8)$$

It follows from (2.4.3, 2.4.4, 2.4.7) that

$$u_1(1) = \frac{\Gamma(c)\Gamma(c - a - b)}{\Gamma(c - a)\Gamma(c - b)},$$

$$u_2(1) = \frac{\Gamma(2 - c)\Gamma(c - a - b)}{\Gamma(1 - a)\Gamma(1 - b)},$$

$$u_3(1) = \frac{\Gamma(1 - c)\Gamma(a + b + 1 - c)}{\Gamma(a + 1 - c)\Gamma(b + 1 - c)},$$

$$u_4(1) = \frac{\Gamma(1 - c)\Gamma(1 + c - a - b)}{\Gamma(1 - a)\Gamma(1 - b)}.$$

With the help of computations which involve simple trigonometric identities and the reflection formula for the gamma function (see (1.14)), the elements of the connection matrix take the form

$$S_{11}^{(10)} = \frac{\Gamma(c)\Gamma(c - a - b)}{\Gamma(c - a)\Gamma(c - b)},$$

$$S_{21}^{(10)} = \frac{\Gamma(2 - c)\Gamma(c - a - b)}{\Gamma(1 - a)\Gamma(1 - b)},$$

$$S_{22}^{(10)} = \frac{\Gamma(c)\Gamma(a + b - c)}{\Gamma(a)\Gamma(b)},$$

$$S_{12}^{(10)} = \frac{\Gamma(2 - c)\Gamma(a + b - c)}{\Gamma(a + 1 - c)\Gamma(b + 1 - c)}. \qquad (2.4.9)$$

The elements of the matrix $S^{(\infty 0)}$ are computed in a similar way. It follows from (2.4.7) that, if $a > b$, then

$$\lim_{z \to \infty} (-z)^a u_1(z) = \frac{\Gamma(c)\Gamma(b - a)}{\Gamma(c - a)\Gamma(b)},$$

$$\lim_{z \to \infty} (-z)^a u_2(z) = \frac{\Gamma(2 - c)\Gamma(c - a - b)}{\Gamma(1 - a)\Gamma(1 - b)},$$

$$u_3(1) = \frac{\Gamma(1 - c)\Gamma(a + b + 1 - c)}{\Gamma(a + 1 - c)\Gamma(b + 1 - c)},$$

$$u_4(1) = \frac{\Gamma(1 - c)\Gamma(1 + c - a - b)}{\Gamma(1 - a)\Gamma(1 - b)}.$$

As a result we obtain the needed matrix elements

$$S_{11}^{(\infty 0)} = \frac{\Gamma(c)\Gamma(b-a)}{\Gamma(c-a)\Gamma(b)}, \qquad S_{12}^{(\infty 0)} = \frac{\Gamma(c)\Gamma(a-b)}{\Gamma(a)\Gamma(c-b)},$$

$$S_{21}^{(\infty 0)} = \frac{\Gamma(c)\Gamma(a+b-c)}{\Gamma(a)\Gamma(b)}, \qquad S_{22}^{(\infty 0)} = \frac{\Gamma(2-c)\Gamma(a+b-c)}{\Gamma(a+1-c)\Gamma(b+1-c)}.$$

$$(2.4.10)$$

2.4.2 Connection relations for solutions of confluent hypergeometric equations

Local solutions can be constructed either at the regular singularity $z = 0$ or at the irregular singularity $z = \infty$. In the latter case, there are two solutions related to the Stokes line $\arg z = 0$, two solutions related to the Stokes line $\arg z = \pi$, and two solutions related to the Stokes line $\arg z = -\pi$. All these solutions have corresponding integral representations. As a result, connection matrices can be constructed in the same manner as in the previous subsection. However, solutions at the irregular singularity may also been related to Thomé solutions (formal asymptotic series). Connection matrices in terms of corresponding asymptotic series are much more instructive for numerical needs.

The principal point is whether asymptotic expansions of the solution at the Stokes line correspond uniquely to the solution itself. In the case of recessive solutions, the fact has long been known (see, for instance, [96], [43], [115]) . However, recent studies in the so-called resurgent analysis [124], [14], [8] show that, under a certain definition of asymptotic series, it is also valid for dominant solutions. To prove this rigorously, even in our simple cases, exceeds our purposes. Hence, here is given a visual but by no means a rigorous computation of some connection relations based on confluence processes. Other relations are given without proof in a table.

The starting point is a relation for hypergeometric functions:

$$F(a, b; c; z) = \frac{\Gamma(c)\Gamma(c-a-b)}{\Gamma(c-a)\Gamma(c-b)} F(a, b; a+b+1-c; 1-z)$$

$$+ \frac{\Gamma(c)\Gamma(a+b-c)}{\Gamma(a)\Gamma(b)} (1-z)^{c-a-b}$$

$$\times F(c-a, c-b; c-a-b+1; 1-z). \qquad (2.4.11)$$

In more convenient notation, eqn (2.4.11) can be rewritten according to (2.4.2) as

$$F(a, b; c; z) = \frac{\Gamma(c)\Gamma(c-a-b)}{\Gamma(c-a)\Gamma(c-b)} z^{-a}$$

$$\times F\left(a, a-c+1; a+b+1-c; 1-\frac{1}{z}\right)$$

$$+ \frac{\Gamma(c)\Gamma(a+b-c)}{\Gamma(a)\Gamma(b)}(1-z)^{c-a-b}z^{a-c}$$

$$\times F\left(c-a, 1-a; c-a-b+1; 1-\frac{1}{z}\right)$$

$$= A(b)f_1(b; z) + B(b)f_2(b; z), \tag{2.4.12}$$

where

$$A(b) = \frac{\Gamma(c)\Gamma(c-a-b)}{\Gamma(c-a)\Gamma(c-b)}b^{-a},$$

$$B(b) = \frac{\Gamma(c)\Gamma(a+b-c)}{\Gamma(a)\Gamma(b)}b^{c-a}.$$

Then the definition of the functions $f_1(b; z)$ and $f_2(b; z)$ is evident. According to the confluence principle, functions at the left-hand side and at the right-hand side of eqn (2.4.12) tend to particular solutions of the confluent hypergeometric equation at $b \to \infty$. First we need to find the limits $\lim_{b\to\infty} A(b)$ and $\lim_{b\to\infty} B(b)$. It is possible only under the additional suppositions that either $\pi - \epsilon > \arg(b) > \epsilon$ or $-\pi +\epsilon < \arg(b) < -\epsilon$. Assume that the latter case holds. Keeping the leading terms of asymptotics of the gamma function (see the Stirling formula (1.8)) and using the definition of the exponential base

$$\lim_{b\to\infty}\left(1+\frac{1}{b}\right)^b = e,$$

we arrive at the limiting expressions

$$\lim_{b\to\infty} A(b) = \frac{\Gamma(c)}{\Gamma(c-a)}e^{i\pi a},$$

$$\lim_{b\to\infty} A(b) = \frac{\Gamma(c)}{\Gamma(a)}.$$

The coefficients of the series expansion of the function

$$F\left(a, a-c+1; a+b+1-c; 1-\frac{1}{z}\right) = \sum_0^\infty h_k(1-z^{-1})^k$$

obey the recurrence relation

$$(k+1)(k+a+b+1-c)h_{(k+1)} = (k+1)(k+a+1-c)h_k,$$

which, at $b \to \infty$, converts eqn (2.1.20) to the recurrence relation for a confluent hypergeometric function $\Psi(a; c; z)$. By analogous considerations, we obtain that the coefficients r_k of the series expansion

$$F\left(c-a, 1-a; c-a-b+1; 1-\frac{1}{z}\right) = \sum_0^\infty r_k(1-z^{-1})^k$$

obey the recurrence relation (2.1.22) at $b \to \infty$. As a result, we arrive at the following

Table 2.4.1. Connection formulae for confluent hypergeometric functions

Actual solution	Ray or sector	Thomé solutions
$\Psi(a; c; z)$	$\arg(z) = 0$ $|\arg(z)| < 2\pi - \epsilon$	$y_1^{\{1;2\}}(a; c; z)$
$\Psi(a; c; z)$	$\arg(z) = 2\pi$	$y_1^{\{1;2\}}(z) + ie^{i\pi(c-a)}\sin[\pi(c-a)]\times$ $\dfrac{\Gamma(c-a)}{\Gamma(a)}y_2^{\{1;2\}}(z)$
$\Phi(a; c; z)$	$\arg(z) = 0$	$\dfrac{\Gamma(c)}{\Gamma(c-a)}\cos(\pi a)y_1^{\{1;2\}}(z) + \dfrac{\Gamma(c)}{\Gamma(a)}y_2^{\{1;2\}}(z)$
$\Phi(a; c; z)$	$\pi - \epsilon > \arg(z) > \epsilon$	$\dfrac{\Gamma(c)}{\Gamma(c-a)}e^{i\pi a}y_1^{\{1;2\}}(z) + \dfrac{\Gamma(c)}{\Gamma(a)}y_2^{\{1;2\}}(z)$

Connection formulae for parabolic cylinder functions

Actual solution	Ray or sector	Thomé solutions
$D_\nu(z)$	$\arg(z) = 0$ $\arg(z) < \pi/2 - \epsilon$	$y_1(a; z)$
$D_\nu(z)$	$\arg(z) = \pm\pi/2$	$y_1(a; z) \pm e^{i\pi(\nu+1)}\dfrac{\sqrt{2\pi}}{2\Gamma(-\nu)}y_2(a; z)$
$D_\nu(z)$	$\pi - \epsilon > \arg(z) > \pi/2 + \epsilon$	$y_1(a; z) + e^{i\pi(\nu+1)}\dfrac{\sqrt{2\pi}}{\Gamma(-\nu)}y_2(a; z)$
$D_\nu(z)$	$\arg(z) = \pi$	$\cos(\pi\nu)e^{-i\pi\nu}y_1(a; z)+$ $e^{i\pi(\nu+1)}\dfrac{\sqrt{2\pi}}{\Gamma(-\nu)}y_2(a; z)$
$V_\nu(z)$	$\arg(z) = 0$	$y_2(a; z)$

Connection formulae for Airy functions

Actual solution	Ray or sector	Thomé solutions
$2\sqrt{\pi}\,\mathrm{Ai}(z)$	$\arg(z) = 0$ $|\arg(z)| < 2\pi/3 - \epsilon$	$y_1^{\{;5/2\}}(z)$
$2\sqrt{\pi}\,\mathrm{Ai}(z)$	$\arg(z) = \pm 2\pi/3$	$y_1^{\{;5/2\}}(z) \pm \dfrac{i}{2}y_2^{\{;5/2\}}(z)$
$2\sqrt{\pi}\,\mathrm{Ai}(z)$	$4\pi/3 - \epsilon > \arg(z) > 2\pi/3 - \epsilon$	$y_1^{\{;5/2\}}(z) + iy_2^{\{;5/2\}}(z)$
$\sqrt{\pi}\,\mathrm{Bi}(z)$	$\arg(z) = 0$	$y_2^{\{;5/2\}}(z)$
$\sqrt{\pi}\,\mathrm{Bi}(z)$	$2\pi/3 - \epsilon > \arg(z) > \epsilon$	$y_2^{\{;5/2\}}(z) + \dfrac{i}{2}y_1^{\{;5/2\}}(z)$
$\sqrt{\pi}\,\mathrm{Bi}(z)$	$\arg(z) = 2\pi/3$	$\dfrac{3}{4}y_2^{\{;5/2\}}(z) + \dfrac{i}{2}y_1^{\{;5/2\}}(z)$

connection formula for solutions of ChE valid for $\pi - \epsilon > \arg(z) > \epsilon$:

$$\Phi(a; c; z) = \frac{\Gamma(c)}{\Gamma(c-a)} e^{i\pi a} y_1^{\{1;2\}}(a; c; z = \infty, z)$$

$$+ \frac{\Gamma(c)}{\Gamma(a)} y_2^{\{1;2\}}(a; c; z = \infty, z). \qquad (2.4.13)$$

By choosing parameter b to be in the upper complex half plane, the connection formula for the lower half plane of z is obtained similar to eqn (2.4.13). It is only needed to substitute $e^{i\pi a}$ for $e^{-i\pi a}$. On the real axis, we take the average of both expressions (see [115]).

Further use of connection matrices for hypergeometric functions (see (2.4.9, 2.4.10)) leads to other connection formulae for solutions of ChE. A list of connection formulae between actual solutions of the confluent hypergeometric equation (confluent hypergeometric function) and Thomé solutions of this equation (which in turn uniquely correspond to a proper actual solution) is given in Table 2.4.1.

In the same way, connection formulae related to other equations belonging to the hypergeometric class can be constructed. However, a simpler way is to use explicit expression for solutions of these equations in terms of solutions of ChE. Namely, if we use eqn (2.1.33), we arrive at a list of connection formulae between actual solutions of the biconfluent hypergeometric equation (parabolic cylinder functions) and Thomé solutions of this equation which is given in Table 2.4.1. Connection formulae for modified Bessel functions simply follow from connection formulae for hypergeometric functions (see (2.1.51–2.1.52)). Finally comes a list of connection formulae between actual solutions of the reduced biconfluent hypergeometric equation (Airy functions) and Thomé solutions of this equation. It is given in Table 2.4.1.

2.5 Polynomial solutions

2.5.1 Introduction

Three of the equations of hypergeometric class, namely the hypergeometric equation, the confluent hypergeometric equation, and the biconfluent hypergeometric equation, generate equations belonging to three sets of orthogonal polynomials. These three sets are called classical orthogonal polynomials and consist of Jacobi, Laguerre, and Hermite polynomials. For historical reasons, and according to applications, the equations for these polynomials and their standardizations are slightly different from what we have considered in our previous studies. Hence we first study polynomials which we call hypergeometric polynomials and which by no means are conventional objects in mathematical physics. Their importance lies in the fact that they inherit the known facts about the hypergeometric function expounded above. Further, the properties of hypergeometric polynomials extend to the much better known Jacobi polynomials and to their specialized cases, namely to Legendre and Gegenbauer polynomials. The confluence process allows us to construct the corresponding formulae for Laguerre polynomials and Hermite polynomials.

It is also important to stress that the above-mentioned sets of polynomials are uniquely related to sets of eigenfunctions generated by appropriate boundary conditions for the seed equation. The eigenvalues of these boundary problems are expressed explicitly in terms of singular parameters of the equation. Mostly, those properties of the polynomials are studied which are important from the point of view of boundary problems.

Throughout this section, for the sake of convenience, it is assumed that the parameters take real values. However, analytic continuation can always be used for complex values of parameters.

2.5.2 Polynomial solutions of the hypergeometric equation

Suppose that the parameter a is a negative integer $-n$ and the parameter b is replaced by $n + b$ in the series expansion for the hypergeometric function. In addition, the conditions $b > c > 1$ are supposed to be fulfilled. Then the series (2.1.8) truncates, and polynomials of order n are obtained as a result. We call these polynomials *hypergeometric polynomials* and denote them by $P_n^{\{1,1;1\}}(b; c; z)$. Neither the notion nor its notation is conventional in the literature. However, they both will be used as starting points for more conventional considerations. We have

$$P_n^{\{1,1;1\}}(b; c; z) = F(-n, n + b; c; z)$$

$$= \sum_0^n (-1)^k z^k \frac{n!\Gamma(c)\Gamma(k + n + b)}{k!(n - k)!\Gamma(n + b)\Gamma(k + c)}. \tag{2.5.1}$$

Hypergeometric polynomials $P_n^{\{1,1;1\}}(b; c; z)$ satisfy the slightly modified hypergeometric differential equation

$$z(1 - z)y''(z) + (c - (b + 1)z)y'(z) + n(b + n)y(z) = 0. \tag{2.5.2}$$

The values of the hypergeometric polynomials at the points $z = 0$ and $z = 1$ can easily be computed:

$$P_n^{\{1,1;1\}}(b; c; 0) = 1,$$

$$P_n^{\{1,1;1\}}(b; c; 1) = (-1)^n \frac{\Gamma(c)\Gamma(n + b + 1 - c)}{\Gamma(n + c)\Gamma(b + 1 - c)}. \tag{2.5.3}$$

The latter formula follows from the known value of the hypergeometric function at unity (see (2.3.18)) and the following relation for the gamma function:

$$\frac{\Gamma(c - b)}{\Gamma(c - b - n)} = (c - b - 1) \ldots (c - b - n) = \frac{\Gamma(n + b + 1 - c)}{\Gamma(b + 1 - c)}.$$

Often the coefficient κ_n of the leading power in the polynomial is needed:

$$P_n^{\{1,1;1\}}(b; c; z) \sim \kappa_n z^n \quad \text{as } z \to \infty. \tag{2.5.4}$$

It can also be obtained from eqn (2.5.1):

$$\kappa_n = (-1)^n \frac{\Gamma(c)\Gamma(2n+b)}{\Gamma(n+b)\Gamma(n+c)}. \tag{2.5.5}$$

The hypergeometric polynomials may be expressed in an explicit form as a result of differentiation of an elementary function.

Theorem 2.3 *The following formula (Rodriguez formula) is valid for hypergeometric polynomials.*

$$P_n^{\{1,1;1\}}(b; c; z) = \frac{\Gamma(c)}{\Gamma(n+c)} z^{1-c} (1-z)^{c-b} \frac{d^n}{dz^n} [(z(1-z))^n z^{c-1} (1-z)^{b-c}]$$

$$= \frac{\Gamma(c)}{\Gamma(n+c)} (\rho(z))^{-1} \frac{d^n}{dz^n} [(z(1-z))^n \rho(z)], \tag{2.5.6}$$

$$\rho(z) = z^{c-1} (1-z)^{b-c}.$$

Here the function $\rho(z)$ is called the weight function. □

Proof: The auxiliary function $\phi(z) = \rho(z)(z(1-z))^n$ satisfies the differential equation

$$z(1-z)\phi'(z) = (n+c-1-z(2n+b-1))\phi(z).$$

Consider another auxiliary function $\psi(z)$ which is the nth derivative of the first one:

$$\psi(z) = D^n \phi(z), \qquad D^n = \frac{d^n}{dz^n}.$$

Differentiating the equation for $\psi_n(z)$ $n+1$ times, we obtain the equation for $\psi(z)$:

$$z(1-z)\psi''(z) + (2-c-z(3-b))\psi'(z) + (n+1)(1-n-b)\psi(z) = 0, \tag{2.5.7}$$

which is actually a hypergeometric equation. It corresponds to the GRS

$$\begin{pmatrix} 1 & 1 & 1 \\ 0 & 1 & \infty \\ 0 & 0 & n+1 \\ c-1 & b-c & 1-b+n \end{pmatrix}. \tag{2.5.8}$$

It is easily seen from eqn (2.5.8) that, under the s-homotopic transformation

$$y(z) = z^{1-c} (1-z)^{c-b} \psi(z),$$

the equation for the function $y(z)$ coincides with the one for the hypergeometric polynomials. Compare the values of $y(z)$ and the hypergeometric polynomials at zero:

$$y(0) = \frac{\Gamma(n+c)}{\Gamma(c)}.$$

This completes the proof. □

The equation for the hypergeometric polynomials can be transformed to the self-adjoint form by the substitution

$$v_n(b, c, z) = z^{(c-1)/2}(1-z)^{(b-c)/2}P_n^{\{1,1;1\}}(b; c; z) = (\rho(z))^{1/2}P_n^{\{1,1;1\}}(b; c; z).$$
(2.5.9)

Quasipolynomials $v_n(b, c, z)$ satisfy the equation

$$\frac{d}{dz}z(1-z)\frac{d}{dz}v_n(b, c, z) - \frac{z(1-z)}{4}\left(\frac{(1-c)^2}{z^2}\right.$$

$$\left. + \frac{(c-b)^2}{(1-z)^2}\right)v_n(b; c; z) + \lambda_n v_n(b; c; z) = 0,$$
(2.5.10)

where

$$\lambda_n = n(n+b-c) + c(c-b)/2 + (c-1)/2.$$
(2.5.11)

As a result, the quasipolynomials $v_n(b; c; z)$ constitute an infinite set of orthogonal functions on the interval [0, 1].

Theorem 2.4 *Hypergeometric polynomials constitute an infinite set of orthogonal polynomials on* [0, 1] *having the weight function* $\rho(z) = z^{c-1}(1-z)^{b-c}$.

$$\int_0^1 v_n(b; c; z)v_m(b; c; z)\,dz$$

$$= \int_0^1 P_n^{\{1,1;1\}}(b; c; z)P_m^{\{1,1;1\}}(b; c; z)\,dz = N^2(b, c)\delta_{nm}$$
(2.5.12)

with δ_{nm} *being the Kronecker symbol.* □

It is important to know the normalization coefficient $N_n^2(b, c)$:

$$N_n^2(b, c) = \int_0^1 v_n^2(b; c; z)dz = \int_0^1 \rho(z)[P_n^{\{1,1;1\}}(b; c; z)]^2\,dz$$

$$= \frac{\Gamma(c)}{\Gamma(n+c)}\int_0^1 P_n^{\{1,1;1\}}(b; c; z)\frac{d^n}{dz^n}[z^{n+c-1}(1-z)^{n+b-c}]\,dz$$

$$= \frac{\Gamma^2(c)\Gamma(2n+b)}{\Gamma(n+b)\Gamma^2(n+c)}\int_0^1 [z^{n+c-1}(1-z)^{n+b-c}]\,dz$$

$$= \frac{\Gamma^2(c)\Gamma(n+b-c+1)}{(2n+b)\Gamma(n+b)\Gamma(n+c)}.$$
(2.5.13)

The Rodrigues formula for one of the multipliers, n-fold integration by parts, and evaluation of the last integral by means of the beta function, i.e.

$$\int_0^1 [z^{n+c-1}(1-z)^{n+b-c}]\,dz = \frac{\Gamma(n+c)\Gamma(n+b-c+1)}{\Gamma(2n+b+1)},$$
(2.5.14)

were used.

The conventional integral representation for the hypergeometric function with an integral over $(1, 1)$ (see $(2.3.16)$) is not valid for hypergeometric polynomials because, at the chosen values of parameters, the integral diverges. However, the Cauchy theorem along with the Rodrigues formula leads to another type of representation:

$$P_n^{\{1,1;1\}}(b; c; z) = z^{1-c}(1-z)^{c-b} \frac{(n+1)!\,\Gamma(c)}{2\pi i \Gamma(n+c)} \oint \frac{\zeta^{n+c-1}(1-\zeta)^{n+b-c}}{(\zeta - z)^{n+1}}\, d\zeta. \tag{2.5.15}$$

This representation, for instance, is valid for $0 < z < 1$ with cuts on the complex ζ plane along $]-\infty, 0]$ and $[1, +\infty[$ and integration anticlockwise around the point $\zeta = z$.

Beyond the Rodrigues formulae, recurrence relations are widely used in order to find explicit or numerical representations of the hypergeometric polynomials. The corresponding formulae can be easily obtained from eqns $(2.2.5)$–$(2.2.6)$

$$z P_n^{\{1,1;1\}}(b; c; z) = -\frac{n(b-c)}{(b+n)(b+n-1)} P_{n-1}^{\{1,1;1\}}(b; c; z)$$

$$= \frac{c(b-n-1)+2nb}{(b+n+1)(b+n-1)} P_n^{\{1,1;1\}}(b; c; z)$$

$$- \frac{b(n+c)}{(b+n)(b+n+1)} P_{n+1}^{\{1,1;1\}}(b; c; z), \tag{2.5.16}$$

$$z(1-z)\,[P_n^{\{1,1;1\}}(b; c; z)]' = -\frac{nb(b-c)}{(b+n)(b+n-1)} P_{n-1}^{\{1,1;1\}}(b; c; z)$$

$$- \frac{nb(2c+n-b-1)}{(b+n+1)(b+n-1)} P_n^{\{1,1;1\}}(b; c; z)$$

$$+ \frac{nb(n+c)}{(b+n)(b+n+1)} P_{n+1}^{\{1,1;1\}}(b; c; z). \tag{2.5.17}$$

The difference to eqns $(2.2.5)$–$(2.2.6)$ is that the recurrence relations for hypergeometric polynomials are truncated from the left side, satisfying the initial conditions

$$P_0^{\{1,1;1\}}(b; c; z) = 1,$$

$$P_{-1}^{\{1,1;1\}}(b; c; z) = 0. \tag{2.5.18}$$

The second condition is a formal definition for $P_{-1}^{\{1,1;1\}}(b; c; z)$.

2.5.3   Jacobi polynomials

Jacobi polynomials differ from hypergeometric polynomials by

(i) a different interval of consideration,
(ii) a different definition of parameters,
(iii) a different normalization.

These are of course not the 'inner' properties of polynomials. Hence, the formulae need only a simple recomputation. We substitute a new independent variable by $z \mapsto (1+z)/2$ and new parameters according to $c - 1 := \alpha$ and $b - c := \beta$, with $\alpha > 0$ and $\beta > 0$, in eqn (2.5.12):

$$(1 - z^2)y''(z) + (\beta - \alpha - (\alpha + \beta + 2)z)y'(z) + n(n + \alpha + \beta + 1)y(z) = 0. \tag{2.5.19}$$

The corresponding GRS reads

$$\begin{pmatrix} 1 & 1 & 1 \\ -1 & 1 & \infty \\ 0 & 0 & -n \\ -\beta & -\alpha & n + \alpha + \beta + 1 \end{pmatrix}. \tag{2.5.20}$$

Polynomial solutions of this equation are called *Jacobi polynomials* and are denoted by $P_n^{(\alpha,\beta)}$. Their standardization is usually taken as

$$P_n^{(\alpha,\beta)}(-1) = (-1)^n \frac{\Gamma(\beta + n + 1)}{n!\Gamma(\beta + 1)},$$
$$P_n^{(\alpha,\beta)}(1) = \frac{\Gamma(\alpha + n + 1)}{n!\Gamma(\alpha + 1)}. \tag{2.5.21}$$

It implies the following relation between hypergeometric polynomials and Jacobi polynomials:

$$P_n^{(\alpha,\beta)}(z) = \frac{(-1)^n \Gamma^2(\alpha + n + 1)\Gamma(\beta + 1)}{n!\Gamma^2(\alpha + 1)\Gamma(\beta + n + 1)}. \tag{2.5.22}$$

As a result, Theorem 2.3 can be reformulated as follows.

Theorem 2.5 *The Jacobi polynomials constitute an orthogonal set of polynomials on the interval $(-1, 1)$ having the weight function $\rho(z) = (1 - z)^\alpha(1 + z)^\beta$ and the normalization*

$$N_n^2(\alpha, \beta) = \int_{-1}^1 \rho(z)[P_n^{(\alpha,\beta)}(z)]^2 \, dz$$
$$= \frac{2^{\alpha+\beta+1}\Gamma(n + \alpha + 1)\Gamma(n + \beta + 1)}{(2n + \alpha + \beta + 1)n!\Gamma(n + \alpha + \beta + 1)}. \tag{2.5.23}$$

□

The chosen standardization simplifies the Rodrigues formula for Jacobi polynomials:

$$P_n^{(\alpha,\beta)}(z) = \frac{1}{(-2)^n n!} z^{-\beta}(1 - z)^{-\alpha} \frac{d^n}{dz^n}[(1 - z^2)^n(1 + z)^\beta(1 - z)^\alpha]$$
$$= \frac{1}{(-2)^n n!}(\rho(z))^{-1} \frac{d^n}{dz^n}[(1 - z^2)^n \rho(z)]. \tag{2.5.24}$$

This yields the integral representation

$$P_n^{(\alpha,\beta)}(z) = \frac{1}{(2)^{n+1}\pi i}(\rho(z))^{-1} \oint \frac{\rho(\zeta)(1-\zeta^2)^n}{(\zeta-z)^{n+1}} \, d\zeta. \tag{2.5.25}$$

After introducing a new standardization, the recurrence relations (2.5.16–2.5.18) for hypergeometric polynomials generate corresponding recurrence relations for Jacobi polynomials:

$$\begin{aligned}
z P_n^{(\alpha,\beta)}(z) &= \frac{2(n+1)(n+\alpha+\beta+1)}{(2n+\alpha+\beta+1)(2n+\alpha+\beta+2)} P_{n+1}^{(\alpha,\beta)}(z) \\
&\quad - \frac{\alpha^2-\beta^2}{(2n+\alpha+\beta)(2n+\alpha+\beta+2)} P_n^{(\alpha,\beta)}(z) \\
&\quad + \frac{2(n+\alpha)(n+\beta)}{(2n+\alpha+\beta+1)(2n+\alpha+\beta)} P_{n-1}^{(\alpha,\beta)}(z),
\end{aligned} \tag{2.5.26}$$

$$\begin{aligned}
(1-z^2)[P_n^{(\alpha,\beta)}(z)]' &= (n+\alpha+\beta+1) \\
&\quad \times \Bigg[-\frac{2n(n+1)}{(2n+\alpha+\beta+1)(2n_\alpha+\beta+2)} P_{n+1}^{(\alpha,\beta)}(z) \\
&\quad + \frac{2n(\alpha-\beta)}{(2n+\alpha+\beta)(2n_\alpha+\beta+2)} P_n^{(\alpha,\beta)}(z) \\
&\quad + \frac{2(n+\alpha)(n+\beta)}{(2n+\alpha+\beta)(2n_\alpha+\beta+1)} P_{n-1}^{(\alpha,\beta)}(z) \Bigg].
\end{aligned} \tag{2.5.27}$$

Relations (2.5.26–2.5.27) are often used for the numerical evaluation of Jacobi polynomials.

2.5.4 Specializations of Jacobi polynomials

By specialization of Jacobi polynomials is meant their grouping into sets of polynomials which arise by an appropriate specialization of the parameters α, β. The following specializations of Jacobi polynomials are studied:

(a) Gegenbauer polynomials,

(b) Legendre polynomials,

(c) Chebyshev polynomials.

Gegenbauer polynomials C_n^ν—or, as they are often called, *ultraspherical polynomials*—are obtained by setting $\alpha = \beta = \nu - 1/2$ with $\nu > \frac{1}{2}$ in eqn (2.5.19), resulting in

$$(1-z^2)y''(z) - (2\nu+1)zy'(z) + n(n+2\nu)y(z) = 0. \tag{2.5.28}$$

Equation (2.5.28) corresponds to the GRS

$$
\begin{pmatrix}
1 & 1 & 1 \\
-1 & 1 & \infty \\
0 & 0 & -n \\
1/2 - \nu & 1/2 - \nu & n + 2\nu
\end{pmatrix}.
\tag{2.5.29}
$$

In order to avoid complications in formulating the following results, the value $\nu = \frac{1}{2}$ will be excluded in the forthcoming formulae. This value corresponds to the case when Gegenbauer polynomials are specialized to Legendre polynomials. The standardization of C_n^ν is taken as

$$
C_n^\nu(1) = \frac{\Gamma(n + 2\nu)}{n!\,\Gamma(2\nu)}.
\tag{2.5.30}
$$

Equation (2.5.28) does not change under the transformation $z \mapsto -z$. Hence the Gegenbauer polynomials are even or odd according to whether the number n is even or odd. The chosen standardization yields the following expression for Gegenbauer polynomials in terms of Jacobi polynomials:

$$
C_n^\nu(z) = \sigma_n P_n^{(\nu-1/2,\nu-1/2)}(z), \quad \text{with } \sigma_n = \frac{\Gamma(n + 2\nu)\Gamma(\nu + 1/2)}{\Gamma(2\nu)\Gamma(n + \nu + 1/2)}.
\tag{2.5.31}
$$

Relation (2.5.31) leads to the Rodrigues formula

$$
C_n^\nu(z) = \sigma_n \frac{1}{(-2)^n n!} (\rho(z))^{-1} \frac{d^n}{dz^n}[(1 - z^2)^n \rho(z)]
$$

$$
\text{with } \rho(z) = (1 - z^2)^{\lambda - \frac{1}{2}}.
\tag{2.5.32}
$$

Theorem 2.5 has the following analogue.

Theorem 2.6 *The Gegenbauer polynomials constitute an orthogonal set of polynomials on the interval $[-1, 1]$ having the weight function $\rho(z) = (1 - z^2)^{\nu - \frac{1}{2}}$ and the normalization*

$$
N_n^2(\nu) = \int_{-1}^{1} \rho(z)[C_n^\nu(z)]^2 \, dz
$$

$$
= \frac{2^{2\nu-1}\Gamma^2(\nu + \frac{1}{2})}{n!(n + \nu)\Gamma^2(2\nu)}.
\tag{2.5.33}
$$

□

Practical computations of Gegenbauer polynomials are mostly established by means of recurrence relations which follow from (2.5.26–2.5.27):

$$
(n + 1)C_{n+1}^\nu(z) = 2(n + \nu)z C_n^\nu(z) - (n + 2\nu - 1)C_{n-1}^\nu(z),
\tag{2.5.34}
$$

$$
(1 - z^2)\frac{d}{dz}C_n^\nu(z) = -nz C_n^\nu(z) + (n + 2\nu - 1)C_{n-1}^\nu(z).
\tag{2.5.35}
$$

Legendre polynomials $P_n(z)$ arise as specializations of Jacobi polynomials by $\alpha = \beta = 0$ or as specialization of Gegenbauer polynomials by $\nu = \frac{1}{2}$. They are regular solutions of the differential equation

$$((1 - z^2)y'(z))' + n(n + 1)y(z) = 0 \tag{2.5.36}$$

at the two regular singularities $z_1 = -1$ and $z_2 = 1$. The standardization is taken as

$$P_n(1) = 1, \tag{2.5.37}$$

which implies

$$P_n(z) = C_n^{\frac{1}{2}}(z) = P_n^{(0,0)}(z).$$

Equation (2.5.36) corresponds to the GRS

$$\begin{pmatrix} 1 & 1 & 1 \\ -1 & 1 & \infty \\ 0 & 0 & -n \\ 0 & 0 & n+1 \end{pmatrix}. \tag{2.5.38}$$

The Rodrigues formula in the case of Legendre polynomials changes for

$$P_n(z) = \frac{1}{(2)^n n!} \frac{d^n}{dz^n}[(z^2 - 1)^n]. \tag{2.5.39}$$

This means that the weight function is unity. The Rodrigues formula implies the integral representation, also called Schlaefli integral,

$$P_n(z) = \frac{1}{(2)^{n+1}\pi i} \oint_\gamma \frac{(\zeta^2 - 1)^n}{(\zeta - z)^{n+1}} d\zeta, \tag{2.5.40}$$

where γ is a contour encircling anticlockwise the point $\zeta = z$.

In the case of the Legendre polynomials, we study one more object related to sets of orthogonal polynomials—the generating function. The function $G(z, t)$ of two variables is called the *generating function* for a set of polynomials $P_n(z)$ (not necessarily being Legendre polynomials) if it can be expanded in a Taylor series of the form

$$G(z, t) = \sum_{j=0}^{\infty} P_j(z)t^j \tag{2.5.41}$$

with coefficients $P_j(z)$. Assume that $P_j(z)$ are Legendre polynomials. For sufficiently small t, we have

$$\sum_{j=0}^{\infty} \left(\frac{\zeta^2 - 1}{2(\zeta - z)}\right)^j t^j = \left(1 - \frac{t(\zeta^2 - 1)}{2(\zeta - z)}\right)^{-1}.$$

Combining this with eqns (2.5.40–2.5.41), we arrive at the expression

$$G(z,t) = \sum_{j=0}^{\infty} P_j(z)t^j = \frac{-1}{\pi i} \oint_{\gamma'} \frac{d\zeta}{t\zeta^2 - 2\zeta + (2z - t)},$$

where the contour γ' encircles the root of the denominator that is nearest to zero at small t. Calculating the integral with the help of residue theory, we get the final expression:

$$G(z,t) = (1 - 2zt + t^2)^{-\frac{1}{2}}. \tag{2.5.42}$$

The generating functions can be constructed for Jacobi polynomials and for Gegenbauer polynomials as well, but the corresponding formulae are more complicated than eqn (2.5.42), and their derivation is more laborious.

The main theorem for Legendre polynomials reads as follows.

Theorem 2.7 *The Legendre polynomials constitute an orthogonal set of polynomials on the interval* $[-1, 1]$ *having the normalization*

$$N_n^2 = \frac{2}{2n + 1}. \tag{2.5.43}$$

$\square$

Legendre polynomials usually are computed with the help of the three-term recurrence relation (cf. (2.5.26))

$$(n + 1)P_{n+1}(z) = (2n + 1)zP_n(z) - nP_{n-1}(z), \tag{2.5.44}$$

while the derivatives of the polynomials are got by the recurrence relations

$$(1 - z^2)\frac{d}{dz}P_n(z) = -n(zP_n(z) - P_{n-1}(z)). \tag{2.5.45}$$

Chebyshev polynomials are specializations of Jacobi polynomials having the weight function $\rho(z) = (1 - z^2)^{\frac{1}{2}}$. Chebyshev polynomials of the first kind, $T_n(z)$, relate to the sign $-$, while Chebyshev polynomials of the second kind $U_n(z)$ relate to the sign $+$. All formulae for these polynomials can be easily obtained from their representation in terms of Gegenbauer polynomials:

$$T_n(z) = \frac{n}{2}C_n^0(z), \qquad U_n(z) = C_n^1(z). \tag{2.5.46}$$

2.5.5 Laguerre polynomials

One possible way of presenting the Laguerre polynomials is to study the confluent hypergeometric equation in the same manner as was done for hypergeometric polynomials. The other way is to use the confluence process. Here we set out both approaches. *Laguerre polynomials* $L_n^\alpha(z)$ arise by setting $a = -n$ and $c = \alpha + 1$ in the confluent hypergeometric equation and in the Taylor expansion for the confluent hypergeometric function. However, the standardization of these polynomials differs in the literature. In most textbooks on special functions (see e.g. [40, 5, 96]), the following standardization is taken:

$$L_n^\alpha(z) = \frac{(\alpha+1)_n}{n!}, \tag{2.5.47}$$

where $(\bullet)_n$ is the Pochhammer factorial, while in books on physics (see e.g. [93, 72]) another notation and another standardization is taken.

Laguerre polynomials satisfy the differential equation

$$zy''(z) + (\alpha + 1 - z)y'(z) + ny(z) = 0. \tag{2.5.48}$$

Equations (2.5.47–2.5.45) along with eqn (2.1.17) imply the following representation of the Laguerre polynomials in terms of the confluent hypergeometric function:

$$L_n^\alpha(z) = \frac{(\alpha+1)_n}{n!}\Phi(-n, \alpha + 1; z). \tag{2.5.49}$$

The weight function for the Laguerre polynomials can be obtained as a result of a confluence process in the expression for the weight function related to hypergeometric polynomials:

$$\lim_{b\to\infty} z^{c-1}\left(1 - \frac{z}{b}\right)^{b-c} = z^{c-1}\mathrm{e}^{-z}.$$

In view of the chosen standardization, the Rodrigues formula reads

$$L_n^\alpha(z) = \frac{1}{n!\rho(z)}\frac{\mathrm{d}^n}{\mathrm{d}z^n}z^n\rho(z), \qquad \rho(z) = z^\alpha \mathrm{e}^{-z}. \tag{2.5.50}$$

The main theorem for the Laguerre polynomials is stated as follows.

Theorem 2.8 *The Laguerre polynomials constitute an orthogonal set of polynomials on the interval* $[0, \infty]$ *having the weight function* $\rho(z) = z^\alpha \mathrm{e}^{-z}$ *and the normalization*

$$N_n^2 = \frac{1}{n!}\int_0^\infty L_n^\alpha(z)\frac{\mathrm{d}^n}{\mathrm{d}z^n}z^{n+\alpha}\mathrm{e}^{-z}\,\mathrm{d}z$$

$$= \int_0^\infty z^{n+\alpha}\mathrm{e}^{-z}\,\mathrm{d}z = \frac{\Gamma(n+\alpha+1)}{n!}. \tag{2.5.51}$$

$\square$

Practical recurrent computations of Laguerre polynomials use the relations

$$(n+1)L_{n+1}^\alpha(z) + (z - 2n - \alpha - 1)L_n^\alpha(z) + (n+\alpha)L_{n-1}^\alpha(z) = 0. \tag{2.5.52}$$

2.5.6 Hermite polynomials

Hermite polynomials $H_n(z)$ arise as polynomial solutions of the biconfluent hypergeometric equation (see (2.1.23)) for $a = -n$. The series (2.1.26) terminates for this value of parameter. The Hermite polynomials are either even or odd functions in accordance with n being even or odd. The Rodrigues formula for Hermite polynomials reads

$$H_n(z) = \frac{(-1)^n}{\rho(z)} \frac{d^n}{dz^n} \rho(z), \qquad \rho(z) = e^{-z^2}. \tag{2.5.53}$$

The expression for the weight function $\rho(z)$ can be obtained with the help of a confluence process from the expression for the weight function related to Laguerre polynomials. The leading coefficient in the polynomial is obtained from the asymptotic behaviour $H_n(z) \sim 2^n z^n$.

Theorem 2.9 *The Hermite polynomials constitute an orthogonal set of polynomials on the interval $[-\infty, \infty]$ having the weight function $\rho(z) = e^{-z^2}$ and the normalization*

$$N_n^2 = (-1)^n \int_{-\infty}^{\infty} H_n(z) \frac{d^n}{dz^n} e^{-z^2} \, dz = 2^n \int_{-\infty}^{\infty} e^{-z^2} \, dz = 2^n n! \sqrt{\pi}. \tag{2.5.54}$$

□

The following recurrence relation holds:

$$H_{n+1}(z) - 2z H_n(z) + 2n H_{n-1}(z) = 0. \tag{2.5.55}$$

Since the biconfluent hypergeometric equation can be obtained from the confluent hypergeometric equation under the substitution $z \mapsto z^2$, Hermite polynomials can be expressed in terms of Laguerre polynomials.

3

THE HEUN CLASS OF EQUATIONS

3.1 A classification scheme

3.1.1 The Heun equation

The Heun class of linear second-order ODEs is generated by the Heun equation—the Fuchsian equation with four singularities. The Heun equation was first studied by K. Heun [51]. A hundred years later, a centennial workshop on the Heun equation was organized by A. Ronveaux, A. Seeger, and W. Lay, where achievements in the theory were discussed [113]. Attention was mainly devoted to the confluent Heun equations first presented in details in articles [33, 34]. A result of the conference was the appearence of the first book on Heun equations in 1995 [107].

The general form of this equation is characterized by twelve generally complex parameters, of which four define the locations of the singularities, seven are the characteristic exponents of the singularities (one is lacking, according Fuchs's theorem) and one is the nonlocal accessory parameter. If, by means of a Möbius transformation, three of the singularities are fixed at the points $z_0 = 0$, $z_1 = 1$, $z_4 = \infty$, respectively, and three of the characteristic exponents take zero values, then the number of parameters is reduced to six.

Four confluent equations arise from the general Heun equation by means of different confluence processes. Five reduced confluent equations appear as a result of weak confluence processes. If elementary singularities are regarded as special types of singularities, then ten more equations are to be added to the Heun class. The full classification scheme is given in Tables 1.3.3 and 3.1.1.

The standard canonical natural form of Heun's equation reads

$$L_z^{\{1,1,1;1\}}(a, b; c, d; t)y(z) - \lambda y(z) :=$$
$$\big[z(z-1)(z-t)D^2 + \{c(z-1)(z-t) + dz(z-t)$$
$$+ (a+b+1-c-d)z(z-1)\}D + (abz - \lambda)\big]y(z) = 0. \tag{3.1.1}$$

Local characteristic properties of this equation are exhibited by the corresponding GRS

$$\begin{pmatrix} 1 & 1 & 1 & 1 & \\ 0 & 1 & t & \infty & ; z \\ 0 & 0 & 0 & a & ; \lambda \\ 1-c & 1-d & c+d-a-b & b & \end{pmatrix}.$$

Other standard forms of eqn (3.1.1) may be found in Table 3.1.1.

Table 3.1.1. Standard forms of Heun equation

c	$L_z^{\{1,1,1;1\}}y(z) =$ $\lambda y(z), z_1 = 0,$ $z_2 = 1, z_3 = t$	$L_z^{\{1,1,1;1\}}(a, b; c_j; t) = r(z)D^2$ $\sum_{j=1}^{3} c_j(z - z_{j+1})(z - z_{j+2})D + abz,$ $j = j \ (\text{mod } 3)$	$\sum_{j=1}^{3} c_j = a + b + 1$ $r(z) = \prod_{j=1}^{3}(z - z_j)$
c	$\tilde{L}_z^{\{1,1,1;1\}}y(z) =$ $\lambda(r(z))^{-1}y(z)$	$\tilde{L}_z^{\{1,1,1;1\}}(a, b; c_j; t) = D^2 +$ $\sum_{j=1}^{3} \frac{c_j}{z - z_j}D + \frac{abz}{r(z)}$	$\tilde{L} = L/r(z)$
n	$N_z^{\{1,1,1;1\}}w(z) =$ $\tilde{\lambda}(r(z))^{-1}w(z)$	$N_z^{\{1,1,1;1\}}(a, b; c_j; t) = D^2 +$ $\frac{1}{4}\sum_{j=1}^{3}\frac{1 - (1 - c_j)^2}{(z - z_j)^2} + \frac{1}{r(z)} \times$ $\left(ab - \sum_{i,j=1,i\neq j}^{3} c_i c_j/4\right)z$	$y = \prod_{j=0}^{3}(z - z_j)^{-c_j/2}w$ $\tilde{\lambda} = \lambda - \sum_{j=1}^{3} c_j c_{j+1} z_{j+2}$
s-a	$M_z^{\{1,1,1;1\}}v(z) =$ $\hat{\lambda}v(z)$	$M_z^{\{1,1,1;1\}}(a, b; c_j; t) =$ $Dr(z)D - \frac{r(z)}{4}\sum_{j=1}^{3}\frac{(1 - c_j)^2}{(z - z_j)^2} +$ $\left(ab - \left(\sum_{i,j=1,i\neq j}^{3}\frac{c_i c_j}{4} - \frac{3}{2}\right)\right)z$	$w = r(z)^{1/2}v$ $\hat{\lambda} = \tilde{\lambda} + \sum_{j=1}^{3}\frac{z_j}{2}$

Local solutions of eqn (3.1.1) may be constructed in a vicinity of each of the four regular singularities. The standard solution of eqn (3.1.1), valid for

$$c \neq 0, -1, -2, \ldots, \qquad |z| < \min(|t|, 1),$$

is the Heun function

$$y_1^{\{1,1,1;1\}}(z) = y_1^{\{1,1,1;1\}}(a, b; c, d; t; \lambda; z = 0, z), \qquad y_1^{\{1,1,1;1\}}(0) = 1.$$

It is one of the two Frobenius solutions at zero, defined by the power series

$$y_1^{\{1,1,1;1\}}(z) = \sum_0^{\infty} g_k^{\{1,1,1;1\}} z^k. \tag{3.1.2}$$

The coefficients $g_k^{\{1,1,1;1\}}$ are obtained from the three-term recurrence relation generated under substitution of (3.1.2) into eqn (3.1.1):

$$t(k + 1)(k + c)g_{k+1}^{\{1,1,1;1\}}$$
$$- \{k[(k - 1 + c)(1 + t) + dt + a + b + 1 - c - d] + \lambda\}g_k^{\{1,1,1;1\}}$$
$$+ (k - 1 + a)(k - 1 + b)g_{k-1}^{\{1,1,1;1\}} = 0, \qquad g_0^{\{1,1,1;1\}} = 1.$$

The other linear independent particular solution of Heun's equation is given by the function

$$y_2^{\{1,1,1;1\}}(a,b;c,d;t;\lambda;z=0,z) =$$

$$z^{1-c}y_1^{\{1,1,1;1\}}(a+1-c,b+1-c;2-c,d;t;\lambda';z=0,z),$$

$$\lambda' = \lambda + (a+b+1-c)(1-c).$$

A modified form of (3.1.1) with more symmetrical locations of the singularities is

$$L_z^{\{1,1,1;1\}}(a,b;c,d;t)y(z) - \lambda y(z) =$$

$$\{(z^2-1)(z-t)D^2 + [c(z-1)(z-t)$$

$$+ d(z+1)(z-t) + (a+b+1-c-d)(z^2-1)]D.$$

$$+ (abz-\lambda)\}\,y(z) = 0.$$

The parameters a,b,c,d,t,λ characterizing (3.1.1) can be divided into three groups:

(1) Local (dimensionless) parameters a, b, c, d, determining the characteristic exponents of the Frobenius solutions at the singularities.

(2) A scaling parameter t, determining the location of one of the singularities.

(3) An accessory parameter λ. It usually serves as a spectral parameter.

Under the conditions $c = d = c+d-a-b = 1/2$, eqn (3.1.1) has three elementary singularities at $z=0$, $z=1$, and $z=t$ and is known as the algebraic Lamé equation (see [71]).

3.1.2 Confluent Heun equations

Among the confluent equations belonging to Heun's class first comes the singly confluent equation (CHE)

$$L_z^{\{1,1;2\}}(a;c,d;t;)y(z) + \lambda y(z) := \left[z(z-1)D^2\right.$$

$$\left. + \{-tz(z-1)+c(z-1)+dz\}D + (-taz+\lambda)\right]y(z) = 0. \qquad (3.1.3)$$

It is generated from eqn (3.1.1) by a confluence process of the singular points at $z=t$ and at $z=\infty$. The corresponding limiting processes are set out in Table 3.1.2. The number of parameters under a confluence process is decreased by one. There are several standard forms of the CHE.

The corresponding GRS to eqn (3.1.3) reads

$$\begin{pmatrix} 1 & 1 & 2 & & \\ 0 & 1 & \infty & ;z & \\ 0 & 0 & a & ;\lambda & \\ 1-c & 1-d & c+d-a & & \\ & & 0 & & \\ & & t & & \end{pmatrix}. \qquad (3.1.4)$$

Table 3.1.2.   Standard forms of confluent Heun equation

	HE $\mapsto$ CHE		
	$t := \epsilon^{-1},$ $b := p\epsilon^{-1},$ $\epsilon \to 0,\ p := t$	$-\epsilon L^{\{1,1,1;1\}} \to L^{\{1,1;2\}}$	
c	$(L_z^{\{1,1;2\}} + \lambda)\times$ $y(z) = 0$	$L_z^{\{1,1;2\}}(a; c_j; t) = r(z)D^2 +$ $\left(\sum_{j=1}^{2} c_j(z - z_{j+1})\right) - tr(z)\times$ $D - atz, \quad j = j\,(\mathrm{mod}\,2)$	$z_1 = 0,\ z_2 = 1,$ $c_1 = c,\ c_2 = d$ $r(z) = \prod_{j=1}^{2}(z - z_j)$
c	$(\tilde{L}_z^{\{1,1;2\}} + \lambda(r(z))^{-1})\times$ $y(z) = 0$	$\tilde{L}_z^{\{1,1;2\}}(a; c_j; t) = D^2 +$ $\left(\sum_{j=1}^{2}\frac{c_j}{z - z_j} - t\right)D - \frac{atz}{r(z)}$	$\tilde{L} = L/r(z)$
n	$(N_z^{\{1,1;2\}} + \tilde{\lambda}(r(z))^{-1})\times$ $w(z) = 0$	$N_z^{\{1,1;2\}}(a; c_j; t) = D^2 -$ $\frac{t^2}{4} + \sum_{j=1}^{2}\frac{1 - (1 - c_j)^2}{4(z - z_j)^2} -$ $\frac{1}{r(z)}\left(a - \sum_{j=1}^{2}\frac{c_j}{2}\right)tz$	$y = \prod_{j=1}^{2}(z - z_j)^{-c_j/2}\cdot e^{tz/2}w$ $\tilde{\lambda} = \lambda + \sum_{j=1}^{2}\frac{c_j c_{j+1}}{4} +$ $t\sum_{j=1}^{2}\frac{c_j z_{j+1}}{2}$
s-a	$(M_z^{\{1,1;2\}} + \hat{\lambda})\times$ $v(z) = 0$	$M_z^{\{1,1;2\}}(a; c_j; t) = DrD -$ $\frac{1}{4}r(z)\left(t^2 + \sum_{j=1}^{2}\frac{(1 - c_j)^2}{(z - z_j)^2}\right) -$ $\left(ab - \sum_{j=1}^{2}\frac{c_j}{2}\right)tz$	$w = r^{1/2}v$ $\hat{\lambda} = \tilde{\lambda} - 1/2$

In (3.1.3)–(3.1.4), the notation λ is kept for the accessory parameter, and notations a, c, d refer to the local parameters. The scaling parameter t now determines not the location of the singularity but the location of the turning points. This will be discussed below.

In applications, the following modified self-adjoint form of the CHE often appears. In this form, regular singularities are located at the points $z = \pm 1$:

$$\frac{\mathrm{d}}{\mathrm{d}z}(z^2 - 1)\frac{\mathrm{d}}{\mathrm{d}z}v(z) + \left(-t^2(z^2 - 1) + 2taz - \nu - \frac{m^2 + s^2 + 2msz}{z^2 - 1}\right)v(z) = 0.$$

(3.1.5)

Equation (3.1.5) is called the generalized spheroidal equation (GSE). It is generated by (3.1.1) under the substitution $z \mapsto 2z - 1$. From a physical point of view, eqn (3.1.5) is a Schrödinger equation the potential of which consists of two Coulomb and two centrifugal terms. If the latter are equal, i.e. if $s = 0$, then the GSE specializes to the

Coulomb spheroidal equation

$$\frac{d}{dz}(z^2 - 1)\frac{d}{dz}v(z) + \left(-t^2(z^2 - 1) + 2taz - v - \frac{m^2}{z^2 - 1}\right)v(z) = 0. \quad (3.1.6)$$

If $a = 0$, then equation (3.1.6) simplifies to the spheroidal equation

$$\frac{d}{dz}(z^2 - 1)\frac{d}{dz}v(z) + \left(-t^2(z^2 - 1) - v - \frac{m^2}{z^2 - 1}\right)v(z) = 0. \quad (3.1.7)$$

For $m = \frac{1}{2}$, eqn (3.1.7) reduces to the Mathieu equation

$$\frac{d}{dz}(z^2 - 1)\frac{d}{dz}v(z) + \left(-t^2(z^2 - 1) - v - \frac{1}{4(z^2 - 1)}\right)v(z) = 0.$$

Further modifications of the CHE will be studied in Chapter 4. Local solutions of eqn (3.1.3) may be constructed either in the vicinities of the regular singularities at $z = 0$ and $z = 1$ or in the vicinity of the irregular singularity $z = \infty$. In the vicinity of the point $z = 0$, a standard solution of the CHE may be presented as a Taylor series valid for $c \neq 0, -1, -2, \ldots$ and $|z| < 1$:

$$y_1^{\{1,1;2\}}(a; c, d; t; \lambda; z = 0, z) = \sum_0^\infty g_k^{\{1,1;2\}} z^k. \quad (3.1.8)$$

The coefficients $g_k^{\{1,1;2\}}$ are obtained from the three-term recurrence relation which is generated by substituting the series (3.1.8) into eqn (3.1.3):

$$(k+1)(k+c)g_{k+1}^{\{1,1;2\}} - [k[(k-1+c)(1+t) + d + t] - \lambda]g_k^{\{1,1;2\}}$$
$$+ t(k-1+a)g_{k-1}^{\{1,1;2\}} = 0, \qquad g_0^{\{1,1;2\}} = 1. \quad (3.1.9)$$

Another way to obtain (3.1.9) is to perform a confluence process in eqn (3.1.3). The second standard particular solution in a vicinity of zero is the function

$$y_2^{\{1,1;2\}}(a, b; c, d; t; \lambda; z = 0, z) = z^{1-c} y_1^{\{1,1;2\}}(a + 1 - c; 2 - c, d; t; \lambda'; z = 0, z),$$
$$\lambda' = \lambda + (a + 1 - c)(1 - c).$$

The second solution of the CHE may be constructed as a local solution at infinity (Thomé solution) valid as a recessive solution for $\operatorname{Re} t > 0$:

$$y_1^{\{1,1;2\}}(a; c, d; t; \lambda; z = +\infty, z) = z^{-a} \sum_0^\infty h_k^{\{1,1;2\}} z^k \quad (3.1.10)$$

with the following three-term recurrence relation for the coefficients $h_k^{\{1,1;2\}} z^k$:

$$t(k+1)h_{k+1}^{\{1,1;2\}} + [(k+a)(k+a+1+t-d-c) + \lambda]h_k^{\{1,1;2\}}$$
$$+ (k+a-c)(k-1+a)h_{k-1}^{\{1,1;2\}} = 0, \qquad h_0^{\{1,1;2\}} = 1.$$

Confluence processes of singularities of the CHE may be carried out in two ways: either the regular singularity at $z = 1$ and the irregular singularity at infinity coalesce, generating an irregular singularity at infinity the s-rank of which is $R = 3$, or two regular singularities at $z = 0$ and $z = 1$ coalesce generating an irregular singularity at zero the s-rank of which is $R = 2$.

In the former case, the biconfluent Heun equation (BHE) arises as a result:

$$L_z^{\{1;3\}}(a; c; t)y(z) + \lambda y(z) :=$$
$$\left[zD^2 + (-z^2 - tz + c)D + (-az + \lambda) \right] y(z) = 0. \qquad (3.1.11)$$

It corresponds to the GRS

$$\begin{pmatrix} 1 & 3 & \\ 0 & \infty & ; z \\ 0 & a & ; \lambda \\ 1-c & c+1-a & \\ & 0 & \\ & t & \\ & 0 & \\ & 1 & \end{pmatrix}.$$

Other standard forms of equation (3.1.11) and the related confluence processes are given in Table 3.1.3. Equation (3.1.11) arises on studying the rotating oscillator in quantum mechanics, scattering problems on paraboloids etc.

A standard solution of the BHE in the vicinity of zero valid for $c \neq 0, -1, -2, \ldots$ is the function

$$y_1^{\{1;3\}}(a; c; t; \lambda; z = 0, z) := \sum_0^\infty g_k^{\{1;3\}} z^k. \qquad (3.1.12)$$

The coefficients $g_k^{\{1;3\}}$ obey the three-term recurrence relation

$$(k + 1)(k + c)g_{k+1}^{\{1;3\}} + (tk + \lambda)\, g_k^{\{1;3\}}$$
$$- (k - 1 + a)g_{k-1}^{\{1;3\}} = 0, \qquad g_0^{\{1;3\}} = 1. \qquad (3.1.13)$$

The recessive Thomé solution at plus infinity reads

$$y_1^{\{1;3\}}(a; c; t; \lambda; z = +\infty, z) := z^{-a} \sum_0^\infty h_k^{\{1;3\}} z^{-k} \qquad (3.1.14)$$

with the following three-term recurrence relation for the coefficients $h_k^{\{1;3\}}$:

$$(k + 1)h_{k+1}^{\{1;3\}} + [t(k + a) + \lambda]\, h_k^{\{1;3\}}$$
$$+ (k + a - c)(k - 1 + a)h_{k-1}^{\{1;3\}} = 0, \qquad h_0^{\{1;3\}} = 1.$$

Table 3.1.3. Standard forms of biconfluent Heun equation

	CHE $\mapsto$ BHE		
	$z := \epsilon z, \; t := -\epsilon^{-2},$ $d := p\epsilon^{-1} + \epsilon^{-2},$ $\epsilon \to 0, \; p := t$	$-\epsilon L^{\{1,1;2\}} \to L^{\{1;3\}}$	
c	$\left(L_z^{\{1;3\}} + \lambda\right) y(z) = 0$	$L_z^{\{1;3\}}(a; c; t) = zD^2 +$ $(-z^2 - tz + c)D - az$	
c	$\left(\tilde{L}_z^{\{1;3\}} + \dfrac{\lambda}{z}\right)y(z) = 0$	$\tilde{L}_z^{\{1;3\}}(a; c; t) = D^2 +$ $\left(-z - t + \dfrac{c}{z}\right)D - a$	$\tilde{L} = L/z$
n	$\left(N_z^{\{1;3\}} + \dfrac{\tilde\lambda}{z}\right)y(z) = 0$	$N_z^{\{1;3\}}(a; c; t) = D^2 + \dfrac{1-(1-c)^2}{4z^2} -$ $\dfrac{1}{4}((z+t)^2 - 2(c+1) + 4a)$	$y = e^{z^2/4 + tz/2} \cdot z^{-c/2} w$ $\tilde\lambda = \lambda - ct/2$
s-a	$\left(M_z^{\{1;3\}} v(z) + \hat\lambda\right)v(z) = 0$	$M_z^{\{1;3\}}(a; c; t) = DzD - \dfrac{(1-c)^2}{4z} -$ $\dfrac{z}{4}((z+t)^2 - 2(c+1) + 4a)$	$w = z^{1/2} v$ $\hat\lambda = \tilde\lambda$

Table 3.1.4. Standard forms of doubly confluent Heun equations

	CHE $\mapsto$ DHE		
	$z := \epsilon^{-1}z, \; t := \epsilon,$ $c := c - p\epsilon^{-1},$ $d := p\epsilon^{-1},$ $\epsilon \to 0, \; p := t$	$-\epsilon L^{\{1,1;2\}} \to L^{\{2;2\}}$	
c	$\left(L_z^{\{2;2\}} + \lambda\right)y(z) = 0$	$L_z^{\{2;2\}}(a; c; t) = z^2 D^2 +$ $(-z^2 + t + cz)D - az$	
c	$\left(\tilde{L}_z^{\{2;2\}} + \dfrac{\lambda}{z^2}\right)y(z) = 0$	$\tilde{L}_z^{\{2;2\}}(a; c; t) = D^2 +$ $\left(-1 + \dfrac{t}{z^2} + \dfrac{c}{z}\right)D - \dfrac{a}{z}$	$\tilde{L} = L/z^2$
n	$\left(N_z^{\{2;2\}} + \dfrac{\tilde\lambda}{z^2}\right) = 0$	$N_z^{\{2;2\}}(a; c; t) = D^2 - \left(\dfrac{1}{2} - \dfrac{t}{2z^2}\right)^2 -$ $\dfrac{a - c/2}{z} + \dfrac{t(1 - c/2)}{z^3}$	$y = e^{z/2 + t/2z} \cdot z^{-c/2} w$
s-a	$\left(M_z^{\{2;2\}} + \hat\lambda\right)v(z) = 0$	$M_z^{\{2;2\}}(a; c; t) = Dz^2 D - \left(\dfrac{z}{2} - \dfrac{t}{2z}\right)^2 -$ $(a - c/2)z + \dfrac{t(1 - c/2)}{z}$	$w = zv$

The second variant of a confluence process described above leads to the doubly confluent Heun equation (DHE)

$$L_z^{\{2;2\}}(a, c, t)y(z) + \lambda y(z) := \left(z^2 D^2 + (-z^2 + cz + t)D\right.$$
$$+ (-az + \lambda))y(z) = 0. \tag{3.1.15}$$

The corresponding GRS reads

$$\begin{pmatrix} 2 & 2 & \\ 0 & \infty & ; z \\ 0 & a & ; \lambda \\ 2-c & c-a & \\ 0 & 0 & \\ -t & 1 & \end{pmatrix}.$$

We define two Thomé solutions of the DHE. One of these solutions is recessive as $z \to +\infty$:

$$y_1^{\{2;2\}}(a; c; t; \lambda; z = +\infty, z) := z^{-a}\sum_0^\infty h_k^{\{2;2\}} z^{-k} \tag{3.1.16}$$

with the following three-term recurrence relation for the coefficients $h_k^{\{2;2\}}$:

$$- (k+1)h_{k+1}^{\{2;2\}} + [k(k+1+2a-c) + \lambda + a(a+1-c)]h_k^{\{2;2\}}$$
$$- t(k-1+a)h_{k-1}^{\{2;2\}} = 0, \qquad h_0^{\{2;2\}} = 1.$$

The other Thomé solution which is recessive as $\mathrm{Re}\, t > 0$ may be constructed in the form

$$y_1^{\{2;2\}}(a; c; t; \lambda; z = +0, z) := \sum_0^\infty f_k^{\{2;2\}} z^k. \tag{3.1.17}$$

The coefficients $f_k^{\{2;2\}}$ obey the three-term recurrence relation

$$t f_{k+1}^{\{2;2\}} + [k(k+c-1) + \lambda]f_k^{\{2;2\}}$$
$$- (k-1-a)f_{k-1}^{\{2;2\}} = 0, \qquad f_0^{\{2;2\}} = 1.$$

The last equation arising under a strong confluence process is the triconfluent Heun equation (THE). It has one irregular point at infinity, which is a result of a confluence process of all four regular singularities of the starting HE. It is characterized by the s-multisymbol $\{; 4\}$:

$$L_z^{\{;4\}}(a; p;)y(z) + \lambda y(z) := (D^2 + (-z^2 - t)D + (-az + \lambda))y(z) = 0. \tag{3.1.18}$$

Table 3.1.5. Standard forms of triconfluent Heun equations

BHE $\mapsto$ THE	
$z := \epsilon^{-1}z + \epsilon^{-3}$, $t := -2\epsilon^{-3}$, $c := -\epsilon^{-6} - p\epsilon^{-2}$, $\epsilon \to 0, \ p := t$	$\epsilon L^{\{1;3\}} \to L^{\{;4\}}$
c $(L_z^{\{;4\}} + \lambda)y(z) = 0$	$L_z^{\{;4\}}(a; t) = D^2 + $ $(-z^2 - t)D - az$
n $(N_z^{\{;4\}} + \tilde{\lambda})w(z) = 0$	$M_z^{\{;4\}}(a; t) = D^2 - $ $y = e^{z^3/6 + tz/2}w$ $\dfrac{1}{4}(z^2 + t)^2 - az$

The corresponding GRS reads

$$
\begin{pmatrix}
4 & \\
\infty & ; z \\
a & ; \lambda \\
2 - a & \\
0 & \\
t & \\
0 & \\
0 & \\
0 & \\
1 &
\end{pmatrix}.
$$

The normal form of eqn (3.1.18) is known in the literature as the differential equation describing the quantum quartic oscillator. The Thomé solution, which is recessive at plus infinity, may be constructed as

$$
y_1^{\{;4\}}(a; t; \lambda; z = +\infty, z) := z^{-a} \sum_0^\infty h_k^{\{;4\}} z^{-k} \tag{3.1.19}
$$

with the following three-term recurrence relation for the coefficients $h_k^{\{;4\}}$:

$$
(k + 1)h_{k+1}^{\{;4\}} + [k(k - 1 + 2a) + \lambda + a(a + 1)]\, h_k^{\{;4\}}
$$
$$
+ t(k - 1 + a)h_{k-1}^{\{;4\}} = 0, \qquad h_0^{\{;4\}} = 1.
$$

3.1.3   Reduced confluent Heun equations

In the following, the reduced confluent equations belonging to Heun's class are studied. These equations may be considered as a result of a weak confluence process.

There are five equations of this sort. All of them have at least one singularity with a half-integer s-rank. In the following, only the equations are exhibited. Solutions which are mainly Thomé solutions in the form of subnormal asymptotic series are not studied.

First comes the reduced singly confluent Heun equation (RCHE)

$$(L_z^{\{1,1;3/2\}}(a; c, d; t) + \lambda)y(z) := (z(z-1)D^2$$
$$+ (c(z-1) + dz)D + (-tz + \lambda))y(z) = 0. \qquad (3.1.20)$$

It appears by a weak confluence process, i.e. when the singularities at $z = t$ and at $z = \infty$ coalesce according to

$$t \mapsto \epsilon^{-1}, \; b \mapsto (t/\epsilon)^{1/2}, \; a \mapsto (t/\epsilon)^{1/2}, \; \lambda \mapsto -\lambda\epsilon^{-1}, \; \epsilon \to \infty$$
$$\implies \quad (-\epsilon)(L_z^{\{1,1,1;1\}})(a, b; c, d; t) \mapsto (L_z^{\{1,1;3/2\}})(c, d; t).$$

The corresponding GRS reads

$$\begin{pmatrix} 1 & 1 & 3/2 & \\ 0 & 1 & \infty & ; z \\ 0 & 0 & (c+d)/2 - 1/4 & ; \lambda \\ 1-c & 1-d & (c+d)/2 - 1/4 & \\ & & \sqrt{t} & \end{pmatrix}.$$

From eqn (3.1.20), the reduced biconfluent Heun equation (RBHE) may be obtained. It has an irregular singularity at infinity, the s-rank of which is $R(\infty) = 5/2$:

$$(L_z^{\{1;5/2\}}(c; t) + \lambda)y(z) := (zD^2 + cD + (-z^2 - tz + \lambda))y(z) = 0. \qquad (3.1.21)$$

The corresponding GRS reads

$$\begin{pmatrix} 1 & 5/2 & \\ 0 & \infty & ; z \\ 0 & 1/4 - c/2 & ; \lambda \\ 1-c & 1/4 - c/2 & \\ & t/2 & \\ & 0 & \\ & 1 & \end{pmatrix}.$$

If, instead of a strong confluence process at zero, a weak one takes place, then we obtain the reduced doubly confluent, Heun equation (RDHE). However, preliminary transformation of the equation is needed. Namely, from the function $y(z)$ in (3.1.3), we transform to the function $\check{y}(z)$:

$$y = \prod_{j=1}^{2} (z - z_j)^{-c_j/2} \check{y}.$$

Table 3.1.6. Standard forms of the reduced confluent Heun equation

HE $\mapsto$ RCHE

	$a := b := p^{1/2}\epsilon^{-1},$ $t := \epsilon^{-2},\ \epsilon \to 0,$ $p := t$	$-\epsilon L^{\{1,1,1;1\}} \to L^{\{1,1;3/2\}}$	

c $\quad \left(L_z^{\{1,1;3/2\}} + \lambda\right)y(z) = 0$

$\qquad L_z^{\{1,1;3/2\}}(c_j;t) = r(z)D^2 + \left(\sum_{j=1}^{2} c_j(z - z_{j+1})\right)D - tz,$

$\qquad j = j \pmod 2$

$\qquad z_1 = 0,\ z_2 = 1,$
$\qquad c_1 = c,\ c_2 = d$
$\qquad r(z) = \prod_{j=1}^{2}(z - z_j)$

c $\quad \left(\tilde{L}_z^{\{1,1;3/2\}} + \dfrac{\lambda}{r(z)}\right)y(z) = 0$

$\qquad \tilde{L}_z^{\{1,1;3/2\}}(c_j;t) = D^2 + \left(\sum_{j=1}^{2}\dfrac{c_j}{z - z_j}\right)D - \dfrac{tz}{r(z)}$

$\qquad \tilde{L} = L/r(z)$

n $\quad \left(N_z^{\{1,1;3/2\}} + \dfrac{\nu}{r(z)}\right)w(z) = 0$

$\qquad N_z^{\{1,1;3/2\}}(c_j;t) = D^2 + \sum_{j=1}^{2}\dfrac{1 - (1 - c_j)^2}{4(z - z_j)^2} - \dfrac{tz}{r(z)}$

$\qquad y = \prod_{j=1}^{2}(z - z_j)^{-c_j/2}w$
$\qquad \nu = \lambda + \sum_{j=1}^{2}\dfrac{c_j(c_{j+1} + 2tz_{j+1})}{4}$

s-a $\quad \left(M_z^{\{1,1;3/2\}} + \mu\right)v(z)$

$\qquad M_z^{\{1,1;3/2\}}(c_j;t) = DrD - \dfrac{r(z)}{4}\sum_{j=1}^{2}\dfrac{(1 - c_j)^2}{(z - z_j)^2} - tz$

$\qquad w = r^{1/2}v$
$\qquad \mu = \nu - 1/2$

The CHE transforms to the equation

$$r(z)\left(\check{L}_z^{\{1,1;2\}}(a;c_j;t) + \check{\lambda}\right)\check{y}(z) = \left(D^2 - tD + \right.$$

$$\sum_{j=1}^{2}\dfrac{1 - (1 - c_j)^2}{4(z - z_j)^2} + t\sum_{j=1}^{2}\dfrac{c_j}{2(z - z_j)} - \dfrac{at}{z - 1} + \check{\lambda}\left.\right)\check{y}(z) = 0.$$

From this equation, by means of a weak confluence process, we obtain

$$(L_z^{\{3/2;2\}}(a,t) + \lambda)y(z) := \left(z^2 D^2 - z^2 D + (-az - t/z + \lambda)\right)y(z) = 0.$$

$$(3.1.22)$$

The corresponding GRS reads

$$\begin{pmatrix} 3/2 & 2 & \\ 0 & \infty & ; z \\ 3/4 & a & ; \lambda \\ 3/4 & -a & \\ \sqrt{t} & 0 & \\ & 1 & \end{pmatrix}.$$

Table 3.1.7. Standard forms of the reduced biconfluent Heun equation

CHE $\mapsto$ RBHE			
$\tilde{L}_z^{\{1,1;2\}} \mapsto \hat{L}_z^{\{1,1;2\}}$	$\hat{L}_z^{\{1,1;2\}}(a;c;t) = D^2 + \dfrac{c}{z}D-$		$y = (z-1)^{-d/2}\cdot e^{t/2}\hat{y}$
$\hat{L}_z^{\{1,1;2\}}\hat{y}(z) = \dfrac{\lambda\hat{y}(z)}{z}$	$\dfrac{t^2}{4} + \dfrac{1-(1-d)^2}{4(z-1)^2} + \dfrac{ct}{2z} +$		
	$\dfrac{(-a+d/2)t}{z-1}$		
$z := \epsilon^2 z,\ t := 1,$ $4a := p\epsilon^{-4} + \epsilon^{-6},$ $d := \epsilon^{-3}\epsilon \to 0,$ $p := t$	$\epsilon^{-4}\hat{L}^{\{1,1;2\}} \to \tilde{L}^{\{1;5/2\}}$		
c	$L_z^{\{1;5/2\}}y(z) = \lambda y(z)$	$L_z^{\{1;5/2\}}(c;t) = zD^2 + cD-$ $z^2 - tz$	
c	$\tilde{L}_z^{\{1;5/2\}}y(z) = \dfrac{\lambda y(z)}{z}$	$\tilde{L}_z^{\{1;5/2\}}(c;t) = D^2+$ $\dfrac{c}{z}D - z - t$	$\tilde{L} = L/z$
n	$M_z^{\{1;5/2\}}w(z) = \dfrac{\mu w(z)}{z}$	$M_z^{\{1;5/2\}}(c;t) = D^2-$ $z - t + \dfrac{1-(1-c)^2}{4z^2}$	$y = z^{-c/2}w$ $\mu = \lambda$
s-a	$N_z^{\{1;5/2\}}v(z) = vv(z)$	$N_z^{\{1;5/2\}}(c;t) = DzD-$ $z^2 - tz - \dfrac{(1-c)^2}{4z}$	$w = z^{1/2}v$ $v = \mu$

The doubly reduced doubly confluent Heun equation reads

$$(L_z^{\{3/2;3/2\}}(t) + \lambda)y(z) = (z^2D^2 + (-z + \lambda - t/z))y(z) = 0. \qquad (3.1.23)$$

It corresponds to the GRS

$$\begin{pmatrix} 3/2 & 2 & \\ 0 & \infty & ;z \\ 3/4 & 1/4 & ;\lambda \\ 3/4 & -1/4 & \\ \sqrt{t} & 1 & \end{pmatrix}.$$

Eventually comes the reduced triconfluent Heun equation

$$(L_z^{\{;7/2\}}(t) + \lambda)y(z) := \left(D^2 + (-z^3 - tz + \lambda)\right)y(z) = 0 \qquad (3.1.24)$$

Table 3.1.8. Standard forms of the reduced doubly confluent Heun equation

CHE $\mapsto$ RDHE		
$z := \epsilon^{-2}z,\ t := \epsilon^2,\ d := -c,$ $c := c - p\epsilon^{-1}, c := (2p)^{1/2}\epsilon^{-1},$ $\epsilon \to 0,\ p := t$	$\epsilon^{-4}\check{L}^{\{1,1;2\}} \to L^{\{3/2;2\}}$	
c	$(L_z^{\{3/2;2\}} + \lambda)y(z)$	$L_z^{\{3/2;2\}}(a;t) = z^2D^2 - z^2D - az - \dfrac{t}{z}$	
c	$\tilde{L}_z^{\{3/2;2\}}y(z) = \dfrac{\lambda y(z)}{z^2}$	$\tilde{L}_z^{\{3/2;2\}}(a;t) = D^2 - D - \dfrac{t}{z^3} - \dfrac{a}{z}$	$\tilde{L} = L/z^2$
n	$\left(N_z^{\{3/2;2\}} + \dfrac{\nu}{z^2}\right)w(z) = 0$	$N_z^{\{3/2;2\}}(a;t) = D^2 - \dfrac{1}{4} - \dfrac{t}{z^3} - \dfrac{a}{z}$	$y = e^{z/2}w$
s-a	$(M_z^{\{3/2;2\}} + \mu v(z)) = 0$	$M_z^{\{3/2;2\}}(a;t) = Dz^2D - \dfrac{z^2}{4} - az - \dfrac{t}{z}$	$w = zv$

Table 3.1.9. Standard forms of DRDHE

RCHE $\mapsto$ DRDHE		
$z := -\epsilon^{-2}z,\ t := -p\epsilon^2,$ $c := 2^{1/2}\epsilon^{-1}, \epsilon \to 0,\ p := t$	$\epsilon^{-4}M^{\{1,1;3/2\}} \to M^{\{3/2;3/2\}}$	
n	$\left(N_z^{\{3/2;3/2\}} + \dfrac{\nu}{z^2}\right)w(z) = 0$	$N_z^{\{3/2;3/2\}}(t) = D^2 - \dfrac{t}{z} - \dfrac{1}{z^3}$
s-a	$(M_z^{\{3/2;3/2\}} + \mu)w(z) = 0$	$M_z^{\{3/2;3/2\}}(t) = Dz^2D - tz - \dfrac{1}{z}$

Table 3.1.10. Standard forms of the reduced triconfluent Heun equation

BHE $\mapsto$ RTHE		
$z := \epsilon^{-4}z - \epsilon^{-10},\ t := 3\epsilon^{-10} - p\epsilon^2,$ $\lambda := 3\epsilon^{-20} - p\epsilon^{-8} - \nu\epsilon^{-2}, c := 2\epsilon^{-15} + 1$ $\epsilon \to 0,\ p := t,\ \nu := \lambda$	$\epsilon^{-8}N^{\{1;5/2\}} \to N^{\{;7/2\}}$	
n	$(N_z^{\{;7/2\}} + \nu)w(z)$	$N_z^{\{;7/2\}}(t) = D^2 - z^3 - tz$

having the GRS

$$
\begin{pmatrix}
7/2 & \\
\infty & ; z \\
3/4 & ; \lambda \\
3/4 & \\
t & \\
0 & \\
0 & \\
1 &
\end{pmatrix}.
$$

3.2 Types of solutions

3.2.1 Solutions of the Heun equation

Several types of solutions of second-order linear homogeneous ODEs can be distinguished as the most important for physical applications. These are: (1) polynomial solutions, (2) eigenfunctions, (3) path-multiplicative (Floquét) solutions, (4) local solutions (of Frobenius and Thomé type), (5) scattering-problem solutions. A preliminary discussion can be found in Section 1.4. Here we give a more refined treatment starting from the Heun equation.

First, local solutions at regular singularities are studied. A canonical form of the equation can be taken as basic. One of the Frobenius solutions at the singularity $z = 0$ has been introduced and studied in the previous section. It was denoted by $y_1^{\{1,1,1;1\}}(a, b; c, d; t; \lambda; z = 0, z)$. Another notation for this function introduced in the book [107] is $Hl(t, \lambda; a, b, c, d; z)$. Since there are four singularities and two Frobenius solutions at each singularity, the total number of local solutions for a given equation is eight. However, it is clear that the transformation

$$
y(z) = (z - z_j)^{\rho_2(z_j)} u(z), \tag{3.2.1}
$$

where z_j is an arbitrary regular singularity and $\rho(z_j)$ is the nonzero characteristic exponent at this singularity, preserves the canonical form of the equation. This means that, with the help of (3.2.1), it is possible to construct 24 solutions located at singularities at finite points. On the other hand, the singularity at infinity can be exchanged with any finite singularity, which increases the total number of local solutions to 72. Of course, there are many simple relations between these solutions. The simplest way to produce these solutions is to use the generalized Riemann scheme. What is important is that all the mentioned solutions can be expressed in terms of the function $y^{\{1,1,1;1\}}(a, b, c, d; t; \lambda; z)$ defined in the previous section. As the argument of the function serves in general a bilinear function of z and all parameters beyond t can be altered. In the case of the Heun equation, in contrast to the hypergeometric equation, there are—in general—no explicit expressions for the connection matrices binding solutions at two different singularities, at least not in terms of gamma function.

The eigenfunctions $y_n(a, b, c, d; t; z)$ will be studied as solutions of an appropriate CTCP. An explicit study of the problem is given in Section 3.6. Here, we give only a draft. For simplicity, it is assumed that the parameters a, b, c, d, t are real and moreover that $t < 0$. It will be also assumed that $c > 0$, $d > 0$.

The eigenvalue problem for Heun equation on the interval $[0, 1]$ can be posed by the boundary conditions

$$|y(0)| < \infty, \qquad |y(1)| < \infty. \tag{3.2.2}$$

The function $\omega(z)$ arising in the transformation of the Heun equation to its self-adjoint form is expressed as

$$w(z) = \omega(z)y(z), \qquad \omega(z) = z^{(c-1)/2}(1-z)^{(d-1)/2}(z-t)^{(e-1)/2}. \tag{3.2.3}$$

The eigenfunctions $y_n(a, b, c, d; t; z)$ obey the orthogonality condition with the weight function $\omega^2(z)$:

$$\int_0^1 \omega^2(z)y_n(z)y_m(z)\mathrm{d}z = 0 \qquad \text{for } n \neq m. \tag{3.2.4}$$

Eigenfunctions can be studied on other intervals, for instance, $[a, 0]$. All sets of eigenfunctions related to Heun equation are known as Heun functions.

The sign attached to the spectral parameter in the equation allows, as is conventional, that $\lambda_n \to \infty$ as $n \to \infty$. This is ensured by a proper choice of the sign of the parameter t. The chosen notation corresponds to $t < 0$. The asymptotic behaviour of eigenvalues is studied in the next sections. Heun functions can be computed in several ways. A general approach is presented in Section 3.6. Here, a specific method valid for the Heun equation only (not for confluent cases) is presented.

Expand Heun functions in series of hypergeometric polynomials (see (2.5.1)):

$$y_n(a, b, c, d; t; z) = \sum_{j=1}^{\infty} g_n P_n(c+d-1; c; z). \tag{3.2.5}$$

In order to justify this form of polynomial expansion, we need to write down a differential equation (cf. (2.5.2)) satisfied by the nth polynomial:

$$K_n P_n(c+d-1; c; z) = (z(1-z)D^2 + (c - (c+d)z)D$$
$$+ n(n+c+d-1))P_n(c+d-1; c; z)$$
$$= 0. \tag{3.2.6}$$

The characteristic exponents of the Heun equation and eqn (3.2.6) at the singular points $z = 0$ and $z = 1$ coincide. Moreover, the operator $L_z^{\{1,1,1;1\}}$ related to the

Heun equation, in terms of the operator K_n and the operators T and Z introduced in Section 2.5, can be expressed as

$$L_z^{\{1,1,1;1\}} = (t - z)K_n - (a + b + 1 - c - d)T$$
$$+ (ab + n(n + d + c - 1)Z - t(n + d + c - 1). \qquad (3.2.7)$$

This means that the operator $L_z^{\{1,1,1;1\}}$ acts on the functions in the right hand-side of eqn (3.2.6) as a linear combination of the operators T and Z. It follows from eqns (2.5.16–2.5.17) that coefficients of expansion (3.2.5) satisfy the three-term recurrence relation

$$\alpha_j g_{j+1} + (\beta_j - \lambda)g_j + \gamma_j g_{j+1} = 0. \qquad (3.2.8)$$

Explicit expressions for the coefficients α_j, β_j, γ_j in eqn (3.2.8) require rather tiresome calculations. They can be found in [107].

$$\alpha_j = -\frac{(j + 1)(j + c + d - a)(j + c + d - b)(j + d)}{(2j + c + d)(2j + c + d + 1)},$$

$$\beta_j = \frac{1}{2}\left(\frac{(c - d)[j(j + c + d - 1)(2a + 2b - c - d) + (c + d - 2)ab]}{(2j + c + d - 2)(2j + c + d)}\right.$$

$$\left. + (1 - 2t)j(j + c + d - 1) + ab\right),$$

$$\gamma_j = -\frac{(j + a - 1)(j + b - 1)(j + c + d - 2)(j + c - 1)}{(2j + c + d - 2)(2j + c + d - 3)}. \qquad (3.2.9)$$

Only for $\lambda = \lambda_n$ is the solution of eqn (3.2.8) recessive, leading to a convergence of series (3.2.5). A more detailed study shows that, in this case, the series converges in the interior of the ellipse the foci of which are at $z = 0$ and at $z = 1$ and which has the singular point $z = t$ on the boundary. Expansion (3.2.5) was proposed by Svartholm [125].

The question arises whether, under additional assumptions on the parameters of Heun's equation, some of the eigenfunctions turn out to be polynomials, as in the case of the hypergeometric equation. Clearly, in this latter case, the eigenfunction is holomorphic also in the vicinity of $z = t$. The necessary condition is that one of the parameters a and b is a nonpositive integer. Since any of these parameters can be chosen, it is possible to put

$$a = -n.$$

Under this assumption, exactly n eigenfunctions, corresponding to n eigenvalues, can be constructed. It is easy to find the first eigenvalue and the first eigenfunction by substituting the polynomial of corresponding degree with undetermined coefficients in Heun's equation. For instance, if $a = 0$, then

$$y_1 = 1, \qquad \lambda_1 = 0.$$

In the general case, an algebraic equation of nth order needs to be solved. However, there are possibilities of explicit solutions for special values of coefficients.

Floquét solutions (they are also called path-multiplicative solutions) in the case of Heun equations also can be related only to two singularities, for instance, $z = 0$, $z = 1$. Suppose that γ is a simple closed contour encircling these points anticlockwise. Moreover suppose that there is a smooth parametrization of this contour:

$$\gamma : z = \varphi(s) \quad \text{for } s \in [0, 1], \qquad \varphi(0) = \varphi(1) = z_0.$$

A Floquét solution is defined as a solution satisfying

$$y(\varphi(1)) = e^{2\pi\sigma i} y(\varphi(0)). \tag{3.2.10}$$

The value σ is called a *Floquét exponent* or *path exponent* or *characteristic exponent*. It depends on the parameters of the equation.

Example: Let $\sigma = k$, with $k \in \mathbb{Z}$. Then the studied functions $y(z)$ are holomorphic in a domain D containing singularities $z = 0$ and $z = 1$ and hence, under condition $c > 1$ and $d > 1$, are eigenfunctions of the corresponding eigenvalue problem. $\square$

We denote functions $y(z)$ satisfying the Heun equation and the condition (3.2.10) by

$$Hp(a, b, c, d; t; \lambda; z).$$

Theoretically they can be constructed as a series

$$Hp(a, b, c, d; t; \lambda; z) = \sum_{-\infty}^{\infty} c_n f_{\sigma+n}(z), \tag{3.2.11}$$

where the functions $f_{\sigma+n}(z)$ in (3.2.11) have the same path-multiplicative property (3.2.10) and the same behaviour at the singularities $z = 0$, $z = 1$. Schmidt [110] proposed to take

$$f_{\sigma+n}(z) = z^{\sigma} F(-\sigma, 1 - c - \sigma; 2 - c - d - 2\sigma; z)$$

as such functions. Several conjectures about the convergence of the corresponding series (3.2.11) are known (see [110, 107]). However, in the general case, the form of the dependence of the path exponents σ on the parameters of the Heun equation is lacking, and hence expansions (3.2.11) cannot be used for practical numerical calculations.

3.2.2 Confluent cases of the Heun equation

First, the case of the singly confluent Heun equation is studied. Two basic local solutions—one at the regular singularity $z = 0$ and the other at the irregular singularity $z = \infty$—have already been introduced in the previous section. It is necessary to add three more solutions related to the regular singularities and three more solutions at infinity (different characteristic exponents and different Stokes lines). It is possible to enlarge this number by interchanging the regular singularities and by s-homotopic transformations preserving the canonical forms of the equation. However, it does not really lead to any new important results.

Considering the eigenvalue problem, we get a new situation when compared to the one in the case of the Heun equation. Namely there are two different eigenvalue problems corresponding to intervals $[0, 1]$ and $[1, \infty[$ respectively. The case $]-\infty, 0]$ is similar to $[1, \infty[$. It is assumed that $c > 1$, $d > 1$, $t > 0$. Eigenfunctions related to $[0, 1]$ are called angular and the ones related to $[1, \infty[$ are called radial. The first ones can be denoted as $y_n^{(a)}(a; c, d; t; z)$ and the second ones as $y_n^{(r)}(a; c, d; t; z)$. Both sets of eigenfunctions are orthogonal on the corresponding intervals with the weight function

$$\omega^2(z), \qquad \omega(z) = z^{(c-1)/2}(1-z)^{(d-1)/2}e^{pz/2},$$

i.e.

$$\int_0^1 \omega^2(z)y_n^{(a)}(z)y_m^{(a)}(z)\,\mathrm{d}z = 0 \quad \text{for } n \neq m,$$

$$\int_1^\infty \omega^2(z)y_n^{(r)}(z)y_m^{(r)}(z)\,\mathrm{d}z = 0 \quad \text{for } n \neq m. \tag{3.2.12}$$

The asymptotic behaviour of eigenvalue curves at large values of the parameter t is studied in Section 3.5. A numerical method of computing eigenfunctions and eigenvalues in the radial case is discussed in Section 3.6.

Angular eigenfunctions can be constructed in a way similar to (3.2.5). Namely

$$y_n^{(a)}(a; c, d; t; z) = \sum_{j=1}^\infty h_n P_n(c + d - 1; c; z). \tag{3.2.13}$$

Further computations are reminiscent of (3.2.6–3.2.8) and lead once again to the three-term recurrence relation

$$\sigma_j h_{j+1} + (\rho_j - \lambda)h_j + \kappa_j h_{j+1} = 0. \tag{3.2.14}$$

Coefficients σ_j, ρ_j, κ_j in eqn (3.2.14) can be found by using a confluence process.

$$\sigma_j = -t\frac{(j+1)(j+c+d-a)(j+d)}{(2j+c+d)(2j+c+d+1)},$$

$$\rho_j = -t\frac{(c-d)\,[j(j+c+d-1)+a(c+d-2)/2]}{(2j+c+d-2)(2j+c+d)}$$
$$\quad + j(j+c+d-1) - at/2,$$

$$\gamma_j = t\frac{(j+a-1)(j+c+d-2)(j+c-1)}{(2j+c+d-2)(2j+c+d-3)}. \tag{3.2.15}$$

Polynomial solutions can occur only if $a = -n$. For this value of the parameter a, the number of polynomials is at most n. Eigenvalues λ are calculated as roots of an nth order algebraic equation. For this values of λ, the series (3.2.13) terminates. Several specific cases are discussed in the book [69].

Radial eigenfunctions related to the interval $[1, \infty[$ can be expanded in a series of confluent hypergeometric functions. A more detailed study can be found in the book [107].

3.3 Integral equations and integral relations

3.3.1 Introduction

Integral representations for the special functions of the hypergeometric class, as was seen above, enable further simplification of the features of these functions. However, there is no hope of obtaining such representations in terms of elementary functions for the special functions belonging to the Heun class. Moreover, even representations in terms of special functions belonging to the hypergeometric class do not exist, since this would mean that there are representations with double integrals in terms of elementary functions.

Still, there are integral relations binding special functions belonging to the Heun class. The kernels of these relations either are expressed in terms of elementary functions or also include the special functions of the hypergeometric class. In the case of eigenfunctions, often a Fredholm equation may be obtained. The number of possible integral relations is large. Here we restrict ourselves to those which to our mind are the most important with respect to applications. Among these applications, we mention the possibility of comparing the values of the functions or their asymptotic behaviour at different singularities, of carrying out simple estimates of the functions, of evaluating the integrals with Heun class special functions, and of studying the monodromy properties.

For our goals we use the same methods as used in Chapter 2 for the hypergeometric equations, namely theorems, lemmas, etc. Several formulae have been obtained in collaboration with A. Kazakov [61], [62]. Further results may be found in [39], [4], [91], [2], [111].

3.3.2 Integral equations

The starting point in the case of each equation studied below will be a canonical form, although the final results have a better presentation for the self-adjoint form.

First, the integral relation is sought in the form

$$y(z) = \mu \int_C A(z, \zeta) y^*(\zeta) \, d\zeta = Ay^*, \qquad (3.3.1)$$

where $y(z)$ is an eigensolution of the two-point connection problem related to one of the equations belonging to the Heun class, and $y^*(\zeta)$ is a a proper solution of the adjoint equation. The first step in the analysis of the problem is to check some commutation rules. For a rigorous study, it is necessary to verify that (i) the integrals converge, (ii) all constant terms arising from integration by parts vanish, (iii) we get the required asymptotic behaviour of the studied functions at singularities. All these demands can be fulfilled under appropriate restrictions on the parameters and on the

integration path. They will be checked at the end of this section only for one studied
equation.

As a standard integration path, we take an interval $[T_1, T_2]$ (finite or infinite) on
the real axis.

The following differential operators, related to different types of equations belonging to the Heun class, are studied: $L_z^{(\vec{j})}$, where $\vec{j}$ denotes the s-multisymbol. The
corresponding equation itself may be written as

$$L_z^{(\vec{j})} y(z) + \lambda y(z) = 0. \tag{3.3.2}$$

By means of a transformation discussed in Section 1.3, operators $L_z^{(\vec{j})}$ are transformed
to self-adjoint operators

$$M_z^{(\vec{j})} = [G_z^{(\vec{j})}]^{-1} L_z^{(\vec{j})} G_z^{(\vec{j})} = M_z^{(\vec{j})*} \tag{3.3.3}$$

corresponding to the equation

$$\left(M_z^{(\vec{j})} + \tilde{\lambda}\right) w(z) = 0, \qquad w(z) = \left(G_z^{(\vec{j})}\right)^{-1} y(z). \tag{3.3.4}$$

It should be noted that

$$M_z^{(\vec{j})} = [G^{(\vec{j})}(z)](L_z^{(\vec{j})})^* [G_z^{(\vec{j})}]^{-1}, \qquad w(z) = [G_z^{(\vec{j})}] y^*(z). \tag{3.3.5}$$

The functions $G^{\{\vec{j}\}}(z)$ are given in the tables presented in Section 3.1. However, for
the sake of convenience, we rewrite the most important of them.

$$G^{\{1,1,1;1\}}(z) = z^{(1-c)/2}(1-z)^{(1-d)/2}(z-t)^{(1-e)/2}, \tag{3.3.6}$$

$$G^{\{1,1;2\}}(z) = z^{(1-c)/2}(1-z)^{(1-d)/2} \exp(tz/2), \tag{3.3.7}$$

$$G^{\{1;3\}}(z) = z^{(1-c)/2} \exp(z^2/4 + tz/2), \tag{3.3.8}$$

$$G^{\{2;2\}}(z) = z^{(1-c)/2} \exp(z/2 + t/(2z)). \tag{3.3.9}$$

The function $G^{\{\vec{j}\}}(z)$ for other multisymbols we omit for brevity. Explicit expressions
are used whenever they are needed. As a consequence of (3.3.1, 3.3.3), we have

$$w(z) = \mu \int_{T_1}^{T_2} K(z, \zeta) w(\zeta) \, d\zeta = \mu K w, \tag{3.3.10}$$

where

$$K(z, \zeta) = [G(z)]^{-1} A(z, \zeta) [G(\zeta)]^{-1}.$$

The following lemma will be used in the subsequent computations.

Lemma 3.1 *Suppose that (3.3.1) holds and that the function $A(z, \zeta)$ is a solution of the partial differential equation*

$$(L_z - L_\zeta)A(z, \zeta) = 0. \tag{3.3.11}$$

Let $y^*(\zeta)$ be a proper solution of $((L_\zeta^{(\vec{j})})^* + \lambda^*)y(\zeta)^* = 0$. Then $y(z)$ is a proper solution of (3.3.2). □

Proof: The proof follows from formal computations

$$Ly(z) = \int_C v(\zeta)L_z A(z, \zeta)\,d\zeta = \int_C v(\zeta)(L_z - L_\zeta)A(z, \zeta)\,d\zeta$$

$$+ \int_C A(z, \zeta)L_\zeta^* v(\zeta)\,d\zeta. \tag{3.3.12}$$

It is assumed that constant terms arising from integration by parts vanish. □

Besides this, one can seek solutions of (3.3.11) by the method of separation of variables. In this book, we present the most simple solutions of (3.3.11).

First, the kernel $A(z, t)$ in (3.3.1) is taken as

$$A(z, \zeta) = A(\xi), \qquad \xi = z + \zeta. \tag{3.3.13}$$

Then, the operator in partial derivatives $L_z^{\{\vec{i}\}} - L_\zeta^{\{\vec{i}\}}$ is factorized for equations belonging to the Heun class as

$$L_z^{\{1,1;2\}} - L_\zeta^{\{1,1;2\}} = (z - \zeta)((\xi - 1)D_\xi^2$$
$$+ (t(\xi - 1) + c + d)D_\xi + ta), \tag{3.3.14}$$

$$L_z^{\{1;3\}} - L_\zeta^{\{1;3\}} = (z - \zeta)\left(D_\xi^2 - (t + \xi)D_\xi - a\right), \tag{3.3.15}$$

$$L_z^{\{2;2\}} - L_\zeta^{\{2;2\}} = (z - \zeta)\left(\xi D_\xi^2 + (t - \xi)D_\xi - a\right), \tag{3.3.16}$$

$$L_z^{\{;4\}} - L_\zeta^{\{;4\}} = -(z - \zeta)(\xi D_\xi + a), \tag{3.3.17}$$

$$L_z^{\{1,1;3/2\}} - L_\zeta^{\{1,1;3/2\}} = (z - \zeta)((\xi - 1)D_\xi^2 + (c + d)D_\xi - t), \tag{3.3.18}$$

$$L_z^{\{1;5/2\}} - L_\zeta^{\{1;5/2\}} = (z - \zeta)(D_\xi^2 - (\xi + t)), \tag{3.3.19}$$

$$L_z^{(7)} - L_\zeta^{(7)} = (z - \zeta)(\xi D_\xi^2 + \rho D_\xi - 1). \tag{3.3.20}$$

The presented equalities are proved by direct substitution.

Another possibility is to take the kernel $A(z, t)$ in the form

$$A(z, \zeta) = A(\eta), \qquad \eta = z\zeta. \tag{3.3.21}$$

Then the differential operators

$$L_z^{\{\vec{i}\}} - L_\zeta^{\{\vec{i}\}}$$

may be factorized as

$$L_z^{\{1,1,1;1\}} - L_\zeta^{\{1,1,1;1\}} = (z - \zeta)\left(\eta(\eta - t)D_\eta^2\right.$$
$$\left. + ((a + b + 1)\eta - c\zeta)D_\eta + ab\right), \quad (3.3.22)$$

$$L_z^{\{1,1;2\}} - L_\zeta^{\{1,1;2\}} = (z - \zeta)\left(\eta D_\eta^2 - (t\eta + c)D_\eta - ta\right), \quad (3.3.23)$$

$$L_z^{\{1;3\}} - L_\zeta^{\{1;3\}} = -(z - \zeta)\left(\eta D_\eta^2 + (\eta + c)D_\eta + a\right), \quad (3.3.24)$$

$$L_z^{\{2;2\}} - L_\zeta^{\{2;2\}} = -(z - \zeta)\left((\eta - t)D_\eta + a\right), \quad (3.3.25)$$

$$L_z^{\{1,1;3/2\}} - L_\zeta^{\{1,1;3/2\}} = -(z - \zeta)\left(\eta D_\eta^2 + cD_\eta + t\right). \quad (3.3.26)$$

We turn from eqns (3.3.2) to eqns (3.3.4) and to Sturm–Liouville problems related to these equations. Instead of the integral relation (3.3.1), an integral equation of the Fredholm type is sought in the form

$$w(z) = \lambda \int_{T_1}^{T_2} K^{\{\vec{j}\}}(z, t)w(t)\, dt. \quad (3.3.27)$$

As follows from (3.3.3) and (3.3.4),

$$K^{\{\vec{j}\}}(z, t) = A^{\{\vec{j}\}}(z, t)\left[G^{\{\vec{j}\}}(z)\right]^{-1}\left[G^{\{\vec{j}\}}(t)\right]^{-1}. \quad (3.3.28)$$

Suppose that the set of eigenfunctions of eqn (3.3.27) is the same as the set of eigenfunctions generated on the interval $[T_1, T_2]$ by one of eqns (3.3.4). Then the following necessary condition arising on integration by parts in (3.3.11) should be fulfilled:

$$R(t)[w(t)D_t K(z, t) - K(z, t)D_t w(t)] \mid_{T_1}^{T_2} = 0. \quad (3.3.29)$$

Since eqn (3.3.29) depends on z as a parameter, the points T_1 and T_2 should be singularities of the differential equation (unless some additional symmetry takes place in the corresponding equation).

In the following, several theorems are formulated. All of them can be proved in the same manner. An example of a proof is given in the next section. The basic items of these proofs are: (i) convergence of the related integrals, (ii) validity of the condition (3.3.29). However, we should mention that the theorems formulated below do not exhaust all possible cases.

Theorem 3.1 *Assume that following conditions for the parameters of the Heun equation are fulfilled: $c > 1$, $d > 1$, and $t < 0$. Then the boundary conditions*

$$\mid w(0) \mid < \infty, \qquad \mid w(1) \mid < \infty, \quad (3.3.30)$$

for the self-adjoint form of CHE (see (3.3.4)), generate a Sturm–Liouville problem on the interval $[0, 1]$ having an infinite set of corresponding eigenfunctions, denoted

$w_n(z)$, and of corresponding eigenvalues denoted $\tilde{\lambda}_n$. This set of eigenfunctions coincides with the set of a eigenfunctions of a Fredholm-type integral equation with Hermitian kernel:

$$w(z) = \mu \int_0^1 \left[G^{\{1,1,1;1\}}(z) G^{\{1,1,1;1\}}(\zeta) \right]^{-1} F(a,b,c,z\zeta/a) w(\zeta) \, d\zeta, \quad (3.3.31)$$

where $F(a,b;c;z)$ is the hypergeometric function and $G^{\{1,1,1;1\}}(z)$ is defined in (3.3.6). □

This result is a particular case of a more general integral equation obtained in [39]. In this particular case of Heun's equation, all boundary problems related to different intervals do not differ in essence.

Integral equations related to confluent cases differ not only in the kernel but also in the interval of consideration.

Theorem 3.2 *Suppose that the parameters in the confluent Heun equation satisfy the conditions $c > 1$, $d > 1$. Then boundary conditions (3.3.30) generate for the self-adjoint form of the CHE (3.1.3) a Sturm–Liouville problem on the interval $[0, 1]$ with an infinite set of corresponding eigenfunctions, denoted by $w_n(z)$, and of corresponding eigenvalues, denoted by λ_n. This set of eigenfunctions coincides with the set of eigenfunctions of two different Fredholm-type integral equations having Hermitian kernels:*

$$w(z) = \mu \int_0^1 \left[G^{\{1,1;2\}}(z) G^{\{1,1;2\}}(\zeta) \right]^{-1} \Phi(a,c,-tz\zeta) w(\zeta) \, d\zeta, \quad (3.3.32)$$

$$w(z) = \mu \int_0^1 \left[G^{\{1,1;2\}}(z) G^{\{1,1;2\}}(\zeta) \right]^{-1} \Phi(a,c+d,-t(z+\zeta-1)) w(\zeta) \, d\zeta. \quad (3.3.33)$$

Here $G^{\{1,1;2\}}(z)$ is defined by (3.3.7) and $\Phi(a,c;z)$ is the confluent hypergeometric function. □

Two more integral equations arise when, instead of the interval $[0, 1]$, the interval $[1, \infty[$ is taken. For this case one more condition is needed, namely $t > 0$, and the boundary condition at infinity should be posed. The integral equations read

$$w(z) = \lambda \int_1^\infty \left[G^{\{1,1;2\}}(z) G^{\{1,1;2\}}(\zeta) \right]^{-1} \Psi(a,c,-tz\zeta) w(\zeta) \, d\zeta,$$

$$w(z) = \mu \int_1^\infty \left[G^{\{1,1;2\}}(z) G^{\{1,1;2\}}(\zeta) \right]^{-1} \Psi(a,c+d,-t(z+\zeta-1)) w(\zeta) \, d\zeta.$$

The confluent hypergeometric function Φ is substituted for the function Ψ—another solution of the confluent hypergeometric equation. Particular cases of eqns (3.3.32) have been studied in [4] in relation to the two-centres Coulomb problem.

Next comes BHE.

120 SPECIAL FUNCTIONS

Theorem 3.3 *Suppose that the parameters c and t satisfy*

$$t > 0, \qquad c > 1.$$

Let $w_n(z)$ be eigenfunctions generated by the boundary conditions

$$|\, w(0)\, | < \infty, \qquad |\, w(\infty)\, | < \infty, \tag{3.3.34}$$

for the BHE (3.1.11) on the interval $[0, \infty[$. Then $w_n(z)$ also are eigenfunctions of the following Fredholm, type integral equation:

$$w(z) = \mu \int_0^\infty \left[G^{\{1;3\}}(z) G^{\{1;3\}}(\zeta) \right]^{-1} \Phi(a/2, 1/2; \infty, (z + \zeta - t)^2/2) w(\zeta)\, d\zeta.$$

$$(3.3.35)$$

□

The other possible integral equation is related to the boundary problem at the interval $]-\infty, 0]$. The integral equation (3.3.35) is a particular case of the equation obtained in [91].

For arbitrary values of parameters in eqn (3.3.24), we do not get a corresponding integral equation. However, if the parameter a is a nonpositive integer, then the following integral equation holds:

$$w(z) = \mu \int_0^\infty \left[G^{\{1;3\}}(z) G^{\{1;3\}}(\zeta) \right]^{-1} \Psi(a, c; -z\zeta) w(\zeta)\, d\zeta. \tag{3.3.36}$$

Here Ψ is a confluent hypergeometric function which under our assumption reduces to a (Laguerre) polynomial.

In the case of the DHE two kernels are possible.

Theorem 3.4 *Assume that $t > 0$. Then the boundary conditions (3.3.34) posed for the self-adjoint form of the DHE (3.1.15) generate a set of eigenfunctions $w_n(z)$ which simultaneously are eigenfunctions of the two Fredholm-type integral equations*

$$w(z) = \mu \int_0^\infty \left[G^{\{2;2\}}(z) G^{\{2;2\}}(\zeta) \right]^{-1} (z\zeta + t)^{-a} w(\zeta)\, d\zeta, \tag{3.3.37}$$

$$w(z) = \mu \int_0^\infty \left[G^{\{2;2\}}(z) G^{\{2;2\}}(\zeta) \right]^{-1} \Psi(a, c, z + \zeta) w(\zeta)\, d\zeta. \tag{3.3.38}$$

□

In the case of the THE, there is no integral equation having kernels which are equivalent to a boundary problem for the self-adjoint form of the THE with corresponding conditions at $\pm\infty$. The same statement is also true for the reduced THE.

For studying reduced confluent equations, we first write the equations themselves, since they can slightly differ from the conventional notations. The reduced CHE is taken in the form

$$L_z^{\{1,1;3/2\}} y(z) + \lambda y(z) :=$$
$$z(z-1)y''(z) + (c(z-1) + dz)y'(z) + (-tz + \lambda)y(z) = 0. \qquad (3.3.39)$$

The factor $G^{\{1,1;3/2\}}(z)$ is

$$G^{\{1,1;2\}}(z) = z^{(1-c)/2}(1-z)^{(1-d)/2}.$$

Factorization according to (3.3.13) leads to

$$L_z^{\{1,1;3/2\}} - L_\zeta^{\{1,1;3/2\}} = (z-\zeta)((\xi-1)D_\xi^2 + (c+d)D_\xi - t),$$

while factorization according to (3.3.21) leads to

$$L_z^{\{1,1;3/2\}} - L_\zeta^{\{1,1;3/2\}} = -(z-\zeta)(\eta D_\eta^2 + cD_\eta - t).$$

As a result we get the following theorem.

Theorem 3.5 *Assume that $c > 1$, $d > 1$, $t > 0$. Then the boundary conditions (3.3.34) posed for the self-adjoint form of the reduced CHE (eqn (3.3.39)) generate a set of eigenfunctions $w_n(z)$ which, simultaneously, are eigenfunctions of the Fredholm-type integral equations*

$$w(z) = \mu \int_1^\infty \left[G^{\{1,1;3/2\}}(z)G^{\{1,1;3/2\}}(\zeta) \right]^{-1}$$
$$\times (z + \zeta - 1)^{(1-c-d)/2} K_{1-c-d}(2\sqrt{t(z+\zeta-1)})w(\zeta)\,d\zeta. \qquad (3.3.40)$$

Boundary conditions (3.3.30) for the same equation generate a set of eigenfunctions $w_n(z)$ which, simultaneously, are eigenfunctions of the Fredholm-type integral equations

$$w(z) = \mu \int_0^1 \left[G^{\{1,1;3/2\}}(z)G^{\{1,1;3/2\}}(\zeta) \right]^{-1} (z\zeta)^{(1-c)/2} I_{c-1}(2\sqrt{t(z\zeta)})w(\zeta)\,d\zeta,$$
$$(3.3.41)$$

where $I_\nu(x)$ is the modified Bessel function and $K_\nu(x)$ is the MacDonald function (see Section 2.1). $\square$

In the case of the reduced BHE, the use of the standard canonical form leads to the following integral equation.

Theorem 3.6 *Assume that $c > 1$. Then the boundary conditions (3.3.34) posed for the self-adjoint form of the reduced BHE (eqn (3.3.7)) generate a set of eigenfunctions $w_n(z)$ which simultaneously are eigenfunctions of the Fredholm-type integral equations*

$$w(z) = \mu \int_0^\infty (z\zeta)^{(c-1)/2} \mathrm{Ai}(z + \zeta - t)w(\zeta)\,d\zeta, \qquad (3.3.42)$$

where $\mathrm{Ai}(z)$ is the standard solution of the Airy equation decreasing at $+\infty$ (see Section 2.1). $\square$

Equation (3.3.42) was proposed for the first time in [2].

Theorem 3.7 *Assume that $t > 0$. Then the boundary conditions (3.3.34) posed at the interval $[0, \infty[$ for the self-adjoint form of the reduced DHE (eqn (3.3.8)) generate a set of eigenfunctions $w_n(z)$ which, simultaneously, are eigenfunctions of two Fredholm-type integral equations:*

$$w(z) = \mu \int_0^\infty (zt)^{\rho/2-1} \exp(t/2(1/z + 1/t))(z + \zeta)^{(1-c)/2} K_{1-c}\left(2\sqrt{z+\zeta}\right) w(\zeta) \, d\zeta,$$

$$\tag{3.3.43}$$

$$w(z) = \mu \int_0^\infty (zt)^{c/2-1} \exp\left(t/2(1/z + 1/\zeta)\right) \exp\left(z\zeta/t\right) w(\zeta) \, d\zeta \tag{3.3.44}$$

where $K_\nu(x)$ is the MacDonald function. □

3.3.3 An example of a proof

As pointed out above, the proofs of the formulated conjectures are similar. They include checking whether the involved integrals converge and whether all constant terms arising from integration by parts vanish.

As an example, the proof for the case of eqn (3.3.44) is given. The self-adjoint form of (3.3.8) reads

$$Kw = \frac{d}{dz}z^2\frac{d}{dz}w(z) + \left(-\frac{t^2}{4z^2} - \frac{t(1-c/2)}{z} + \frac{c(1-c/2)}{2} + \lambda - z\right) w(z) = 0. \tag{3.3.45}$$

It is easy to obtain the local behaviour of solutions of (3.3.31) at zero and at infinity:

$$w_{z\to+0}^{(1)} = e^{t/2z}z^{c/2-1}(1 + O(z)),$$

$$w_{z\to+0}^{(2)} = e^{-t/2z}z^{-c/2+1}(1 + O(z)),$$

$$w_{z\to+\infty}^{(3)} = e^{-2\sqrt{z}}z^{-3/4}(1 + O(z^{-1})),$$

$$w_{z\to+\infty}^{(4)} = e^{2\sqrt{z}}z^{-3/4}(1 + O(z^{-1})). \tag{3.3.46}$$

Only the solutions $w^{(1)}$ and $w^{(3)}$ satisfy boundary conditions (3.3.34) under the assumed condition $t < 0$. The eigenvalues λ_n are roots of the equation arising when the Wronskian of solutions $w^{(1)}$ and $w^{(3)}$ is set to be zero. Standard theorems about the spectrum of singular self-adjoint differential operators state the existence of an infinite set of eigenvalues satisfying the condition $\lambda_n \to \infty$ as $n \to \infty$. The asymptotic estimates (3.3.46) for the eigenfunctions show that, for any $z \in [0, \infty[$, the integral in the right-hand side of (3.3.44) is absolutely and uniformly convergent and defines a function $U(z)$ which is regular on the ray $]0, \infty[$. The estimates of the behaviour of the integral for small and for large values of z give the same behaviour of $U(z)$ as the corresponding solutions $w_n(z)$ of the differential operator.

The final step is to prove that $U(z)$ satisfies the same differential equation as $w(z)$. By direct differentiation, it can be proved that $L^{\{2;3/2\}}G = GK$, where $K = K^*$ is defined in eqn (3.3.45) and $G(z) = z^{1-c/2}\exp(-t/(2z))$. Applying the differential operator from (3.3.45) to the function $U(z)$, satisfying (3.3.44), we obtain

$$K_z U(z) = \mu \int_0^\infty \left[K_z G^{-1}(z)\exp(z\zeta/t) \right] G^{-1}(\zeta)w(\zeta)\,d\zeta$$

$$= \mu \int_0^\infty G^{-1}(z) \left[L_z^{\{2;3/2\}}\exp(z\zeta/t) \right] G^{-1}(\zeta)w(\zeta)\,d\zeta$$

$$= \mu \int_0^\infty G^{-1}(z) \left[L_\zeta^{\{2;3/2\}}\exp(z\zeta/t) \right] G^{-1}(\zeta)w(\zeta)\,d\zeta$$

$$= \mu \int_0^\infty G^{-1}(z) \left[K_\zeta G^{-1}(\zeta)\exp(z\zeta/t) \right] w(\zeta)]\,d\zeta$$

$$= \mu G^{-1}(z) \left[\zeta^2 \left(w(\zeta)\frac{d}{d\zeta}(G^{-1}(\zeta)\exp(z\zeta/t) - G^{-1}(\zeta)\exp(z\zeta/t)w'(\zeta) \right) \right]_0^\infty$$

$$+ \mu \int_0^\infty G^{-1}(z)[G^{-1}(\zeta)\exp(z\zeta/t)]K_\zeta w(\zeta)]\,d\zeta = 0. \qquad (3.3.47)$$

All differentiations under the integral sign and all constant terms arising from integration by parts vanish due to estimates of the function $w(z)$ in (3.3.46). Since the function $U(z)$ satisfies the same differential equation as $w_n(z)$, and the same boundary conditions, it must be proportional to $w_n(z)$. The corresponding proportional coefficient is μ_n.

The eigenvalues of the integral equation μ_n can be calculated with the help of the formula

$$\mu_n = \left[\int_0^\infty \zeta^{c/2-1}\exp(t/2\zeta)w_n(\zeta)\,d\zeta \right]^{-1} \lim_{z\to 0}\left[z^{1-c/2}\exp(-t/2z)w_n(z) \right].$$

$$(3.3.48)$$

The proposed list of integral equations can be enlarged by several means. First, more complicated kernels than presented above can be used. Secondly, when the parameters of the initial differential equation satisfy additional conditions, further integral equations can be proposed.

3.3.4 Integral relations

First, we consider the integral relations of the form

$$y(z) = S^{\{1,1,1,;1\}}\hat{y}(\zeta) = \int_C (z-\zeta)^\sigma \hat{y}(\zeta)\,d\zeta, \qquad (3.3.49)$$

where $y(z)$ and $\hat{y}(\zeta)$ are eigenfunctions of some boundary-value problems for Heun's equation but with different sets of parameters $\{a,\ b,\ c,\ d,\ t,\ \lambda\}$ and $\{a',\ b',\ c',\ d',\ t',\ \lambda'\}$ respectively.

Lemma 3.2 *The equation*

$$\left(L_z^{\{1,1,1;1\}} - \hat{L}_\zeta^{\{1,1,1;1\}*}\right)(z-t)^{-\sigma} = 0 \qquad (3.3.50)$$

is satisfied if the above-mentioned parameters are related according to

$$(\sigma - a)(\sigma - b) = 0, \qquad c' = c + 1 - \sigma, \qquad d' = d + 1 - \sigma,$$
$$a' = 2 - \sigma, \qquad b' = -2\sigma + a + b + 1,$$
$$\lambda' = \lambda + (-\sigma + 1)(a + b + 1 - d + dt + ct) + \sigma(\sigma - 1)(t + 1). \qquad (3.3.51)$$

$\square$

Proof: The proof is based on a direct application of the operators and a further nullification of coefficients standing in front of different powers of z and ζ. $\square$

There are two roots of eqn (3.3.51) for σ. However, because of the symmetry of the Heun equation with respect to an interchange of the parameters a and b, both roots lead to the same result. Hence, without lack of generality, the parameters of the Heun equation for the function $v(z)$ in (3.3.49) can be chosen as

$$c' = c + 1 - a, \qquad d' = d + 1 - a,$$
$$a' = 2 - a, \qquad b' = b + 1 - a,$$
$$\lambda' = \lambda + a(a - 1)(t + 1) - (a - 1). \qquad (3.3.52)$$

If the integral transform (3.3.49) is applied once again, then we get initial values for the coefficients of Heun's equation. Hence

$$(S^{(0)})^2 = I \qquad (3.3.53)$$

where I is the unity operator and (3.3.49) is an involution transformation. It is worth mentioning that the kernel of the transformation $(S^{(0)})^2$ is a Cauchy kernel. With the help of the general relation (3.3.49), it is possible to obtain different particular formulae. We are interested only in formulae for the eigenfunctions.

Theorem 3.8 *Suppose that $t < 0$, $d > a$, $c > 1$, $a + b > c + d$, $a < 1$. Let $\{v_n(s)\}$ be the set of eigenfunctions for the Sturm–Liouville problem on the interval $[0, 1]$ related to the Heun equation (3.1.1) with parameters a', b', c', d', λ', defined in (3.3.52). ($v_n(s)$ are bounded at $z = 0$ and at $z = 1$). Then the relation*

$$y_n(z) = S^{(0)} v_n = \int_0^1 (z - s)^{-a} v_n(s) \, ds \qquad (3.3.54)$$

generates the set of eigenfunctions $y_n(z)$ related to the Heun equation with parameters a, b, c, d, λ on the interval $[t, 0]$. ($y_n(z)$ is bounded at $z = 0$ and at $z = 1$). $\square$

Proof: The integral in (3.3.54) is convergent if $z \in]t, 0[$ and is bounded at $z = t$. It is also bounded at $z = 0$, since $a < 1$. The constant terms arising from integration by parts vanish due to the behaviour of the solutions at the integration endpoints. Hence the functions $y_n(z)$ defined by (3.3.54) are eigenfunctions of the Heun equation on $[t, 0]$. $\square$

Under the substitution $s \mapsto t/s$, the integral relation (3.3.54) transforms to another integral relation having a kernel depending on $z\,s$. Below, we shall alter the integration interval as well.

Theorem 3.9 *Suppose that $t > 1$, $a > c$, $b < 1$, $a + b > c + d$. Let $\{w_n(s)\}$ be the set of eigenfunctions of the Heun equation on the interval $[-\infty, 0]$ with parameters a'', b'', c'', d'', λ'':*

$$a'' = 2 - a, \qquad b'' = 2 - c, \qquad c = 2 - b,$$

$$d'' = b + 2 - c - d, \qquad q'' = q - ad + (2 - c)(a + 1) - \epsilon$$

($w(s)$ is bounded at $-\infty$ and at 0). Then the relation

$$y_n(z) = S^{(0)} w_n(s) = \int_{-\infty}^{0} (1 - \frac{zs}{t})^{-a} w_n(s)\, \mathrm{d}s \qquad (3.3.55)$$

generates $y_n(z)$ on $[1, a]$ ($y_n(z)$ is bounded at $z = 1$ and at $z = t$). $\square$

Although integral relations (3.3.54) and (3.3.55), in essence, coincide, they generate two different sequences of relations for confluent equations. Among these, only the most important ones are given. For other relations, we refer to [61]. By means of a confluence process, we map $L^{\{1,1,1;1\}} \mapsto L^{\{1,1;2\}}$. In terms of eqn (3.3.55), its kernel transforms to an exponent, and we obtain an integral relation for eigenfunctions related to the CHE.

Theorem 3.10 *Assume that $d > 1$, $a < 1$, $t < 0$. Let $\{w_n(s)\}$ be a set of eigenfunctions of the eigenvalue problem on the interval $]-\infty, 0]$ for the CHE (3.1.3) with the parameters $a' = 2 - c$, $c' = 2 - a$, $d' = a + 2 - c - d$, $t' = -t$, $\lambda' = \lambda + d + c - 2 - t$ ($w_n(t)$ is bounded at the points 0 and $-\infty$). Then the relation*

$$y_n(z) = S^{(1)} w_n = \int_{-\infty}^{0} \exp(-tzs) w_n(s)\, \mathrm{d}s \qquad (3.3.56)$$

produces a set of eigenfunctions $y_n(z)$ of the eigenvalue problem on the interval $[0, \infty[$ for the CHE (3.3.8) ($y_n(z)$ is bounded at the points 0 and ∞). $\square$

Theorem 3.11 *Assume that $c < 1$, $b > 1$. Let $\{w_n(s)\}$ be a set of eigenfunctions of the eigenvalue problem on the interval $]-\infty, 0]$ for the BHE (3.1.11)*

$$\left(t D_t^2 + (t^2 + \mu't + c') D_t + a't - q' \right) w(t) = 0, \qquad (3.3.57)$$

with the parameters $a' = 2 - c$, $c' = 2 - a$, $\mu' = -\sigma$, $q' = q + \sigma$ ($w_n(t)$ is bounded at 0 and at ∞). Then the relation

$$y_n(z) = S^{(5)} w_n = \int_{0}^{\infty} \exp(-zs) w_n(s)\, \mathrm{d}s \qquad (3.3.58)$$

produces a set of eigenfunctions $y_n(z)$ of the eigenvalue problem on the interval $[0, \infty[$ for the BHE (3.3.8) ($y_n(z)$ is bounded at the points 0 and ∞). $\square$

Theorem 3.12 *Assume that $c < 1$, $b > 1$. Let $\{w_n(s)\}$ be a set of eigenfunctions of the eigenvalue problem on the interval $]-\infty, 0]$ for the reduced CHE (3.1.20) with the parameters $a' = 2 - c$, $c' = 2 - a$, $\sigma' = -\sigma$, $q' = q + \sigma$ ($w_n(t)$ is bounded at 0 and at $-\infty$). Then the relation*

$$y_n(z) = S^{(5)}w_n = \int_0^{-\infty} \exp(-zs)w_n(s)\,\mathrm{d}s \qquad (3.3.59)$$

produces a set of eigenfunctions $y_n(z)$ of the eigenvalue problem on the interval $[0, \infty[$ for the DHE (3.1.15) ($y_n(z)$ is bounded at the points 0 and ∞). $\square$

Theorem 3.13 *Assume that $c > 1$, $d > 1$, $t > 0$. Let $\{w_n(s)\}$ be a set of eigenfunctions of the eigenvalue problem on the interval $[0, \infty[$ for the DHE (3.1.15) with the parameters $a' = 2 - c$, $c' = 4 - c - a$, $t' = -t$, $\lambda' = \lambda + a + c - 2$ ($w_n(s)$ is bounded at the points 0 and ∞). Then the relation*

$$y_n(z) = \int_{-\infty}^0 \exp(-zs)w_n(s)\,\mathrm{d}s \qquad (3.3.60)$$

produces a set of eigenfunctions $y_n(z)$ of the eigenvalue problem on the interval $[0, 1[$ for the reduced CHE (3.1.20) ($y_n(z)$ is bounded at the points 0 and ∞). $\square$

Theorem 3.14 *Assume that $t > 0$. Let $\{w_n(s)\}$ be a set of eigenfunctions of the eigenvalue problem on the interval $[0, \infty[$ for the reduced DHE (3.1.20) with the parameters $\rho' = 4 - \rho$, $t' = t$, $q' = \rho - 2 - q$ ($w_n(s)$ is bounded at the points 0 and ∞). Then the relation*

$$y_n(z) = \int_0^\infty \exp(-zs/t)w_n(s)\,\mathrm{d}s \qquad (3.3.61)$$

produces a set of eigenfunctions $y_n(z)$ of the eigenvalue problem on the interval $[0, \infty[$ for the reduced DHE (3.1.22) ($y_n(z)$ is bounded at the points 0 and ∞). $\square$

The proposed list of integral relations does not include those relations which arise only for specialized values of the parameters for the equations under consideration. It is possible to obtain relations not only between eigenfunctions but also between other specially chosen solutions. The full list would be too large and therefore is not given here.

In both cases, namely of integral equations and of integral relations, the following suppositions were assumed:

- The kernel, in essence, comprises one multiplier.
- The kernel satisfies an equation belonging either to the hypergeometric class or to the class of elementary functions.

If these suppositions do not hold, then other formulae may be obtained.

3.4 Basic asymptotic formulae for small t

3.4.1 Introduction

In contrast to special functions belonging to the hypergeometric class, of special functions belonging to the Heun class yield a much larger variety of asymptotic formulae. First, these are asymptotic solutions in neighbourhoods of irregular singularities. Here, formal asymptotic expansions have already been exhibited above in this chapter. However, the Stokes multipliers cannot be obtained in an explicit form. There are several publications (see, for instance, [70], [70] [26], [27], [128]) where certain procedures are proposed for numerical calculation of Stokes multipliers in the case of some confluent Heun equations, but a general approach is still lacking. Much more promising is the situation when *critical values* of the scaling parameter t are studied. By these critical values we mean those values when either singularities or turning points of the equation under consideration merge. In the case of the Heun equation, the critical values are $t = 0$, 1, ∞; in the case of the CHE, they are $t = 0$, ∞; in all the other cases of confluent equations, $t = \infty$. On the other hand, semiclassical formulae can be studied arising from an appropriate behaviour of the accessory parameter λ. Here, only results for asymptotics with a scaling parameter are presented that reflect the scientific experience of the authors. This type of asymptotics enables a quantitative explanation of several important physical effects [69]. Mainly, solutions of the central two-point connection problem (singular Sturm–Liouville problem) are studied. The presentation is on a purely formal level, saying that, in every studied case, an asymptotic approximation is constructed which satisfies the equation up to small terms.

3.4.2 Heun equation with nearby singularities

The problem is posed as following. Consider the Heun equation with a slightly altered notation of the scaling parameter $t \mapsto -s$ $(s > 0)$:

$$L_z y(z) = \lambda y(z), \tag{3.4.1}$$

with

$$L_z = z(z-1)(z+s)D^2 + [c(z-1)(z+s) + dz(z+s) + ez(z-1)]D + abz \tag{3.4.2}$$

and

$$D = \frac{\mathrm{d}}{\mathrm{d}z},$$

at small positive values of s. What is the behaviour of the eigenfunctions $y_n(z)$ and the corresponding eigenvalues λ_n of the boundary problem (central two-point connection problem) for eqn (3.4.1)

$$|y(0)| < \infty, \qquad |y(1)| < \infty, \tag{3.4.3}$$

on the interval $[0,\ 1]$ as $s \to 0$?

Several other similar problems can be studied. First, the chosen interval may be $] - \infty, s]$ or $[s, 0]$ or $[s, \infty[$. Secondly, the critical value of the parameter[1] s may be $s = -1$ or $s = \infty$. All these cases can be divided into two groups:

(A) At the critical value of the scaling parameter s, the point $z = -s$ merges with the endpoint of the interval.

(B) The point $z = -s$ merges with the singularity outside the interval.

The problems belonging to the group B are studied by conventional perturbation procedures. The problems belonging to the group A reveal more sophisticated boundary-layer phenomena. The problem (3.4.1–3.4.3) belongs to the group A and is studied below.

It is assumed that a, b, c, d, e are real and (without loss of generality) satisfy the conditions

$$c \geq 1, \qquad d \geq 1, \qquad e \geq 1, \qquad a \geq b, \tag{3.4.4}$$

resulting in the existence of one analytical solution and one unbounded solution at each regular singularity $z = 0$, $z = 1$, $z = -s$ of (3.4.1). Moreover, a and b are supposed not to be integers. Then the Fuchs identity holds:

$$a + b + 1 - c - d - e = 0. \tag{3.4.5}$$

As a consequence of eqns (3.4.5–3.4.4) we have

$$a + b \geq 2, \qquad a + b - d \geq 1. \tag{3.4.6}$$

To study the local behaviour of the solutions of (3.4.1), further auxiliary equations near their singularities GRS are widely used. In the case of eqn (3.4.1) the corresponding GRS reads

$$\begin{pmatrix} 1 & 1 & 1 & 1 & \\ -s & 0 & 1 & \infty & ; z \\ 0 & 0 & 0 & a & \lambda \\ 1-e & 1-c & 1-d & b & \end{pmatrix}. \tag{3.4.7}$$

The principal scheme of further calculations (also used in the next subsection) includes

(i) the use of two different scales (one of which is the boundary-layer scale);

(ii) the presentation of the differential operator as a sum $L = L_1 + sL_2$ and perturbation calculations for the solution;

(iii) the matching of the solutions obtained on different subintervals in the common region of validity.

---

[1] Of course the demand $s > 0$ should be removed.

As a final step the characteristic equation for the eigenvalues is solved by means of successive approximations. The final results include terms of two different types: (i) conventional power series in s; (ii) terms proportional to s^σ, where σ is not an integer.

First, the strictly limiting case is studied at a formal level. Setting $s = 0$ in (3.4.1) and defining $\lambda^0 = \lambda|_{s=0} = \rho_1(\nu)\rho_2(\nu)$, where the unknown quantities $\rho_1(\nu)$ and $\rho_2(\nu)$ satisfy the condition, resulting from the Fuchs relation,

$$\rho_1(\nu) + \rho_2(\nu) = d - a - b, \qquad (3.4.8)$$

we obtain the limiting differential operator

$$L_z \to L_z^0,$$
$$L_z^0 = z\{z(z-1)D^2 - [a+b+1-d - (a+b+1)z]D + ab\},$$
$$L_z^0 y_n^0 = \rho_1(\nu)\rho_2(\nu)y_n^0(z). \qquad (3.4.9)$$

As one of the solutions, eqn (3.4.9) yields the standard hypergeometric function

$$y_n^0 = z^{\rho_1} F(a + \rho_1, b + \rho_1, 1 + \rho_1 - \rho_2; z). \qquad (3.4.10)$$

Unless no specific values of the (integer) parameters are chosen, the other linear independent solution is obtained by the exchange $\rho_1 \mapsto \rho_2$ with $\rho_2 \mapsto \rho_1$ in (3.4.10).

The quantities $\rho_1(\nu)$ and $\rho_2(\nu)$ are parametrized by ν, and this parametrization, referring to the Fuchs condition, may be taken as

$$\rho_1(\nu) = \tau_1(a, b, c, d) + \nu, \quad \rho_2(\nu) = \tau_2(a, b, c, d) - \nu. \qquad (3.4.11)$$

The eigenvalues λ_n are obtained in the above-defined terms by constructing and solving an eigenvalue equation for ν. Splitting the terms of different order of magnitude makes it possible to present the operator L_z in the form

$$L_z = zM_z^0 + sM_z^1,$$
$$M_z^0 = z(z-1)D^2 - (a+b+1-d - (a+b+1)z)D + ab,$$
$$M_z^1 = z(z-1)D^2 + (-c + (d+c)z)D.$$

Along with eqn (3.4.9), an auxiliary equation is studied, namely

$$[zM_z^0 u(z) - \rho_1\rho_2]u(z) = 0. \qquad (3.4.12)$$

Solutions of eqn (3.4.12) bounded at the singularity $z = 1$ are denoted by $u_\nu(z)$. The GRS exhibits the local behaviour of the solutions related to eqn (3.4.12):

$$\begin{pmatrix} 1 & 1 & 1 & \\ 0 & 1 & \infty & ; z \\ \rho_1 & 0 & a & \\ \rho_2 & 1-d & b & \end{pmatrix}. \qquad (3.4.13)$$

The functions $u_\nu(z)$ are expressible in terms of the hypergeometric function (see [96]) as

$$u_\nu(z) = z^{\rho_1} F(a + \rho_1, b + \rho_1, d; 1 - z). \tag{3.4.14}$$

Further on, it will be computed that, in the first approximation, $\nu \sim n$, where $n \in \mathbb{Z}$.

We represent the eigenfunctions $y_n(z)$ of the boundary problem (3.4.1–3.4.3) normalized in accordance with the condition $y_n(0) = 1$ by formal series

$$y_n(z) = \sum_{m=-\infty}^{m=\infty} g_{nm} u_{\nu+m}(z) \quad \text{with } g_{n0} = 1. \tag{3.4.15}$$

This series is valid on the interval $[\epsilon, \ 1]$ where $\epsilon = o(s)$. We denote it by y_n^+.

Another series is needed in a neighbourhood of zero, where a new scale for the independent variable is introduced:

$$z := \zeta \cdot s. \tag{3.4.16}$$

Thus, instead of (3.4.1), we get

$$L_\zeta y(\zeta) = \lambda y(\zeta) \tag{3.4.17}$$

with

$$L_\zeta := \zeta(\zeta + 1)(\zeta s - 1)D^2 + [c(s\zeta - 1)(\zeta + 1)$$
$$+ ds\zeta(\zeta + 1) + e\zeta(s\zeta - 1)]D + abs\zeta, \qquad D = \frac{d}{d\zeta}.$$

The differential operator in eqn (3.4.18) can be split into the sum of two operators:

$$L_\zeta = K_\zeta^0 + s\zeta K_\zeta^1,$$
$$K_\zeta^0 = -\zeta(\zeta + 1)D^2 + [-c - (a + b + 1 - d)\zeta],$$
$$K_\zeta^1 = \zeta(\zeta + 1)D^2 + [d + c + (a + b + 1)\zeta]D + ab.$$

In addition to the auxiliary eqn (3.4.9), a second auxiliary equation valid for $\zeta = O(1)$ can be introduced:

$$(K_\zeta^0 v(\zeta) - \rho_1 \rho_2) v(\zeta) = 0. \tag{3.4.18}$$

The GRS of this equation is given by

$$\begin{pmatrix} 1 & 1 & 1 \\ -1 & 0 & \infty & ; \zeta \\ 0 & 0 & -\rho_1 \\ 1 - e & 1 - c & -\rho_2 \end{pmatrix}. \tag{3.4.19}$$

Solutions of (3.4.18) bounded at zero we denote by $v_\nu(\zeta)$. They may be expressed in terms of the hypergeometric function according to

$$v_\nu(\zeta) = F(-\rho_1, -\rho_2, c; -\zeta). \qquad (3.4.20)$$

The eigenfunctions $y_n(x)$ of the boundary problem (3.4.1 – 3.4.3) are represented in the form of a formal series

$$y_n^-(z) = \sum_{m=-\infty}^{m=\infty} h_{nm} v_{\nu+m}(\zeta). \qquad (3.4.21)$$

This series is assumed to fit the problem at $z = o(s^{-1})$, and we denote it by y_n^-. We should stress that the coefficient h_{n0} is fixed by means of a previous normalization. The procedure of getting the succeeding coefficients g_{nm} and h_{nm} in expansion (3.4.15, 3.4.21) is based on the formal series

$$g_{nm} = \sum_{j=0}^{\infty} s^{|m|+j} g_{nmj}, \qquad (3.4.22)$$

$$h_{nm} = \sum_{j=0}^{\infty} s^{|m|+j} h_{nmj}. \qquad (3.4.23)$$

The eigenvalues $\lambda_n(\nu)$ are expanded in a formal series

$$\lambda_n(\nu) = \sum_{j=0}^{\infty} s^j \lambda_{nj}(\nu). \qquad (3.4.24)$$

The three-term recurrence relation for the coefficients g_{nm} and h_{nm} and the perturbation procedure for obtaining g_{nmj} and h_{nmj} are studied in the paper [82]. On the first step of the recursive procedure, the eigenvalues $\lambda_{n0}(\nu)$ are obtained as

$$\lambda_{n0}(\nu) = \rho_1(\nu)\rho_2(\nu) \qquad (3.4.25)$$

while, for the eigenfunctions $y_n(z)$, we have

$$y_n^+(z) = u_\nu(z), \qquad y_n^-(z) = h_{n0} v_\nu(\zeta). \qquad (3.4.26)$$

Matching the constructed solutions, we obtain formulae corresponding to standard formulae connecting the local solutions of the hypergeometric equation (see 2.4). Changing formulae (2.4.6, 2.4.9) according to the chosen notations in this section, we obtain the asymptotic behaviour of y^+ (bounded at $z = 1$) in the vicinity of $z = 0$:

$$y^+ \sim \frac{\Gamma(d)\,\Gamma(\rho_2(\nu) - \rho_1(\nu))}{\Gamma(b + \rho_2(\nu))\,\Gamma(a + \rho_2(\nu))}\, z^{\rho_1(\nu)}\,[1 + o(z)]$$

$$+ \frac{\Gamma(d)\,\Gamma(\rho_1(\nu) - \rho_2(\nu))}{\Gamma(b + \rho_1(\nu))\,\Gamma(a + \rho_1(\nu))}\, z^{\rho_2\nu}\,[1 + o(z)] \quad \text{as } z \to 0.$$

$$(3.4.27)$$

On the other hand,

$$y^- \sim h_{n0} \frac{\Gamma(c)\,\Gamma(\rho_1(v) - \rho_2(v))}{\Gamma(-\rho_2(v))\,\Gamma(c + \rho_1(v))}\,\zeta^{\rho_1(v)}\left[1 + O(\zeta^{-1})\right]$$

$$+ h_{n0} \frac{\Gamma(c)\,\Gamma(\rho_2(v) - \rho_1(v))}{\Gamma(-\rho_1(v))\,\Gamma(c + \rho_2(v))}\,\zeta^{\rho_2(v)}\left[1 + O(\zeta^{-1})\right] \quad \text{as } \zeta \to \infty.$$

$$(3.4.28)$$

Also, y^- is bounded at $z = 0$.

In the overlapping interval where both asymptotic expansions (3.4.27) and (3.4.28) are valid, the magnitude of z and of ζ might be taken as $z = O(\sqrt{s})$ and $\zeta = O(1/\sqrt{s})$. Therefore it is appropriate to introduce the matching variable ξ according to

$$z = \xi\sqrt{s} \text{ and } \zeta = \xi/\sqrt{s}.$$

From (3.4.27) and (3.4.2) we get

$$y^+ \sim A\xi^{\rho_1(v)} + B\xi^{\rho_2(v)},$$

$$y^- \sim C\xi^{\rho_1(v)} + d\xi^{\rho_2(v)},$$

$$(3.4.29)$$

with the following symmetry relations for the coefficients $A(a, b, d, \rho_1, \rho_2)$, $B(a, b, d, \rho_1, \rho_2)$, $C(c, \rho_1, \rho_2)$, $D(c, \rho_1, \rho_2)$:

$$A(a, b, d, \rho_1, \rho_2) = B(a, b, d, \rho_2, \rho_1), \qquad C(c, \rho_1, \rho_2) = D(c, \rho_2, \rho_1).$$

The coefficients themselves can be easily evaluated from eqns (3.4.27–3.4.28). The matching condition is the zero of the Wronskian of y^- and y^+, which yields

$$AD - BC = 0. \tag{3.4.30}$$

If actual values of A, B, C, D are substituted, the matching condition (3.4.30) results in

$$\frac{\Gamma(d)\,\Gamma(c)\,\Gamma^2(\rho_2(v) - \rho_1(v))}{P(a, b, c, \rho_2)\,\Gamma(-\rho_1(v))}\,s^{\rho_2(v)}\,(1 + O(s)),$$

$$-\frac{\Gamma(d)\,\Gamma(c)\,\Gamma^2(\rho_1(v) - \rho_2(v))}{P(a, b, c, \rho_1)\,\Gamma(-\rho_2(v))}\,s^{\rho_1(v)}\,(1 + O(s)) = 0, \tag{3.4.31}$$

where

$$P(a, b, c, \rho) = \Gamma(a + \rho)\,\Gamma(b + \rho)\,\Gamma(c + \rho).$$

This leads to the characteristic equation for the values of v:

$$\frac{\Gamma^2(\rho_2(v) - \rho_1(v))\,\Gamma(-\rho_2(v))}{\Gamma^2(\rho_1(v) - \rho_2(v))\,\Gamma(-\rho_1(v))}\,\frac{P(a, b, c, \rho_1(v))}{P(a, b, c, \rho_2(v))} = s^{\rho_1(v) - \rho_2(v)}\,(1 + O(s)).$$

$$(3.4.32)$$

Since s is a small parameter, the left-hand side of this equation is also a small quantity. This means that one of the gamma functions in the denominator of the expression in the left-hand side must be large. Suppose that

$$\nu = n + \delta, \qquad \rho_1 = \nu, \qquad \rho_2 = d - a - b - \nu, \qquad (3.4.33)$$

where

$$|\delta| << 1. \qquad (3.4.34)$$

Then the multiplier $\Gamma(-\rho_1)$ in the denominator of the expression in the left-hand side of eqn (3.4.32) is large at small values of δ. Expanding the functions at both sides of (3.4.32) about $\delta = 0$, we obtain

$$\delta = s^{a+b-d+2n} \frac{(-1)^n}{n!} \times G^0(a, b, c, d, n) \qquad (3.4.35)$$

with

$$G^0(a, b, c, d, n) = \frac{1}{\Gamma(-\rho_2^0)} \frac{\Gamma^2(\rho_1^0 - \rho_2^0)}{\Gamma^2(\rho_2^0 - \rho_1^0)} \frac{P(a.b, c, \rho_2^0)}{P(a.b, c, \rho_1^0)}, \qquad (3.4.36)$$

where

$$\rho_1^0 = n, \qquad \rho_2^0 = d - a - b - n. \qquad (3.4.37)$$

Tacitly, we have assumed that in (3.4.32) none of the quantities a, b, c, $a + b - d$ is an integer. The leading term of the eigenvalue asymptotics, according to eqns (3.4.25, 3.4.37), is

$$\lambda_{n0}^0 = n(d - a - b - n). \qquad (3.4.38)$$

The increment $\Delta\lambda_n$ in the eigenvalue λ_n, due to the boundary layer in the neighbourhood of $z = 0$, may be calculated with the help of the formula

$$\Delta\lambda_n = \frac{\partial\lambda}{\partial\nu} \delta + O(\delta^2). \qquad (3.4.39)$$

In the first approximation, we obtain

$$\lambda_{n0} = n(d - a - b - n) \, (1 + O(s))$$

$$- s^{2n+a+b-d}(2n + a + b - d)\frac{(-1)^n}{n!} \times G^0(a, b, c, d, n).$$

$$(3.4.40)$$

According to the condition (3.4.6), the second term is smaller than the first in order of magnitude. However, according to the philosophy of exponential asymptotics, it should be retained. It is rapidly decreasing when the number n is big, and is small from the beginning if $a + b - d$ is large.

As mentioned above, the boundary-layer phenomenon causes a large peak of the eigenfunction at the merging point of the singularities. In order to estimate the height of this peak, we set $A = C$, thus obtaining for the hitherto arbitrary coefficient h_0 the estimate

$$h_0 = \frac{\Gamma(d)\,\Gamma(c+n)\,\Gamma(d-a-b-2n)\,\Gamma(n+a+b-d)}{\Gamma(c)\,\Gamma(d-a-n)\,\Gamma(d-b-n)\,\Gamma(2n+a+b-d)}s^{-n}(1+O(s)).$$

(3.4.41)

The following conjectures may be formulated as a result of our studies. When two regular singularities in Heun's differential equation merge, the corresponding boundary problem is characterized by a boundary-layer phenomenon. Due to this phenomenon, the eigenvalues cannot be expanded in a power series of a small parameter (in our case denoted by s). The reason is that they include terms which exhibit a branch-point behaviour at the origin $s = 0$. These terms get smaller when the number of the eigenvalue increases. A peak near the origin is typical for the corresponding eigenfunctions. These qualitative features are valid if the conditions (3.4.6) hold. At the limiting point $s = 0$, no explicit eigenfunctions exist, although at this point the eigenvalue curves have an explicit limit.

3.5 Large values of the scaling parameter

3.5.1 Introduction

Large values of the scaling parameter relate to what is called WKB asymptotics for the corresponding ODE. However, in this book, a general theory is presented by very specific qualitative phenomena. One of them is the phenomenon of avoided crossings of eigenvalue curves. The other is the absence of the so-called Stokes phenomenon for special cases of the doubly confluent Heun equation.

The first publication to demostrate the phenomenon of avoided crossings of eigenvalue curves for a specific equation of Heun class was the paper by Komarov and Slavyanov [68] on the quantum problem of a two-Coulomb-center system $Z_1\ e\ Z_2$. The system of two Coulomb centers, with different charges Z_1 and Z_2 has been studied at large values of the center separation. It has been shown that the structure of avoided crossings of eigenvalue curves in this model reveals much more specific features than in an arbitrary model.

In [119] a much more general study of avoided crossings was undertaken. It was shown that it is a typical phenomenon for the majority of Heun's equations with confluenced singularities, with exception of the doubly confluent Heun equation. The major statements of this study are the following:

- In the models related to Heun-class equations, there could not be a single avoided crossing for two eigenvalue curves. If there exists one avoided crossing, then there exist a large number of them (according to the largeness of the large parameter) at appropriate (integer) values of the parameter a which controls the phenomenon.
- The locations of avoided crossings constitute a periodic sequence.

The general presentation of asymptotic methods used as investigation tools can be found in [115].

3.5.2 Avoided crossings for the triconfluent equation

We shall start with the triconfluent Heun equation that possesses the least number of parameters, thus revealing the phenomenon in the simplest form. The substitution

$$y(z) = G(z)w(z) = \exp(z^3/6 + tz)w(z)$$

transforms the THE (3.1.18) to a normal (Schrödinger-type) form without the term of the first derivative:

$$N_z^{(;4)}w(z) + \lambda w(z) = (D^2 + (\lambda - (a-1)z - (z^2+t)^2/4))w(z) = 0, \quad (3.5.1)$$

where $N^{\{;4\}} = (G^{\{;4\}})^{-1}L^{\{;4\}}G^{\{;4\}}$. Further on, the notation $\tilde{a} := a - 1$ is used. Two different types of behaviour of the potential

$$V(z) = (z^2+t)^2/4 - \tilde{a}z$$

for large t relative to negative or positive values of this parameter are realizable. At positive values of t, we have a single potential well at zero, and at negative values of t we have two potential wells located in the vicinities of $z = \sqrt{-t}$ and $z = -\sqrt{-t}$. Only the latter case is interesting for our purposes; so we restrict our study to this case. The substitutions

$$t \mapsto -t \quad t > 0, \qquad z \mapsto \sqrt{t}z, \qquad \lambda \mapsto \sqrt{t}\lambda, \qquad t^{3/2} := p$$

leads to an equation with large parameter p and two pairs of close turning points in vicinities of $z = \pm 1$:

$$\tilde{N}_z^{(;4)}w(z) + \lambda w(z) = (D^2 + (p(\lambda - \tilde{a}z) - p^2(z^2-1)^2/4))w(z) = 0. \quad (3.5.2)$$

Each pair can also be regarded as a second-order turning point or as a cluster of turning points.

The comprehensive study of asymptotic solutions of equations with two close turning points has been proposed by one of the authors and can be found in the book [115]. The main technical tool is the construction of a transformation of the independent variable in terms of formal asymptotic series which converts the studied equation to a biconfluent hypergeometric equation. Further on, it will be necessary to match solutions obtained at different potential wells.

At the qualitative level, three different possibilities should be distinguished. In the first case (a), eigenfunctions are concentrated on the right potential well, and eigenvalues up to exponentially small terms are determined from the quantization condition for the right potential well. Exponentially small corrections to eigenvalues are found from the matching condition, corresponding with the exponentially small numerical values of the eigenfunction in the left potential well. The second case (b) is equivalent to the first with the sole difference that the roles of the left and right potential wells are interchanged. The third case (c), which is the most interesting, occurs when

(i) the maximal numerical values of the eigenfunctions in the left and right potential wells are of the same order of magnitude (in terms of a large parameter),

(ii) the quantization conditions are fulfilled in both potential wells simultaneously,

(iii) there exist two eigenvalues with an exponentially small splitting between them. The eigenfunction corresponding to the upper eigenvalue has an additional zero in the subbarrier region.

Further on, we will denote quantities related to the right potential well by an upper index $+$ and to the left by an upper index $-$.

First, power series terms for large values of p for eigenvalues λ are calculated with the help of the quantization condition at the right well. This condition was first formulated by Wentzel [130] and afterwards rigorously proved in [115]. Although valid under a supposition of a single well, this quantization condition enables us to obtain the correct result while neglecting the exponentially small terms in the above-mentioned cases, numbered (i) and (iii).

The eigensolutions concentrated in the right potential well (neglecting exponentially small terms) are presented in the form

$$w^+(t, z) = (u^+(p, z))^{1/2} \exp(-p \int (u^+(p, z))^{-1} \, dz) \qquad (3.5.3)$$

valid for a complete complex neighbourhood of the point $z = 1$. It reflects the fact that this part of asymptotics for eigenfunctions does not to this extent reveal a Stokes phenomenon. It imposes also that the successive approximations for the function $u^+(t, z)$ can have only poles (not branching points) at $z = 1$. The cancelling the branching-point behaviour originated by the multiplier $(u^+(p, z))^{1/2}$ will be discussed below. The use of the function $u^+(z)$ instead of its reciprocal is caused by simpler practical computations.

Substitution of (3.5.3) into (3.5.2) results in the following equation for $u^+(z)$:

$$p^2(1 - (z^2 - 1)^2 u^2/4) + p(\lambda - \tilde{a}z)u^2 + (u''u - u'^2/2)/2 = 0. \qquad (3.5.4)$$

The condition for the function $w(z)$ to be single-valued imposes the quantization condition

$$-p\mathrm{Res}_{z=1}(u^+(p, z))^{-1} = n^+ + 1/2, \qquad (3.5.5)$$

where once again we neglect exponentially small terms. Here n^+ is an integer ($n^+ = 0, 1, \ldots$), which is equal to the number of zeros of the eigensolution $w^+(p, z)$ at the right well, and the term $1/2$ is needed for the sake of cancelling the branching behaviour of the multiplier $(u^+(p, z))^{1/2}$ in (3.5.3). If the eigensolution is exponentially small at the right well, then the relation between the right-hand side and the left-hand side of (3.5.5) still holds, but n^+ is no longer an integer and below will be denoted by ν^+.

The functions $u^+(p, z)$ and $\lambda^+(p)$ are expanded in formal asymptotic series of the form

$$u^+(z, p) = \sum_{k=0}^{\infty} u_k(z) p^{-k}, \tag{3.5.6}$$

$$\lambda^+(p) = \sum_{k=0}^{\infty} \lambda_k p^{-k}. \tag{3.5.7}$$

Successive terms in expansions (3.5.6–3.5.7) are obtained recursively by equating terms of the same order of p in eqns (3.5.4–3.5.5). We should mention here that, in order to obtain λ_k, it is sufficient to know an appropriate number of terms in the Taylor expansions for the $u_j(z)$ ($j = 0, \ldots, k-1$) in the vicinity of $z = 1$.

The first three terms of expansion (3.5.6) are

$$u_0 = \frac{2}{z^2 - 1}, \qquad u_1 = (\lambda_0 - \tilde{a}z)u_0^3/2,$$

$$u_2 = \frac{u_0}{2} \left(\lambda_1 u_0^2 + \frac{3}{4}(\lambda_0 - \tilde{a}z)^2 u_0^4 + \frac{1}{2}\left(u_0'' u_0 - \frac{1}{2}u_0'^2\right) \right). \tag{3.5.8}$$

The corresponding terms of the expansion for the eigenvalues λ^+ are

$$\lambda^+ = \lambda^+(n^+, \tilde{a}) = (2n^+ + 1) + \tilde{a} \tag{3.5.9}$$

$$- \frac{1}{p}\left[\frac{3}{2}((2n^+ + 1)^2 + \tilde{a}(2n^+ + 1)) + \frac{\tilde{a}^2}{4} + \frac{1}{8} \right] + O(p^{-2}).$$

There is no need to recompute the function $u^-(p, z)$ and the eigenvalue λ^- at the left well, namely at the point $z = -1$, since the substitutions

$$z \mapsto -z, \qquad a \mapsto 2 - a \quad \text{or} \quad \tilde{a} \mapsto -\tilde{a}, \tag{3.5.10}$$

leave eqn (3.5.2) unchanged. Therefore there is another set of eigenvalues related to the left potential well which can be obtained with the help of (3.5.9)–(3.5.10):

$$\lambda^- = \lambda^+(n^-, -\tilde{a}) = (2n^- + 1) - \tilde{a} \tag{3.5.11}$$

$$- \frac{1}{p}\left[\frac{3}{2}((2n^- + 1)^2 - \tilde{a}(2n^- + 1)) + \frac{\tilde{a}^2}{4} + \frac{1}{8} \right] + O(p^{-2}).$$

Here, the integers n^- correspond to the number of zeros of the eigensolution $w^-(p, z)$ at the left potential well.

If one tried to draw the curves of eigenvalues on the $(\tilde{a}, \lambda)$ plane, then one would immediately find that there are crossings of these curves (straight lines in the first approximation) at the points with coordinates $\tilde{a} = n^- - n^+$, $\lambda = n^+ + n^- + 1$ (the λ coordinate in the first approximation). It is well known that there are no degenerate eigenvalues for self-adjoint boundary problems for second-order differential equations. Hence the contradiction should be solved by taking into account exponentially small terms.

The solution of eqn (3.5.2) is sought in a more sophisticated form than that of (3.5.3):

$$w(z) = [(\xi^+(p, z))']^{-1/2} D_{v+}(\sqrt{p}\xi^+(p, z)) \tag{3.5.12}$$

where $D_v(t)$ are parabolic cylinder functions (see (2.1.29) with

$$\xi^+(p, z) = \sum_{k=0}^{\infty} \xi_k^+ p^{-k}$$

and

$$v^+ = n^+ + \delta^+.$$

This form of the solution, as we will see later, includes two exponents in the subbarrier region.

For our goals, we need two terms of asymptotic expansions in large values of p for $\xi^+(p, z)$ or, what is more convenient for further computations, for $(\xi^+(p, z))^2$. We have

$$((\xi_0^+)^2)' = 2(z^2 - 1), \quad (\xi_0^+\xi_1^+)' = 2(z^2 - 1)^{-1}[(v_0^+ + 1/2)\xi_0'^2 - \lambda_0^+ + \tilde{a}z] \tag{3.5.13}$$

which yields

$$(\xi_0^+)^2 = 2\left(\frac{z^3}{3} - z\right) + \frac{4}{3}$$

$$+ \frac{1}{p}\left[(2v_0^+ + 1)(2\ln(z + 1) + \ln(z + 2)) + 2\tilde{a}\ln(z + 2)\right] + O(p^{-2}). \tag{3.5.14}$$

Using the asymptotic formula for $D_v(t)$ for large negative values of t, taken from Section 2.4, which is beyond the Poincaré definition of asymptotics but has several other reasonable explanations in modern exponential asymptotics (see [124], [8]), we obtain

$$D_{v+}(\sqrt{p}\xi^+) = \cos(\pi v^+) \exp(-p(\xi^+)^2/4)(-\sqrt{p}\xi^+)^{v^+}[1 + O(p^{-1})]$$

$$+ \frac{\sqrt{2\pi}}{\Gamma(-v^+)} \exp(p(\xi^+)^2/4)(-\sqrt{p}\xi^+)^{-v^+-1}[1 + O(p^{-1})]. \tag{3.5.15}$$

This results in solutions of (3.5.2) for the subbarrier region

$$w^+(p, z) = \cos(\pi v^+)e^{-\frac{p}{6}(z-1)^2(z+2)}\left(\sqrt{\frac{2p}{3}}\right)^{v^+} \frac{(z-1)^{v^+}}{(z+1)^{v^++\tilde{a}+1}}[1 + O(p^{-1})]$$

$$+ \frac{\sqrt{2\pi}}{\Gamma(-v^+)}e^{\frac{p}{6}(z-1)^2(z+2)}\left(\sqrt{\frac{2p}{3}}\right)^{-v^+-1} \frac{(z+1)^{v^++\tilde{a}}}{(z-1)^{v^++1}}[1 + O(p^{-1})]. \tag{3.5.16}$$

The same solution constructed at the left well, denoted by $w^-(p, z)$, reads

$$
w^+(p, z) = \cos(\pi \nu^-) e^{-\frac{p}{6}(z-1)^2(2-z)} \left(\sqrt{\frac{2p}{3}} \right)^{\nu^-} \frac{(z+1)^{\nu^-}}{(z-1)^{\nu^- - \tilde{a} + 1}} [1 + O(p^{-1})]
$$
$$
+ \frac{\sqrt{2\pi}}{\Gamma(-\nu^-)} e^{\frac{p}{6}(z-1)^2(2-z)} \left(\sqrt{\frac{2p}{3}} \right)^{\nu^-} \frac{(z-1)^{\nu^- - \tilde{a}}}{(z+1)^{\nu^- + 1}} [1 + O(p^{-1})].
$$

$$(3.5.17)$$

Matching the solutions w^+ and w^-, we first get the connection formula for ν^+ and ν^-:

$$
\nu^+ - \nu^- = -\tilde{a} \qquad (3.5.18)
$$

as a necessary condition.

Now the three cases (a–c) mentioned near the beginning of this subsection may be considered in connection with (3.5.18) and the quantization conditions at both wells. Suppose that $\tilde{a}$ is not an integer. Then either $\nu^+ = n^+ + \delta^+$, so that eigenvalues λ are obtained from the quantization condition for the right well (3.5.9), or $\nu^- = n^- + \delta^-$, so that eigenvalues λ are obtained from the quantum condition for the left well (3.5.11). We remark that the connection formula (3.5.18) is compatible with the above-mentioned symmetry property for the eigenvalues λ^+ and λ^-.

The exponentially small corrections δ^+ and δ^- are easily found from the condition that the Wronskian of the solutions w^+ and w^- is equal to zero:

$$
W(w^+(p, z), w^-(p, z)) = 0. \qquad (3.5.19)
$$

Formally, in this case, only one of the gamma functions (3.5.16–3.5.17) is evaluated near its poles.

Another possibility appears when parameter $\tilde{a}$ is an integer. Then the quantization conditions are fulfilled at both potential well independently, and, as a result, we have

$$
\nu^+ = n^+ \pm \delta, \qquad \nu^- = n^- \pm \delta. \qquad (3.5.20)
$$

The exponentially small value of δ is obtained from eqn (3.5.19), which turns out to be a quadratic equation at small values of δ and which, after simplification, takes the form

$$
\delta^2 = \frac{e^{-\frac{4p}{3}} \left(\frac{2p}{3} \right)^{(n^- + n^+ + 1)}}{2\pi (n^-)!(n^+)!} [1 + O(p^{-1})]. \qquad (3.5.21)
$$

Note that, this time, both gamma functions in (3.5.16–3.5.17) are evaluated near their poles.

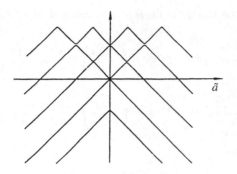

FIG. 3.1. Asymptotic calculation of the phenomenon of avoided crossings.

The splitting of two eigenvalues $\Delta\lambda$ at points of avoided crossings is given by

$$\Delta\lambda = 4 \mid \delta \mid= \frac{4e^{-\frac{2p}{3}} \left(\frac{2p}{3}\right)^{(n^- + n^+ + 1)/2}}{[2\pi(n^-)!(n^+)!]^{1/2}}[1 + O(p^{-1})]. \qquad (3.5.22)$$

The result of the asymptotic calculation is shown in Fig. 3.1. The curves should be compared with the exact calculations in Section 4.4.

3.5.3 Biconfluent Heun equation

In the case of the biconfluent eqn (3.1.11), it is convenient to perform the transformation to the Schrödinger-type form in two steps. First, we convert it to a self-adjoint form by means of a substitution similar to (3.5.2):

$$y(z) \mapsto w(z): \; y(z) = G(z)w(z) = \exp(z^2/4 + tz/2 + (1 - c)\ln z/2)w(z), \qquad (3.5.23)$$

$$M_z^{(1;3)}w(z) = \left(D(zD) + \left[\lambda + \frac{tc}{2} - \left(a - \frac{1+c}{2}\right)z \right.\right.$$
$$\left.\left. - z(z + t)^2/4 - \frac{(1 - c)^2}{4z}\right]\right)w(z) = 0, \qquad (3.5.24)$$

where $M^{(1;3)} = G^{-1}L^{(1;3)}G$. Once again, we study the case when the large parameter p is negative. For this case, we have a typical Coulomb-type behaviour at zero and an oscillator-type well in the vicinity of $z = p$. After carrying out the further substitutions

$$t \mapsto -p \quad (p > 0), \qquad z \mapsto pz,$$

$$\lambda + \frac{pc}{2} \mapsto p\lambda, \qquad a - \frac{1+c}{2} \mapsto \tilde{a}, \qquad w(z) \mapsto z^{-1/2}w(z),$$

eqn (3.5.24) converts to

$$N_z^{(1;3)} w(z) = \left(D^2 + \left[p \frac{\lambda - \tilde{a}z}{z} - p^2(z-1)^2/4 - \frac{1 - (1-c)^2}{4z^2} \right] \right) w(z) = 0.$$

(3.5.25)

Equation (3.5.25) is studied in the same way as eqn (3.5.2), but the sign + labels constructions in the vicinity of the point $z = 1$, i.e. those related to the potential well, and the sign − labels constructions for the Coulomb-type singularity at zero.

The representation (3.5.2) and the quantization condition (3.5.3) impose the first terms of power-type asymptotic expansions for $u^+(p, z)$ and λ^+ in the same way as (3.5.6, 3.5.8):

$$u_0 = \frac{2}{z-1}, \qquad u_1 = (\lambda_0 - \tilde{a}z)\frac{u_0^3}{2z},$$

$$u_2 = \frac{u_0}{2} \left[\lambda_1 \frac{u_0^2}{z} + \frac{3}{8}(\lambda_0 - \tilde{a}z)^2 \frac{u_0^4}{z^2} + (1 - (1-c)^2)\frac{u_0^2}{4z^2} + \frac{5}{16}u_0^4 \right].$$

The corresponding terms of the expansion for the eigenvalues λ^+ are

$$\lambda^+ = \lambda^+(n^+, \tilde{a}) = n^+ + 1/2 + \tilde{a}$$

$$- \frac{1}{p} \left[\frac{3}{2}((n^+ + 1/2)^2 + 2\tilde{a}(n^+ + 1/2)) + \frac{\tilde{a}^2}{2} + \frac{1}{4}c(c-2) \right] + O(p^{-2}).$$

(3.5.26)

where n^+ is a constant indicating the number of zeros of eigensolution at the potential well.

Now we need to construct the same terms in the vicinity of the Coulomb-type potential well at zero. For this, we need to change the sign in the expressions for $u_k(z)$ to the opposite of $u_0(z)$. In order to calculate the successive terms of λ^-, another quantization condition should be used instead of (3.5.5). It takes into account not only the zeros of the eigensolution but also a possible zero or branching point at $z = 0$ according to the value of the parameter c which we further consider to be positive [115]:

$$-p\mathrm{Res}_{z=0}(u^-(p, z))^{-1} = n^- + c/2.$$

(3.5.27)

Here, once again, exponentially small terms are neglected. In (3.5.27), n^- is an integer ($n^- = 0, 1, \ldots$) which is equal to the number of zeros of the eigensolution $w^-(p, z)$ at the left Coulomb-type well. Calculations of the residue give the asymptotic expression for λ^-:

$$\lambda^- = n^- + c/2 + \frac{1}{p} \left[\frac{3}{4}(n^+ + c/2)^2 - \tilde{a}(n^- + c/2)) + \frac{c(c-2)}{4} \right] + O(p^{-2}),$$

(3.5.28)

where n^- is an integer indicating the number of zeros of an eigensolution at the Coulomb-type potential. Unfortunately, here, there is no simple relation connecting λ^- and λ^+ on the basis of symmetries of the equation.

The study of solutions at the right well when two exponents are intended to be retained is completely similar to that given in the previous subsection; so we give only the formulae corresponding to (3.5.13)–(3.5.14):

$$((\xi_0^+)^2)' = 2(z-1), \qquad (\xi_0^+\xi_1^+)' = \frac{2v^+ + 2\tilde{a} + 1}{z}, \tag{3.5.29}$$

which yields

$$(\xi^+)^2 = z(z-2) + 1 + \frac{1}{p}(2v^+ + 2\tilde{a} + 1)\ln z + O(p^{-2}).$$

At the Coulomb-type singularity, an eigensolution $w^-(p, z)$ is sought in the form

$$w^-(p, z) = (\xi'(p, z))^{-1/2}M_{\kappa^-,m}(p\xi^-(p, z)), \tag{3.5.30}$$

where $M_{\kappa,m}(t)$ is the solution of the Whittaker equation bounded at zero with $m = (c-1)/2$, with and κ^- to be found from the matching condition. The first terms of the asymptotic expansion for ξ^- are

$$\xi^-(p, z) = z\left(1 - \frac{z}{2}\right) + \frac{1}{p}\left[\frac{2b - 2\kappa^-}{1-z} - \frac{\kappa^-}{1-z/2}\right] + O(p^{-2}). \tag{3.5.31}$$

For matching the asymptotic expansions, besides (3.5.15), the formula (see Section 2.4)

$$M_{\kappa,m}(p\xi) = \frac{\cos(\pi(m - \kappa + 1/2))}{\Gamma(m - \kappa + 1/2)}e^{-p\xi/2}(p\xi)^\kappa[1 + O(p^{-1})]$$

$$+ \frac{1}{\Gamma(m + \kappa + 1/2)}e^{p\xi/2}(p\xi)^{-\kappa}[1 + O(p^{-1})] \tag{3.5.32}$$

is needed, where both exponents—dominant and recessive—are preserved. Note that, in (3.5.32), a non-standard normalization of the function $M_{\kappa,m}(t)$ was chosen. After having substituted (3.5.31–3.5.32) in (3.5.30), we arrive at an asymptotic expression for $w^-(p, z)$ in the subbarrier region in the form

$$w^-(p, z) = \frac{\cos(\pi(m - \kappa^- + 1/2))}{\Gamma(m - \kappa^- + 1/2)}e^{-p\frac{z}{2}(1 - \frac{z}{2})}(1 - z)^{\tilde{a} - 1/2}$$

$$\times \left(\frac{pz}{1-z}\right)^{\kappa^-}[1 + O(p^{-1})] + \frac{1}{\Gamma(m + \kappa^- + 1/2)}$$

$$\times e^{p\frac{z}{2}(1 - \frac{z}{2})}(1 - z)^{-\tilde{a} - 1/2}\left(\frac{pz}{1-z}\right)^{-\kappa^-}[1 + O(p^{-1})]. \tag{3.5.33}$$

The other asymptotic expression for the function $w^+(p, z)$ reads

$$w^+(p, z) = \cos(\pi v^+) e^{p \frac{z}{2}(1 - \frac{z}{2}) - \frac{p}{4}} z^{v^+ + \tilde{a} + 1/2} (\sqrt{p}(1 - z))^{v^+} [1 + O(p^{-1})]$$

$$+ \frac{\sqrt{2\pi}}{\Gamma(-v^+)} e^{-p \frac{z}{2}(1 - \frac{z}{2}) + \frac{p}{4}} z^{-(v^+ + \tilde{a} + 1/2)} (\sqrt{p}(1 - z))^{-v^+ - 1} [1 + O(p^{-1})].$$

$$(3.5.34)$$

As a necessary condition of matching the two constructed asymptotics, we have

$$v^+ + \tilde{a} + 1/2 = \kappa^-. \tag{3.5.35}$$

As in the previous case of the triconfluent equation, the use of the eigenvalue dispersion eqn (3.5.19) gives rise to two sequences of eigenvalues related to eigensolutions either concentrated at $z = 0$ or concentrated at $z = 1$. The asymptotic formulae for these sequences in neglecting asymptotically small corrections have been obtained already by simpler considerations in (3.5.26) and (3.5.28). Crossings would occur when these sequences intersect in the $(\lambda, \tilde{a})$ plane. However, if $\tilde{a} + (1 - c)/2$ is an integer, then this crossing is suppressed and the phenomenon of avoided crossings appears. Returning to the original parameters of the BHE, this condition may be rewritten as

$$c - a = n^+ - n^-. \tag{3.5.36}$$

The splitting of eigenvalues curves at these points is obtained from (3.5.19) as

$$\delta^2 = \frac{e^{-\frac{p}{2}} p^{(2n^- + n^+ + c + 1/2)}}{\sqrt{2\pi} \Gamma(c) \Gamma(n^- + c)(n^-)!(n^+)!} [1 + O(p^{-1})] \tag{3.5.37}$$

with

$$v^+ = n^+ \pm \delta, \qquad \kappa^- = n^- + \frac{c}{2} \pm \delta.$$

For the sake of illustration, we give some figures from which the reader may get an impression of the phenomenon. They show the exact calculation, the theory of which is given in Sections 1.6 and 3.6, and compare it with the asymptotic calculations, presented above. The reader should also refer to Sections 4.4 and 4.6, where the phenomenon and its suppression or disappearance is shown by numerical means.

Figures[2] 3.2–3.5 show the occurrence and the development of the phenomenon with increasing parameter p. (The method of calculation is developed in Sections 1.6 and 3.6.) Figure 3.6 focuses on a point of avoided crossing for the parameters shown in Fig. 3.5. Figure 3.7 shows the asymptotic calculation of the phenomenon. Figure 3.8 compares the asymptotic calculations given above with the exact numerical results. The full line gives the gap between two curves at a crossing point as a function of p, as a result of the exact numeric calculations. The dashed line gives the same gap as

---

[2]The numerical calculations were carried out by Dipl.-Phys. Karlheinz Bay. The authors thank him for letting them have yet unpublished material.

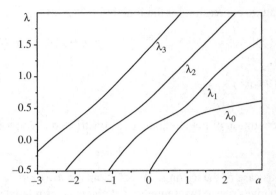

FIG. 3.2. Parameters $p = 10, c = 1$.

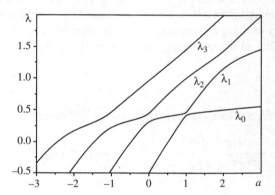

FIG. 3.3. Parameters $p = 20, c = 1$.

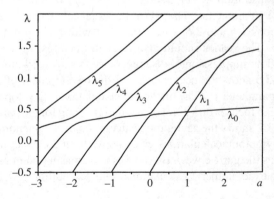

FIG. 3.4. Parameters $p = 30, c = 1$.

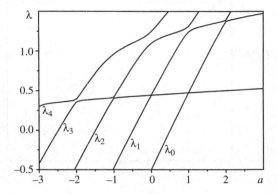

FIG. 3.5. Parameters $p = 40$, $c = 1$.

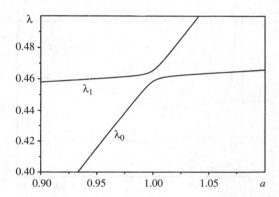

FIG. 3.6. Splitting of the eigenvalue curves at the crossing point.

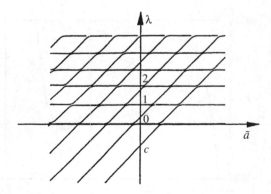

FIG. 3.7. Asymptotic calculation of the phenomenon of avoided crossings.

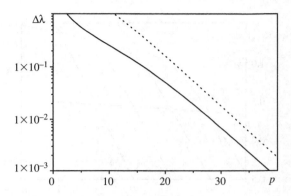

FIG. 3.8. Comparison between asymptotic and numeric calculations of the splitting at the crossing point.

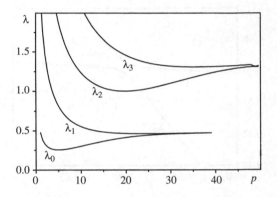

FIG. 3.9. Splitting at the crossing point in dependence on p.

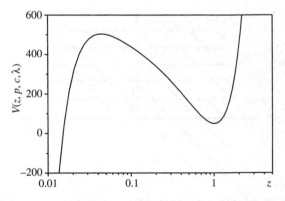

FIG. 3.10. The Potential of the Schrödinger form.

a result of the asymptotic calculations. Figure 3.9 gives the splitting of the curves at the crossing points as function of the parameter p. Figure 3.10 shows the potential of the underlying biconfluent Heun equation in Schrödinger form.

3.5.4 The confluent Heun equation

Since the study of this case is similar to the already studied cases, we give only the final results for the splitting of eigenvalue curves at avoided crossings. All technical details of calculations are discussed in the previous sections.

As a starting point for constructing the asymptotics, the normal form of the CHE can be used in which, for better symmetry properties, the regular singularities are positioned at $z = \pm 1$ and the initial parameters c and d in (3.1.3) are replaced by parameters m and s according to $c = m + s + 1$ and $d = m - s + 1$

$$v''(z) + \left(-p^2 + 2p\frac{\lambda - bz}{1 - z^2} - \frac{m^2 + s^2 + 2msz - 1}{(1 - z^2)^2} \right) v(z) = 0. \qquad (3.5.38)$$

The 'potential' in eqn (3.5.39) can be considered as a sum of two one-dimensional Coulomb potentials. Therefore, as $p \mapsto \infty$ the spectrum of eigenvalues resembles two Coulomb spectra.

For eigensolutions related to the left well, they read

$$\lambda^-(p, \chi, b) = (2\chi - b) + \frac{1}{2p}\left(-2\chi(\chi - b) + \frac{1}{2}(m^2 + s^2 - 1) \right)$$

$$+ \frac{1}{(2p)^2}\left(-\chi(\chi - b)^2 + \chi^2(\chi - b) - \frac{1}{4}(\chi - b)(1 - (m - s)^2) \right.$$

$$\left. - \frac{1}{4}\chi(1 - (m + s)^2) \right)$$

$$+ \frac{1}{(2p)^3}\left(-3\chi^2(\chi - b)^2 - \chi(\chi - b)^3 - \chi^3(\chi - b) \right.$$

$$- \chi(\chi - b) - \frac{1}{2}(\chi - b)^2(1 - (m - s)^2) - \frac{1}{2}\chi^2(1 - (m + s)^2)$$

$$\left. - \chi(\chi - b)(1 - (m^2 + s^2)) - \frac{1}{16}(1 - (m - s)^2)(1 - (m + s)^2) \right)$$

$$+ O\left(\frac{1}{p^4} \right), \qquad (3.5.39)$$

where

$$\chi = \kappa^-. \qquad (3.5.40)$$

An asymptotic formula for eigenvalues $\lambda^+(p, \chi, -b)$ constructed for the right Coulomb-type well is obtained by the substitutions

$$\chi \mapsto \kappa^+, \qquad b \mapsto -b, \qquad m - s \mapsto m + s, \qquad m + s \mapsto m - s$$

in (3.5.41). The relation between κ^- and κ^+, which can be regarded either as the uniqueness of the eigenvalues or a necessary matching condition, is

$$\kappa^+ = \kappa^- - b. \tag{3.5.41}$$

In order to get an equation for the parameter κ^-, we calculate the Wronskian of the asymptotic representation of eigensolutions at the subbarrier region, and equate this Wronskian to zero. The corresponding result reads

$$\frac{1}{\pi} \tan \pi \left(\frac{1}{2} + \frac{m+s}{2} - \kappa^+\right) \cdot \frac{1}{\pi} \tan \pi \left(\frac{1}{2} + \frac{m-s}{2} - \kappa^-\right)$$

$$= \frac{e^{-4p}(4p)^{2(\kappa^+ + \kappa^-)}}{\Gamma\left(\frac{1}{2} + \frac{m+s}{2} + \kappa^+\right)\Gamma\left(\frac{1}{2} + \frac{m-s}{2} + \kappa^-\right)\Gamma\left(\frac{1}{2} + \kappa^+ - \frac{m+s}{2}\right)\Gamma\left(\frac{1}{2} + \kappa^- - \frac{m-s}{2}\right)}$$

$$\times (1 + O(p^{-1})). \tag{3.5.42}$$

The system of two coupled equations (3.5.42)–(3.5.43) can be solved by successive approximations. It leads generally to two sets of solutions. If we neglect exponentially small terms, then they are either

$$\kappa^- = n^- + \frac{m-s}{2} + \frac{1}{2} \tag{3.5.43}$$

or

$$\kappa^+ = n^+ + \frac{m+s}{2} + \frac{1}{2}. \tag{3.5.44}$$

Here n^- and n^+ are integers indexing the eigenvalues corresponding to these two sets. The first set relates to the case when the corresponding eigenfunctions are concentrated near the left end of the interval $]-1, 1[$. The second set relates to the case when the corresponding eigenfunctions are concentrated near the right end.

Avoided crossings of eigenvalue curves in the (b, λ) plane occur under the condition

$$b = n^- - n^+ - s. \tag{3.5.45}$$

The splitting between the eigenvalues curves in this plane can be calculated as

$$\Delta\lambda = \frac{2(4p)^{n^+ + n^- + m + 1} e^{-2p}}{[n^+! \, n^-! \, (n^+ + m + s)! \, (n^- + m - s)!]^{\frac{1}{2}}} (1 + O(p^{-1})). \tag{3.5.46}$$

The asymptotics of the eigensolutions at the points of avoided crossings may be taken as a sum and a difference of the 'local' asymptotic expansions.

3.5.5 Discussion

Here we speculate about the origin and possible links of the phenomenon of avoided crossings. We start with confluent types of hypergeometric equation with unramified singularities at infinity. There are two such equations, namely the confluent hypergeometric equation and the Weber equation for parabolic cylinder functions.

Two formal asymptotic solutions of these equations at the irregular point at infinity are presented in the form

$$y_m(z) = \exp(P_m(z))z^{-\alpha_m}v_m(z) \quad \text{for } m = 1, 2, \tag{3.5.47}$$

where $P_m(z)$ are polynomials, of the first or second order, of the equation, and $v_m(z)$ are formal asymptotic series in powers of z^{-1}. On the radial rays, where the imaginary part of P_m becomes zero, the Stokes phenomenon occurs. This means that the two-dimensional vector that represents an actual solution of the equation in terms of formal asymptotic solutions is multiplied by a matrix. In an appropriate basis, the matrix is triangle. Then it is called Stokes matrix, and the non-diagonal element of this matrix is called Stokes multiplier. It has to be considered in the space of two real variables, one large parameter r—the modulus of the independent variable—and the other parameter φ: the angle between the abscissa and the above-mentioned ray. In this space, on a formal level, the Stokes phenomenon is the occurence of discontinuities of the Stokes multiplier as a function of φ. However, for numerical needs, a further structure can be imported to this space where, instead of formal expansions, finite numbers of terms of these expansions are considered—the larger the number, the larger is r. That leads to Berry's smoothing of the Stokes phenomenon [14].

All these speculations are valid unless $a = \alpha_1 - \alpha_2$ is an integer. In this latter case, the Stokes matrix becomes diagonal and the Stokes phenomenon disappears. One of the solutions for this case is an eigensolution. The phenomenon of the disappearance of the Stokes constant as a function of the parameter a is continuous. It is senseless to investigate the behaviour of the eigenvalues with respect to the parameter a since the eigenvalue is an integer a.

Now comes the Heun class. Formal asymptotic solutions for equations with unramified singularities (which we have studied above) once again may be represented in a form similar to (3.5.48), but the number of parameters at our disposal is much larger. The eigenvalue parameter and the parameter a are distinct. And, besides asymptotics, for large values of the argument z, it is possible to study asymptotics in the large parameter p. So, here, our space of real parameters is four-dimensional, and one can study other discontinuity phenomena in this space. We think that the phenomenon of avoided crossings of eigenvalues λ in the (r, a) space has many features in common with the Stokes phenomenon in the (r, φ) space . Both are discontinuity phenomena when different parts of asymptotics (in the sense of exponential asymptotics) change; both are periodic in the second parameter; both appear to be quite general. For this reason, further study of the phenomena should include estimates for the large terms of the asymptotic expansions in p, the smoothing functions for large but fixed p, the resurgence of terms in some transforms to a new space from the (p, a) space.

3.6   Central two-point connection problems

3.6.1   Introduction

This section exhibits the treatment of the central two-point connection problem (CTCP) for the confluent cases of Heun's differential equation. Reduced confluent Heun equations are not presented, since their treatment is not yet accomplished.

The characteristic feature of our problems is that at least one endpoint of the relevant interval contains an *irregular* singularity of the underlying differential equation, while the other one may be an ordinary point, or a regular or an irregular singularity. All these cases will actually occur.

According to the general procedure outlined in Section 1.6, the first step in our method is the splitting off the asymptotic factors at the relevant singularities. However, we omit this item by starting not with the general forms but with the canonical forms of the confluent cases of Heun's equation. This is done for two reasons. First, the canonical forms, as forms having the least number of parameters, yield the simplest formulae, which is important for the next consideration. Secondly, each of the parameters that occur can be ascribed a specific role that helps the understanding of the CTCPs considerably. In other words: we start *after* the splitting off the asymptotic factors of the particular solutions and, moreover, assuming in the singly-confluent case that one of the characteristic exponents of the non-relevant singularity is zero. Nevertheless, the reader does not need to carry out the necessary transformations for the general forms but can find them in the addendum (see also [79]).

As a consequence, our first step will be the generation of an additional singularity in the biconfluent, triconfluent, and doubly confluent cases where this is necessary.

The second step is a Möbius transformation. Since we start with differential equations in natural form, we need to put the relevant interval between the origin and unity. There is one exception of this rule: in the doubly confluent case, the singularity at the origin is put at minus one, while the one at infinity is put at plus one. This is a natural generalization which is necessary because both of the relevant singularities are of irregular type (cf. Section 1.6).

The crucial point is that the resulting differential equations have one particular solution which is holomorphic at the origin and thus is expandable in a power series. The radius of convergence of this series is precisely unity, and the coefficients obey irregular difference equations of Poincaré–Perron type, the order of which is the sum of the s-ranks of the irregular singularities involved in the boundary problem. Moreover, they all can be solved recursively, and thus have a sufficient number of initial conditions.

As was already discussed in Chapter 1, there exist in all of these cases a fundamental set of asymptotic solutions of the difference equations, called Birkhoff set, consisting in Birkhoff solutions which may be calculated explicitly by means of index-asymptotics and which are given below. As a result, the general solution of these difference equations may be presented in terms of the Birkhoff solutions.

The final result is that the exact eigenvalue condition of the central two-point connection problem for the confluent cases of Heun's differential equation can be expressed by means of the representation of the above-mentioned general solution. As will be shown in Chapter 4, this method not only can be formulated quite uniformly for the whole class of confluent cases of Heun's differential equation but also works with respect to its numerical aspects.

We recall the results of Section 1.6, and our first goal is to add more details to the principal steps of the general procedure shown there.

3.6.2 Differential equations in canonical form

The appropriate canonical forms of the confluent cases of Heun's differential equation for the CTCP are:
the singly confluent case

$$\frac{d^2y}{dz^2} + \left[-p + \frac{c}{z} + \frac{d}{z+1}\right]\frac{dy}{dz} + \left[\frac{apz+\lambda}{z(z+1)}\right]y = 0, \qquad (3.6.1)$$

the biconfluent case

$$\frac{d^2y}{dz^2} + \left[-p(z+1) + \frac{c}{z}\right]\frac{dy}{dz} + \left[-pa + \frac{\lambda}{z}\right]y = 0, \qquad (3.6.2)$$

the triconfluent case

$$\frac{d^2y}{dz^2} + \left[-p(z^2 - 1)\right]\frac{dy}{dz} + [-paz + \lambda]\,y = 0, \qquad (3.6.3)$$

and the doubly confluent case:

$$\frac{d^2y}{dz^2} + \left[\frac{c}{z} - \frac{p(z^2-1)}{z^2}\right]\frac{dy}{dz} + \left[\frac{-paz+\lambda}{z^2}\right]y = 0. \qquad (3.6.4)$$

In all the above-written equations, it is supposed that a, c, d, p, λ are real parameters, and moreover that the parameters p and c satisfy

$$p > 0, \qquad c > 1. \qquad (3.6.5)$$

Confusion should not arise from using the same notation for the parameters in different equations here and below.

For each of the above-mentioned equations, the solution of the CTCP may be stated as follows.
Consider the positive real axis. While approaching the points $z = 0$ and $z = \infty$ from inside the interval $[0, \infty[$ there are two different asymptotic behaviours (called dominant and recessive) of the solutions of each of the equations (3.6.1)–(3.6.4). How do we find the values of the parameter λ for which these differential equations have solutions that behave like the recessive solution while *simultaneously* approaching both points $z = 0$ and $z = \infty$ along the real axis? In fact, these values are the eigenvalues $\lambda = \lambda_i$ of the corresponding eigenvalue problem.

It is assumed that, for all confluent cases, the two relevant points of the CTCP are located at $z_1 = 0$, $z_2 = \infty$. The point $z_1 = 0$ is

(i) an ordinary point of the equation in the triconfluent case,
(ii) a regular singularity in the singly confluent and in the biconfluent cases,
(iii) an irregular singularity in the doubly confluent case.

In the triconfluent case, the point $z = 0$ is an ordinary point of the differential equation. Therefore it would be more natural to extend the interval on which the connection problem is treated from $[0, \infty[$ to $]-\infty, \infty[$. This can be done by applying the procedure given below twice: once for the interval $[0, \infty[$ and once for the interval $]-\infty, 0]$. In the latter case, we need to replace z by $-z$ and p by $-p$. When treating the connection problem on the whole real axis, we speak of the *natural CTCP for the triconfluent case* of Heun's differential equation.

First, we carry out a linear transformations of the dependent variable for each of the equations (3.6.1)–(3.6.4)—that is, an sg-transformation in the biconfluent, triconfluent, and doubly confluent cases:

$$y = (z + 1)^{-a} w. \tag{3.6.6}$$

Writing equations (3.6.1)–(3.6.4) in the non-specified form

$$\frac{d^2 y(z)}{dz^2} + P(z)\frac{dy(z)}{dz} + Q(z)y(z) = 0$$

by means of (3.6.6), we get the differential equation

$$\frac{d^2 w(z)}{dz^2} + \left(P(z) - \frac{2a}{z+1}\right)\frac{dw(z)}{dz} + \left(Q(z) - \frac{aP(z)}{z+1} + \frac{a(a+1)}{(z+1)^2}\right)w(z) = 0. \tag{3.6.7}$$

The crucial difference between these two equations is that the multiplier in front of $w(z)$ in (3.6.7) can be estimated as

$$\tilde{Q}(z) = Q(z) - \frac{aP(z)}{z+1} + \frac{a(a+1)}{(z+1)^2} = O\left(z^{-2}\right) \text{ as } z \to \infty.$$

As a result of this fact, the point $z = -1$ becomes a reducible singularity (see [116], p. 296) in the case of equations (3.6.2)–(3.6.4).

In the following, we apply a Möbius transformation to the dependent variable of each confluent case of eqn (3.6.7) according to

$$x = \frac{z - z_{**}}{z+1}, \tag{3.6.8}$$

where the value of z_{**} is zero in the singly confluent, biconfluent, and triconfluent case and is unity in the doubly confluent case.

Remark: At this point we must mention that there is a situation where a Jaffé transformation for the Heun equation itself makes sense. Suppose that a Heun equation has its singularities at $z = 0, 1, z^*, \infty$. Moreover suppose that we want to solve a CTCP for this Heun equation on the interval $[0, 1]$. If $|z^*| < 1$, where z^*, may be complex, then a Jaffé expansion about $z = 0$ has a radius of convergence that is $\mathcal{R} = |z^*| < 1$ and thus does not reach the other relevant singularity at $z = 1$. In this situation, one has to carry out the Jaffé transformation

$$x = \frac{z\,(1 - z^*)}{z - z^*}. \tag{3.6.9}$$

This transformation has the required features: first, it leaves the relevant singularities unaffected, and, secondly, it makes the unit circle free of non-relevant singularities such that, with a Jaffé expansion, one can solve the CTCP. It leads to a second-order regular difference equation of Poincaré–Perron type, and thus to a three-term recurrence relation. □

In the following, we give the results of the Möbius transformation for each confluent case.

Singly confluent case

$$x(1-x)^2\,\frac{d^2w}{dx^2} + [-px + (c + \{d - 2(a+1)\}x)(1-x)]\,\frac{dw}{dx}$$
$$+ [\lambda - ac + \{a(a+1) - ad\}x]\,w = 0, \tag{3.6.10}$$

biconfluent case

$$x(1-x)^3\,\frac{d^2w}{dx^2} + \left[-px + (1-x)^2\{c - 2(a+1)x\}\right]\frac{dw}{dx}$$
$$+ [\lambda + a(a+1)(1-x)\{x - ca\}]\,w = 0, \tag{3.6.11}$$

triconfluent case

$$(1-x)^4\,\frac{d^2w}{dx^2} + \left[-2px + p - 2(a+1)(1-x)^3\right]\frac{dw}{dx}$$
$$+ \left[\lambda - ap + a(a+1)(1-x)^2\right]w = 0, \tag{3.6.12}$$

and doubly confluent case

$$(1-x^2)^2\,\frac{d^2w}{dx^2} + \left[-8px + 2(1-x^2)\{c - (a+1)(1+x)\}\right]\frac{dw}{dx}$$
$$+ [4\lambda_d + \{a(a+1)(1+x) - 2ac\}(1+x) - 4pa]\,w = 0. \tag{3.6.13}$$

As was already mentioned, the transformations (3.6.6, 3.6.8) taken together we call *Jaffé transformations* in the singly confluent biconfluent, and triconfluent cases, and *Jaffé–Lay transformations* in the doubly confluent case. It was G. Jaffé who recognized in [56] its significance for the CTCP while calculating the spectrum of the ionized hydrogen molecule. The resulting equations (3.6.10–3.6.13) are equations in

Jaffé–Lay form. This form is characterized by the following properties:

(A) At zero there is either an ordinary point or a relevant regular singularity of the equation.

(B) At infinity there is always a regular singularity of the equation (possibly reducible).

(C) Precisely the two relevant singularities lie on or within the unit circle $|x| \leq 1$.

(D) If there is one relevant irregular singularity it lies at $x = 1$; if there are two they lie at $x = 1$ and at $x = -1$.

(E) If there is a relevant regular singularity it lies at $x = 0$.

The differential equations that emerge from Jaffé or Jaffé–Lay transformations in the confluent cases of Heun's equation have the form

$$P_4(x)\frac{d^2 w}{dx^2} + \sum_{i=0}^{3} \Gamma_i x^i \frac{dw}{dx} + \sum_{j=0}^{2} \Delta_j x^j w = 0.$$

$P_4(x)$ is a polynomial of order four, and Γ_i ($i = 0, 1, 2, 3$) and Δ_j ($j = 0, 1, 2, 3$) are parameters which are dependent on the parameters of the initial equation.

Each of the confluent cases (3.6.10)–(3.6.13) admit an expansion of the form

$$w(x) = \sum_{n=0}^{\infty} a_n x^n. \tag{3.6.14}$$

The coefficients a_n obey (irregular) difference equations of Poincaré–Perron type. This results from conditions A and D of the Jaffé–Lay form. The question arises as to what conditions must be imposed on the solutions of the difference equation in order for the function (3.6.14) to be the eigensolution. Formally, the behaviour of the coefficients a_n can be reconstructed from the formula inverse to (3.6.14):

$$a_n = \frac{1}{2\pi i} \int_{\Omega} x^{-n-1} w(x) \, dx.$$

for an appropriate contour Ω (see [10]).

3.6.3   Difference equations and Birkhoff sets

In the following, we list these difference equations for each confluent case.

Singly confluent case:

a_0 arbitrary,

$c\,a_1 + (\lambda - ac)\,a_0 = 0,$

$$\left(1 + \frac{\alpha_1}{n} + \frac{\beta_1}{n^2}\right) a_{n+1} + \left(-2 + \frac{\alpha_0}{n} + \frac{\beta_0}{n^2}\right) a_n$$

$$+ \left(1 + \frac{\alpha_{-1}}{n} + \frac{\beta_{-1}}{n^2}\right) a_{n-1} = 0 \quad \text{for } n \geq 1, \tag{3.6.15}$$

$$\alpha_1 := 1 + c, \qquad \beta_1 := c,$$
$$\alpha_0 := -p + d - c - 2a, \qquad \beta_0 := \lambda - ac,$$
$$\alpha_{-1} := 2a - d - 1, \qquad \beta_{-1} := a(a+1) - d(a-1) - 2a.$$

Biconfluent case:

a_0 arbitrary,

$$c\,a_1 + (\lambda - ca)\,a_0 = 0,$$

$$2(1+c)\,a_2 + (\lambda - ca - p - 2(1+a+c))\,a_1 + (a(a+1) + ca)\,a_0 = 0,$$

$$\left(1 + \frac{\alpha_1}{n} + \frac{\beta_1}{n^2}\right) a_{n+1} + \left(-3 + \frac{\alpha_0}{n} + \frac{\beta_0}{n^2}\right) a_n$$

$$+ \left(3 + \frac{\alpha_{-1}}{n} + \frac{\beta_{-1}}{n^2}\right) a_{n-1} + \left(-1 + \frac{\alpha_{-2}}{n} + \frac{\beta_{-2}}{n^2}\right) a_{n-2} = 0,$$

$$\text{for } n \geq 2, \tag{3.6.16}$$

$$\alpha_1 := 1 + c, \qquad \beta_1 := c,$$
$$\alpha_0 := -p - 2c - 2a + 1, \qquad \beta_0 := \lambda - ac,$$
$$\alpha_{-1} := c + 4a - 5, \qquad \beta_{-1} := a(a+1) + c(a-1) - 4a + 2,$$
$$\alpha_{-2} := -2a + 3, \qquad \beta_{-2} := -a(a+1) + 4a - 2.$$

Triconfluent case:

a_0 and a_1 arbitrary,

$$2\,a_2 + (p - 2(a+1))\,a_1 + ((\lambda - p\,a) + a(a+1))\,a_0 = 0,$$

$$6a_3 + (-8 + 2p - 4(a+1))\,a_2 + (-2p + 6(a+1) + (\lambda - p\,a)$$
$$+ a(a+1))\,a_1 - 2a(a+1)\,a_0 = 0,$$

$$\left(1 + \frac{\alpha_2}{n} + \frac{\beta_2}{n^2}\right) a_{n+2} + \left(-4 + \frac{\alpha_1}{n} + \frac{\beta_1}{n^2}\right) a_{n+1} + \left(6 + \frac{\alpha_0}{n} + \frac{\beta_0}{n^2}\right) a_n$$

$$+ \left(-4 + \frac{\alpha_{-1}}{n} + \frac{\beta_{-1}}{n^2}\right) a_{n-1} + \left(1 + \frac{\alpha_{-2}}{n} + \frac{\beta_{-2}}{n^2}\right) a_{n-2} = 0,$$

$$\text{for } n \geq 2, \tag{3.6.17}$$

$$\alpha_2 := 3, \qquad \beta_2 := 2,$$
$$\alpha_1 := p - 4 - 2(a+1), \qquad \beta_1 := p - 2(a+1),$$
$$\alpha_0 := -2p + 6a, \qquad \beta_0 := \lambda - pa + a(a+1),$$
$$\alpha_{-1} := -6(a+1) + 12, \qquad \beta_{-1} := 2(a+1)(3-a) - 8,$$
$$\alpha_{-2} := 2a - 3, \qquad \beta_{-2} := (a+1)(a-4) + 6.$$

Doubly confluent case:

a_0 and a_1 arbitrary,

$$2 a_2 + 2 (c - a - 1) a_1 + (4(\lambda - p a) + a (a + 1) - 2 c a) a_0 = 0,$$

$$6 a_3 + 4 (c - a - 1) a_2$$
$$\quad + \{-8 p - 2 (a + 1) + 4 (\lambda - p a) + a (a + 1) - 2 c a\} a_1$$
$$\quad + 2 (a (a + 1) - c a) a_0 = 0,$$

$$\left(1 + \frac{\alpha_2}{n} + \frac{\beta_2}{n^2}\right) a_{n+2} + \left(\frac{\alpha_1}{n} + \frac{\beta_1}{n^2}\right) a_{n+1} + \left(-2 + \frac{\alpha_0}{n} + \frac{\beta_0}{n^2}\right) a_n$$

$$+ \left(\frac{\alpha_{-1}}{n} + \frac{\beta_{-1}}{n^2}\right) a_{n-1} + \left(1 + \frac{\alpha_{-2}}{n} + \frac{\beta_{-2}}{n^2}\right) a_{n-2} = 0,$$

for $n \geq 2,$ $\hspace{6cm}$ (3.6.18)

$$\begin{aligned}
\alpha_2 &:= 3, & \beta_2 &:= 2, \\
\alpha_1 &:= 2(c - a - 1), & \beta_1 &:= 2(c - a - 1), \\
\alpha_0 &:= -8p - 2a, & \beta_0 &:= 4(\lambda - pa) + a(a + 1) - 2ca, \\
\alpha_{-1} &:= -2(c - a - 1), & \beta_{-1} &:= -2(a - 1)(c - a - 1), \\
\alpha_{-2} &:= 2a - 3, & \beta_{-2} &:= (a - 1)(a - 2).
\end{aligned}$$

In the following are given the characteristic equations of irregular difference equations of Poincaré–Perron type related to the studied differential equations.

Singly confluent case:

$$(t_s - 1)^2 = 0. \hspace{5cm} (3.6.19)$$

Biconfluent case:

$$(t_b - 1)^3 = 0. \hspace{5cm} (3.6.20)$$

Triconfluent case:

$$(t_t - 1)^4 = 0. \hspace{5cm} (3.6.21)$$

Doubly confluent case:

$$(t_d - 1)^2 (t_d + 1)^2 = 0. \hspace{4cm} (3.6.22)$$

As emerges out from Perron's publications [101]–[103], there is no simple theory of regular difference equations generalizable to irregular ones such that a conclusion could be drawn from the solutions of the characteristic equations as to the ratio of two consecutive terms of the underlying difference equation if it is of irregular type. Therefore we must apply the Birkhoff solutions introduced in Section 1.6: For each of these equations there exist formal solutions, the number of which coincides with the order of the equation. They represent linearly independent particular solutions of the corresponding difference equation asymptotically for $n \to \infty$ [16], [17], [18]. The totality of the Birkhoff solutions of a difference equation is called Birkhoff set [132].

In the following, we write the Birkhoff sets s_m for each of the equations (3.6.15)–(3.6.18).

Singly confluent case:

$$s^m(n) = \exp\left(\gamma_m n^{\frac{1}{2}}\right) n^{r_m}\left[1 + \frac{C_{m1}}{n^{\frac{1}{2}}} + \frac{C_{m2}}{n^{\frac{2}{2}}} + \cdots\right]$$

for $m = 1, 2$. (3.6.23)

Biconfluent case:

$$s^m(n) = \exp\left(\gamma_{m1} n^{\frac{2}{3}} + \gamma_{m2} n^{\frac{1}{3}}\right) n^{r_m}\left[1 + \frac{C_{m1}}{n^{\frac{1}{3}}} + \frac{C_{m2}}{n^{\frac{2}{3}}} + \cdots\right]$$

for $m = 1, 2, 3$. (3.6.24)

Triconfluent case:

$$s^m(n) = \exp\left(\gamma_{m1} n^{\frac{3}{4}} + \gamma_{m2} n^{\frac{1}{2}} + \gamma_{m3} n^{\frac{1}{4}}\right) n^{r_m}\left[1 + \frac{C_{m1}}{n^{\frac{1}{4}}} + \frac{C_{m2}}{n^{\frac{2}{4}}} + \cdots\right]$$

for $m = 1, 2, 3, 4$. (3.6.25)

Doubly confluent case:

$$s^m(n) = \varrho_m^n \exp\left(\gamma_m n^{\frac{1}{2}}\right) n^{r_m}\left[1 + \frac{C_{m1}}{n^{\frac{1}{2}}} + \frac{C_{m2}}{n^{\frac{2}{2}}} + \cdots\right]$$

for $m = 1, 2, 3, 4$. (3.6.26)

The factors in front of the brackets on the right-hand sides of (3.6.23)–(3.6.26) are called the asymptotic factors of the Birkhoff sets (cf. Section 1.6). In the following, we give the coefficients for the asymptotic factors of the Birkhoff sets (3.6.23)–(3.6.26). Singly confluent case:

$$\gamma_1 = 2\, p^{1/2},$$
$$\gamma_2 = -\gamma_1,$$
$$r_1 = r_2 = a - 1 - \frac{c+d}{2} =: r.$$ (3.6.27)

Biconfluent case:

$$\gamma_{m1} = \frac{3}{2} \exp\left(\frac{2\pi i m}{3}\right) p^{1/3},$$

$$\gamma_{m2} = -\frac{3}{2} \exp\left(\frac{4\pi i m}{3}\right) p^{2/3},$$

$$r_1 = r_2 = r_3 = \frac{c - 2a + 4}{3} =: r.$$ (3.6.28)

Triconfluent case:

$$\gamma_{m1} = \frac{4}{3} \exp\left(\frac{2\pi im}{4}\right) p^{1/4},$$

$$\gamma_{m2} = -\frac{1}{2} \exp\left(\frac{4\pi im}{4}\right) p^{1/2},$$

$$\gamma_{m3} = -\frac{19}{24} \exp\left(\frac{6\pi im}{4}\right) p^{3/4},$$

$$r_1 = r_2 = r_3 = r_4 = \frac{a-3}{2} =: r. \tag{3.6.29}$$

Doubly confluent case:

$$\varrho_m = +1 \quad \text{for } m = 1, 2,$$

$$\varrho_m = -1 \quad \text{for } m = 3, 4,$$

$$\gamma_{m1} = \exp(\pi im)(8p)^{1/2} \quad \text{for } m = 1, 2, 3, 4,$$

$$r_1 = r_2 = -1 + a - \frac{c}{2},$$

$$r_3 = r_4 = -2 - \frac{c}{2}. \tag{3.6.30}$$

As one can see from (3.6.27)–(3.6.30), the asymptotic factors of the Birkhoff sets depend only on the parameters a, c, d, p of the differential equations (3.6.1)–(3.6.4) but do not depend upon the accessory parameter λ.

As was explained above, one can see from the discussed paper by Birkhoff and Trjitzinsky that, for each of the difference equations (3.6.15)–(3.6.18) the related Birkhoff solutions represent a complete set of particular solutions for $n \to \infty$. Therefore we can write an asymptotic representation of the general solutions of the difference equations as a linear combination of their Birkhoff solutions:

$$a_n \sim \sum_{m=1}^{j} L_m s^m(n) \quad n \to \infty. \tag{3.6.31}$$

The upper limit of the sum in (3.6.31) is $j = 2$ in the singly confluent case, is $j = 3$ in the biconfluent case, and is $j = 4$ in the triconfluent and in the doubly confluent case. The coefficients L_m depend on the parameters of the differential equations (3.6.1)–(3.6.4) and on the initial data for the solutions, but do not depend on the index variable n.

The crucial point now is the fact that we can distinguish two characteristic behaviours of the solutions of the difference equations (3.6.15)–(3.6.18) for $n \to \infty$: According to the Birkhoff sets (3.6.23)–(3.6.26) and (3.6.27)–(3.6.30), we get—under the rather weak condition $p > 0$ (cf. (3.6.5))—exclusively exponentially increasing and exponentially decreasing solutions. The increasing solutions we call *dominant* and the decreasing ones we call *recessive* (cf. Section 1.6). The CTCP for the confluent cases of Heun's differential equation was formulated in Section 1.6 to be the problem of finding the values of the eigenvalue parameter $\lambda = \lambda_i$ of the differential equation for which the coefficients in front of the dominant solutions of the difference equations (3.6.15)–(3.6.18) in (3.6.31) vanish.

As one can see, the Birkhoff set (3.6.23) for the singly confluent case consists of one exponentially increasing and of one exponentially decreasing asymptotic solution. In the biconfluent case, the Birkhoff set (3.6.24) includes one exponentially increasing and two exponentially decreasing solutions. In the triconfluent case, the Birkhoff set (3.6.25) includes three exponentially increasing and one exponentially decreasing solutions. In the doubly confluent case, the Birkhoff set (3.6.26) includes two exponentially increasing and two exponentially decreasing solutions. The two increasing solutions are both maximum solutions.

3.6.4 The eigenvalue conditions

Our procedure of dealing with the CTCP for the confluent cases of Heun's differential equation results in irregular difference equations of Poincaré–Perron type, the order of which is the sum of the s-ranks of the irregular singularities in the underlying differential equation.

As we have seen in the preceding section, there are specific fundamental systems of the difference equations which are characterized by the index-asymptotic behaviour for $n \to \infty$. These specific systems of particular solutions may be used to formulate the eigenvalue condition.

As one can see from the asymptotic factors of the Birkhoff sets, in the singly confluent and in the biconfluent cases, the dominant solutions comprise only the maximum solution (as defined in Section 1.6) which will be denoted by $s^1(n)$ in (3.6.23)–(3.6.24). The other solutions are recessive ones. The eigenvalue conditions can be stated as follows (cf. Section 1.6). In these two confluent cases of Heun's differential equation, the CTCP is solved when the coefficient in front of maximum solution in (3.6.31) of the resulting difference equation vanishes, i.e. when

$$L_1(\lambda = \lambda_i; p, a, \ldots) = 0.$$

This is a result of our general consideration in Section 1.6. Additionally we state that, in the singly confluent case, one can prove this also by infinite continued-fraction methods. In the biconfluent case, it may be seen from Weierstrass's convergence criterion and Abel's limiting-value theorem.

In the triconfluent case, the dominant solutions contain three particular solutions of the difference equation: one that is real (the maximum solution) and the two complex conjugate ones that have purely imaginary parts to the first asymptotic order. However, as one can see, from the next order they also increase exponentially.

Suppose that we denote the maximum solution as $s^1(n)$ and the two oscillating solutions as $s^2(n)$ and $s^3(n)$. Then the exact eigenvalue condition in the triconfluent case is given by

$$L_1(\lambda = \lambda_i; p, a, \ldots) = L_2(\lambda = \lambda_i; p, a, \ldots) = L_3(\lambda = \lambda_i; p, a, \ldots) = 0$$

in (3.6.4). This can also be proved by applying Weierstrass's convergence criterion and Abel's limiting-value theorem.

The difference equation in the doubly confluent case is of fourth order. However, because of the specific character of the Jaffé–Lay transformation, we have two groups of dominant solutions, both of which consist of one maximum solution, denoted by $s^1(n)$ and $s^3(n)$ in (3.6.26) and (3.6.30). Thus the eigenvalue condition here is given by

$$L_1(\lambda = \lambda_i; p, a, c) = L_3(\lambda = \lambda_i; p, a, c) = 0. \qquad (3.6.32)$$

Obviously, in this case, we need a further parameter in order to meet these two conditions. This is given by the initial parameter a_1 in (3.6.18), which serves as an eigenvalue parameter if we would like to keep a_0 arbitrary for normalizing reasons.

3.6.5 Numerical aspects

The singly confluent, biconfluent, and triconfluent cases
In the singly confluent, biconfluent, and triconfluent cases, the numerical application of the formulae to get eigenvalues or eigenvalue curves is straightforward. One has to apply the recurrence relations (3.6.15)–(3.6.18) in a forward direction. The truncation must not be made before the asymptotic behaviour of the *maximum* solution of the difference equation has been revealed by the corresponding Birkhoff solution. That the maximum solution shows itself is always guaranteed by numerical instabilities, however small the coefficient L_1 in (3.6.31) may be. The last term, thereafter, has to be varied by varying the spectral parameter. The zeros of this variation are the eigenvalues of the CTCP. It is important to mention that the number of dominant solutions of the underlying difference equation does not have any impact on this procedure. In each case, all the dominant solutions are suppressed at once.

A reduction process in the doubly confluent case
The doubly confluent case is the only one where we got two eigenvalue conditions. It is possible to reduce these to only one at the numerical level, as done below.

We consider eqn (3.6.18) in the form

$$A_0(0)a_0 + A_1(0)a_1 + A_2(0)a_2 = 0,$$
$$A_{-1}(1)a_0 + A_0(1)a_1 + A_1(1)a_2 + A_2(1)a_3 = 0,$$
$$A_{-2}(n)a_{n-2} + A_{-1}(n)a_{n-1} + A_0(n)a_n + A_1(n)a_{n+1} + A_2(n)a_{n+2} = 0,$$
$$\text{for } n \geq 2. \qquad (3.6.33)$$

It can also be written in the form of a system of infinitely many linear equations:

$$A \cdot a = 0. \qquad (3.6.34)$$

The exact eigenvalue condition for the characteristic values may be written in the form

$$\det A = 0. \qquad (3.6.35)$$

The system (3.6.34) is truncated at a sufficiently large value number N of n, so that we get a system of $N + 1$ linear equations:

$$A^{(N)} \cdot a^{(N)} = 0 \tag{3.6.36}$$

with

$$A^{(N)} = \begin{bmatrix} A_0(0) & A_1(0) & A_2(0) & 0 & 0 \\ A_{-1}(1) & A_0(1) & A_1(1) & A_2(1) & 0 \\ A_{-2}(2) & A_{-1}(2) & A_0(2) & A_1(2) & A_2(2) \\ 0 & A_{-2}(3) & A_{-1}(3) & A_0(3) & A_1(3) \\ \cdots & \cdots & \cdots & \cdots & \cdots \\ & \cdots & 0 & A_{-2}(N-2) & A_{-1}(N-2) \\ & \cdots & 0 & 0 & A_{-2}(N-1) \\ & \cdots & 0 & 0 & 0 \end{bmatrix}$$

$$\begin{bmatrix} 0 & 0 & \cdots & & \\ 0 & 0 & \cdots & & \\ 0 & 0 & \cdots & & \\ A_2(3) & 0 & \cdots & & \\ \cdots & \cdots & \cdots & \cdots & \\ A_0(N-2) & A_1(N-2) & A_2(N-2) & 0 & 0 \\ A_{-1}(N-1) & A_0(N-1) & A_1(N-1) & A_2(N-1) & 0 \\ A_{-2}(N) & A_{-1}(N) & A_0(N) & A_1(N) & A_2(N) \end{bmatrix},$$

$$a(N) = \begin{bmatrix} a_0 \\ a_1 \\ a_2 \\ \cdot \\ \cdot \\ a_{N-1} \\ a_N \\ a_{N+1} \\ a_{N+2} \end{bmatrix} \tag{3.6.37}$$

Since we have two exponentially increasing Birkhoff solutions in (3.6.26), truncation means that we have to set $a_{N+1} = 0$ and $a_{N+2} = 0$ in the system (3.6.36)–(3.6.37). The number of exponentially decreasing Birkhoff solutions in (3.6.26) is also two. With the initial conditions

$$a_N^{(1)} = 1, \qquad a_{N-1}^{(1)} = 0,$$
$$a_N^{(2)} = 0, \qquad a_{N-1}^{(2)} = 1,$$

we get two linearly independent recessive solutions

$$a_n^{(1)} \text{ and } a_n^{(2)} \quad (n = N, N-1, \ldots 0, -1, -2)$$

by means of numerically stable backward recursions. The general recessive solution of (3.6.36)–(3.6.37) is then given by

$$a_n = K_1 a_n^{(1)} + K_2 a_n^{(2)}$$

with two arbitrary constants K_1 and K_2. The condition (3.6.35) is transformed by means of the truncation and the backward recursion process into

$$\det \begin{pmatrix} a_{-1}^{(1)} & a_{-1}^{(2)} \\ a_{-2}^{(1)} & a_{-2}^{(2)} \end{pmatrix} = 0$$

so that we eventually have to solve a two-times-two determinant in order to calculate the characteristic parameters of the CTCP of the doubly confluent case of Heun's differential equation.

The eigenvalue condition of the doubly confluent case consists in two conditions (3.6.32). This is a consequence of the fact that the two relevant singularities are located on the unit circle; thus two conditions have to be met—one at each of the two endpoints of the relevant interval: at $x = \pm 1$. This situation is compensated by the fact that the solutions are expanded about an ordinary point of the differential equation and not about a (regular) singularity as is done in the other confluent cases. Mathematically, this is expressed by the initial equations of the difference equation (3.6.33), where one can see that it is recursively solvable only when a_0 and a_1 are chosen; thus, there are two initial values. Fixing a_0 by normalizing, we see that we have one further eigenvalue parameter entering the problem in addition to the parameters of the differential equation. It was the aim of our numerical procedure, outlined above to reduce the number of eigenvalue parameters from two to one. This was achieved by the backward recursion procedure, from which we get a_1 (or, what is the same, the ratio a_1/a_0)—playing the role of this second eigenvalue parameter of the CTCP—as a result!

4

APPLICATION TO PHYSICAL SCIENCES

Having presented a theory of linear ordinary second-order differential equations generating the classical and the higher special functions in the preceding chapters, here we give several examples illustrating how the studied equations are involved in physical problems. This comprises their emergence from a separation of variables, central connection problems, confluence processes, and the calculation of exponentially small effects.

It should be mentioned that the examples show the main structural elements of the underlying theory: the singularity analysis, the two-dimensional classification scheme including linear transformations, specialization and confluence processes, the representation of the solutions, and the boundary-value problems.

The first section deals with several basic problems in atomic physics: the hydrogen atom as a trivial example of central two-point connection problems, the Stark effect as an example of exponentially small quantum processes, and the hydrogen-molecule ion, the eigenvalues of which may be calculated by means of the well-known infinite continued fraction method.

In the second section we give an example of coalescence of two non-elementary regular singularities that occurs in the theory of general relativity applied to astrophysics. While calculating the quasi-normal frequencies of black holes in the so-called Kerr limit (for high angular velocity) we encounter the phenomenon that two singularities come together and amalgamate. What is important to understand is that we do not have a confluence process here, since the s-rank of the resulting singularity doesn't change.

In the third section, we show the application of general methods to solid-state physics. First, we treat the famous line-tension model of classical dislocation movement as an example for a central two-point connection problem for Heun's equation. Then we proceed to quantum processes which are dealt with by means of Feynman's path-integral method. It is interesting there that, although the physical background is completely different than before, the mathematical method is analogous, although the differential equation is beyond the Heun equation. As a result, the reader may see here that the methods for treating two-point connection problems may well be generalized as long as the structure of the relevant singularities remains the same. Furthermore we give an example of how the continuous spectrum of a Schrödinger-type equation may be handled.

Double-well potentials are rather familiar in physical applications. The most important example in this respect is the quartic oscillator. In the fourth section, we explicitly give an exact calculation of the spectrum of the quantum quartic oscillator. It is well-known that, if the quartic potential has the form of a double well, then the effect of avoided crossings of eigenvalue curves appears, which is an exponentially small effect with respect to the barrier height. This means that ordinary perturbation theory, which is actually of power type, is not able to take into account this effect. As a consequence, avoided crossings are a test case for our method. The results are given graphically.

In the fifth section, we once again go beyond the theory presented before by showing a rather basic example of a lateral connection problem on the calculation (due to Hill) of the motion of the lunar node and perigee. There are several reasons for this: First, the reader may learn about the second basic type of connection problem, besides the central ones, along with the main ideas for its solution. Secondly, the physical background is interesting, even for those who don't want to go too deeply into the theoretical part. Thirdly, it takes some effort to get the original papers of Hill and go through all of this material.

Therefore we give a more detailed introduction to the background than in the case of the other examples. Moreover, we sketch the consideration that led Hill to introduce infinite determinants into mathematics and stress their application to asymptotic series. The last subsection is devoted to the infinite-determinant method of lateral-connection problems which was given by D. Schmidt [108] and is a modern development of Hill's seminal ideas.

The sixth section is devoted to the effect of suppressing quantum tunneling and the last section deals with a diffusion problem in electron microscopy.

4.1 Problems in atomic and molecular physics

4.1.1 The hydrogen atom

We start with the description of the most simple central two-point connection problem in physics which actually is the calculation of the spectrum of the hydrogen atom. According to the rules of quantum mechanics, we have to consider the Schrödinger equation with the potential $V \sim \frac{1}{r}$ where r is the absolute value of the space vector from the centre of the atom:

$$\Delta \psi + \frac{2m_e}{\hbar^2}\left[\varepsilon + \frac{e^2}{r}\right]\psi = 0.$$

Here m_e is the mass of the electron, and e the electric charge, $\hbar = h/2\pi$, where h is Planck's constant, and ε is the energy parameter.

One has to look for values of the energy parameter ε such that the solution ψ is square-integrable on $\mathbb{R}^3$. Separating the problem in spherical coordinates ϕ, θ, r according to

$$\psi_{lm}(\phi, \theta, r) = Y_{lm}(\phi, \theta) \cdot y_l(r),$$

where $Y_{lm}(\phi, \theta)$ are the spherical harmonics, and normalizing it according to $r \to z$ and $\varepsilon \to E$, yields for the radial functions y (omitting in the following the index) an equation of the form

$$\frac{d^2y}{dz^2} + \frac{2}{z}\frac{dy}{dz} + \left[-\frac{1}{4} + \frac{1/\sqrt{-2E}}{z} - \frac{l(l+1)}{z^2}\right]y = 0$$

$$\text{for } z \in \mathbb{C}, \quad E \in \mathbb{R}, \quad l \in \mathbb{N}. \tag{4.1.1}$$

The physically relevant interval here is $\{z | z \in \mathbb{R}, 0 \le z < \infty\}$. The differential equation is the singly confluent hypergeometric equation with a regular singularity at zero and an irregular singularity of s-rank two at infinity. The last statement may be seen from inverting (4.1.1) by means of $t = 1/z$, resulting in

$$\frac{d^2 y}{dt^2} + \left[-\frac{1/4}{t^4} + \frac{1/\sqrt{-2E}}{t^3} - \frac{l(l+1)}{t^2} \right] y = 0.$$

The *leading* asymptotic behaviour of solutions of (4.1.1) at infinity is given by

$$y_{1,\infty}(z) = \exp(-z/2), \tag{4.1.2}$$

$$y_{2,\infty}(z) = \exp(z/2), \tag{4.1.3}$$

and at zero is given by

$$y_{1,0}(z) = z^l, \tag{4.1.4}$$

$$y_{2,0}(z) = z^{-2l-1}. \tag{4.1.5}$$

The physically relevant solution has to behave like $y_{1,\infty}$, as $z \to \infty$, and also like $y_{1,0}$ as $z \to 0$. In both cases, we mean the asymptotic behaviour along the positive real axis.

The Frobenius solutions at zero are given by

$$y_{1,0}(z) = z^l \sum_{j=0}^{\infty} a_{1j} z^j, \qquad y_{2,0}(z) = z^{-2l-1} \sum_{j=0}^{\infty} a_{2j} z^j,$$

and the Thomé solutions at infinity are given by

$$y_{1,\infty}(z) = \exp(-z/2) z^{n-1} \sum_{j=0}^{\infty} b_{1j} z^j, \qquad y_{2,\infty}(z) = \exp(z/2) z^{-n-1} \sum_{j=0}^{\infty} b_{2j} z^j,$$

with $n = 1/\sqrt{-2E}$. Accordingly, matching the asymptotic behaviour for the connection problem with (4.1.2) and (4.1.4), we look for solutions of the form

$$v(z) = \exp(-z/2) z^l \sum_{j=0}^{\infty} c_j z^j,$$

yielding the first-order difference equation

$$c_{j+1} + f_j c_j = 0 \quad (j = 0, 1, 2, 3, \ldots) \tag{4.1.6}$$

with

$$f_j = \frac{n - l - 1 - j}{j[j + 2l + 3] + 2(l+1)}.$$

Thus

$$f_j \sim -\frac{1}{j} \quad \text{as } j \to \infty. \tag{4.1.7}$$

After having normalized the wave function by choosing a_0, eqn (4.1.6) may be solved recursively. It has two types of solutions, namely trivial and nontrivial ones. The trivial solutions are given by $c_j = 0$ for all $j \geq M$, with $M(E) \in \mathbb{N}_0$. The nontrivial solutions behave like $c_j \sim \frac{1}{j!}$ as $j \to \infty$. As a consequence, we have

$$\sum_{j=0}^{\infty} c_j z^j \sim \exp(z).$$

Thus we may state our final result:

> The eigenvalue condition for the hydrogen atom is the truncation criterion for the difference equation (4.1.6):
> $n - l = j + 1 =: k \in \mathbb{N}.$

As a consequence the eigenvalues of the hydrogen atom are essentially the natural numbers.

4.1.2   The Stark effect on hydrogen

The problem of calculating exponentially small effects is widespread in mathematical physics. Applications in which non-self-adjoint differential equations appear with complex eigenvalues include, for example, diffusion processes in non-equilibrium thermodynamics and quantum mechanics of quasistationary states. An example of the former problem will be given later, while an example of the latter problem in atomic physics is the Stark effect in hydrogen. It has been treated in several articles and books (see e. g. [38]). In the following, we adopt the notation of the approach which was given in [115].

Taking atomic units ($\hbar = e = m_e = 1$) and a Cartesian coordinate system centred in the atom (almost at the proton), the Schrödinger equation may be written as

$$\left\{ \Delta + 2\left[E - \left(F_z - \frac{1}{r} \right) \right] \right\} \psi(\mathbf{r}) = 0.$$

Here, F is the strength of the electric field which is directed along the z axis. Separating the problem in parabolic coordinates ξ, η, ϕ—which are connected to the Cartesian ones via

$$x = \sqrt{\xi \eta} \cos \phi, \qquad y = \sqrt{\xi \eta} \sin \phi, \qquad z = \frac{\xi - \eta}{2}$$

according to

$$\psi(\mathbf{r}) = \sqrt{\xi \eta} V(\xi) U(\eta) \exp(im\phi)$$

we get

$$\frac{d^2 V}{d\xi^2} + \left(\frac{E}{2} + \frac{\beta_1}{\xi} + \frac{F}{4}\xi + \frac{1 - m^2}{4\xi^2} \right) V(\xi) = 0,$$

$$\frac{d^2 U}{d\eta^2} + \left(\frac{E}{2} + \frac{\beta_2}{\eta} + \frac{F}{4}\eta + \frac{1 - m^2}{4\eta^2} \right) U(\eta) = 0. \qquad (4.1.8)$$

Here m is the magnetic quantum number, and the separation constants obey

$$\beta_1 + \beta_2 = 1.$$

As is shown in [115], by means of the transformations

$$\xi \to \xi(-2E/F),$$
$$\eta \to \xi(-2E/F),$$
$$p := (-2E)^{3/2}F^{-1},$$
$$\lambda_1 := \beta_1(-2E)^{-2}F,$$
$$\lambda_2 := \beta_2(-2E)^{-2}F,$$

eqns (4.1.8) admit the forms

$$\frac{d^2V}{d\xi^2} + \left[p^2\left(\frac{\lambda_1}{\xi} - \frac{1}{4} - \frac{\xi}{4}\right) + \frac{1-m^2}{4\xi^2}\right]V(\xi) = 0,$$

$$\frac{d^2U}{d\eta^2} + \left[p^2\left(\frac{\lambda_2}{\eta} - \frac{1}{4} + \frac{\eta}{4}\right) + \frac{1-m^2}{4\eta^2}\right]U(\eta) = 0$$

which are useful in doing parameter asymptotics.

As a result, we may state that dealing with the Stark effect of the hydrogen atom leads to two coupled reduced biconfluent Heun equations, the singularities of which are located at zero and at infinity. While the singularity at zero is regular, the s-rank of the one at infinity is 5/2. To calculate the energy levels is a two-parameter boundary–eigenvalue problem.

4.1.3 The hydrogen-molecule ion

The hydrogen-molecule ion [131] has nine degrees of freedom, namely the coordinates x, y, z of the electron, of mass m, and ξ_1, η_1, ζ_1, ξ_2, η_2, ζ_2 of the two nuclei, both of mass M. The Schrödinger equation of this system is given by

$$\frac{1}{m}\nabla^2_{xyz}\Psi + \frac{1}{M}\nabla^2_{\xi_1\eta_1\zeta_1}\Psi + \frac{1}{M}\nabla^2_{\xi_2\eta_2\zeta_2}\Psi + \frac{8\pi}{h^2}\left(E + \frac{e^2}{r_1} + \frac{e^2}{r_2} + \frac{e^2}{r_{12}}\right)\psi = 0$$

where r_1 and r_2 are the distances of the electron from the nuclei, r_{12} is the distance apart of the nuclei, and E is the energy.

Since m/M is small, we may apply the Born–Oppenheimer approximation [20], getting

$$\frac{1}{m}\nabla^2_{xyz}\Psi + \frac{8\pi}{h^2}\left(E + \frac{e^2}{r_1} + \frac{e^2}{r_2} + \frac{e^2}{r_{12}}\right)\psi = 0.$$

Using prolate spheroidal coordinates ξ, η, ϕ with

$$\xi = \frac{r_1 + r_2}{2c}, \qquad 1 \le \xi < \infty,$$

$$\eta = \frac{r_1 - r_2}{2c}, \qquad -1 \le \eta \le +1,$$

and ϕ, with $0 \le \phi \le 2\pi$, as the azimuth, where c is the distance parameter of the two centres. We get

$$\frac{\partial}{\partial \xi}\left\{ (\xi^2 - 1)\frac{\partial \psi}{\partial \xi} \right\} + \frac{\partial}{\partial \eta}\left\{ (1 - \eta^2)\frac{\partial \psi}{\partial \eta} \right\}\left\{ \frac{1}{\xi^2 - 1} + \frac{1}{1 - \eta^2} \right\}\frac{\partial^2 \psi}{\partial \phi^2}$$

$$+ \frac{8\pi^2 mc^2}{h^2}\left[E(\xi^2 - \eta^2) + \frac{2e^2}{c}\xi \right]\psi = 0. \tag{4.1.9}$$

Separating (4.1.9) in these coordinates according to

$$\psi(\xi, \eta, \phi) = \exp(\pm in\phi)\, X(\xi)\, Y(\eta),$$

we get

$$\frac{d}{d\eta}\left\{ (1 - \eta^2)\frac{dY}{d\eta} \right\} + \left\{ \lambda^2 \eta^2 - \frac{n^2}{1 - \eta^2} + \mu \right\}Y = 0 \quad \text{for } -1 \le \eta \le +1 \tag{4.1.10}$$

as the azimuthal equation and

$$\frac{d}{d\xi}\left\{ (1 - \xi^2)\frac{dX}{d\xi} \right\} + \left\{ \lambda^2 \xi^2 - \kappa\xi - \frac{n^2}{1 - \xi^2} + \mu \right\}X = 0 \quad \text{for } 1 \le \xi < \infty \tag{4.1.11}$$

as the radial equation with which we are concerned in the following. Both equations in (4.1.10) and in (4.1.11) are singly confluent Heun equations. Here, $n \in \mathbb{Z}$ is the angular momentum quantum number, e the electron charge, E is the electronic energy, μ is a constant of separation, and h is Planck's constant. $\lambda^2 = -\frac{8\pi^2 mc^2 E}{h^2}$ and $\kappa = \frac{16\pi^2 mce^2}{h^2}$ are composite constants.

In 1933 it was George Jaffé [56] who discovered—on following a hint from W. Pauli and F. Hund—the exact eigenvalue condition of the hydrogen molecule ion in quantum mechanics. In the following, we give an account of his approach, correcting the misprints which occurred in the article.

In order to solve the radial equation (4.1.11), we carry out what we call a Jaffé transformation:

$$u = \frac{\xi - 1}{\xi + 1}. \tag{4.1.12}$$

This transforms the singular points of (4.1.11) according to the following table:

$$
\begin{array}{ccccc}
\xi : & -\infty & -1 & +1 & +\infty \\
\downarrow & \downarrow & \downarrow & \downarrow & \downarrow \\
u : & 1 & \pm\infty & 0 & 1.
\end{array}
$$

Jaffé (see [56]; obviously there is a misprint in the original article) makes the substitution

$$
X(\xi) = \exp(-\lambda\xi)(\xi - 1)^{n/2}(\xi + 1)^{-\frac{n}{2}+p-1} \cdot y(\xi)
$$

with

$$
y(\xi) = (1 - u)^\rho w(u), \qquad \rho = n + 3 - p,
$$

where we have $p = \frac{\kappa}{2\lambda}$.

The differential equation for $w(u)$ is

$$
(1 - u)^2 u \frac{d^2 w}{du^2} - \{(n - 2\rho - 1)u^2 + (2\lambda + \rho + 1)2u
$$
$$
- (n + 1)\}\frac{dw}{du} + \{\rho(\rho - n)u - v\}w = 0
$$

with

$$
v := \mu + \lambda^2 + (1 - p)(n + 1 + 2\lambda) + 2\lambda n,
$$

which can be solved by means of a power series about the origin:

$$
w(u) = \sum_{m=0}^{\infty} a_m u^m. \tag{4.1.13}
$$

The coefficients $a_m (m = 0, 1, 2 \ldots)$ of these series obey a linear second-order difference equation of Poincaré–Perron type:

$$
a_{m+1} + u_m a_m - v_{m-1} a_{m-1} = 0 \quad (m = 0, 1, 2, \ldots) \tag{4.1.14}
$$

the coefficients u_m and v_m of which are given by

$$
u_m = -\frac{2m^2 + 2m(2\lambda + n + 1 - p) + v}{(m + 1)(m + n + 1)},
$$
$$
v_m = -\frac{(m - p)(m + n - p)}{(m + 1)(m + n + 1)}.
$$

Dividing (4.1.14) by a_m, we get a second-order algebraic equation for a_{m+1}/a_m, the solutions of which are

$$
\frac{a_{m+1}}{a_m} = 1 \pm \sqrt{\frac{4\lambda}{m}}, \tag{4.1.15}
$$

which actually shows the behaviour of the solutions of (4.1.14) for $m \to \infty$. The Gaussian convergence criterion (see [65] on page 297) tells us that (4.1.13) converges at $u = 1$ if and only if the solution of (4.1.14) behaves like

$$\frac{a_{m+1}}{a_m} = 1 - \sqrt{\frac{4\lambda}{m}}.$$

In this case Abel's limiting value theorem (see [65] on page 419) tells us that (4.1.13) represents a function that is continuous as long as one approaches unity from inside the unit circle, and

$$w(u = 1) = \sum_{m=0}^{\infty} a_m \qquad (4.1.16)$$

gives the value of this function there.

In order to fulfill (4.1.3), the parameters κ, λ, μ have to obey the continued-fraction relation

$$F := u_0 + \cfrac{v_0}{u_1 + \cfrac{v_1}{u_2 + \cfrac{v_2}{u_3 + \cfrac{v_3}{u_4 + \cdots}}}} = 0. \qquad (4.1.17)$$

This is a transcendental relation for the parameters κ, λ, μ, where λ is the spectral parameter. As is well known, there is a set of infinitely many triples $\lambda = \lambda_i(\kappa, \mu)$ ($i = 1, 2, 3 \ldots$) that obey (4.1.17), representing the spectrum of the hydrogen molecule ion.

4.2 Teukolsky equations in astrophysics

In 1963 it was Roy Patrick Kerr [63] who found a global stationary axially symmetrical two-parameter solution of Einstein's field equations of gravitation. Suppose that c is the velocity of light in a vacuum, G is the constant of gravitation, and M is the mass of the universe. If we normalize the equations such that $c = G = 2M = 1$, then the metric of this solution in Boyer–Lindquist coordinates (t, r, θ, ϕ) [21] is given by

$$ds^2 = (1 - r/\Sigma)dt^2 \cdot (2ar\sin^2(\theta)/\Sigma)dt d\phi - (\Sigma/\Delta)dr^2 - \Sigma d\theta^2 - \sin^2(\theta)$$
$$\times (r^2 \cdot a^2 + a^2 r \sin^2(\theta)/\Sigma)d\phi^2. \qquad (4.2.1)$$

Here a is the absolute value of the angular momentum, and we have the abbreviations

$$\Sigma = r^2 + a^2 \cos^2(\theta) \qquad (4.2.2)$$

and

$$\Delta = r^2 - r + a^2. \qquad (4.2.3)$$

This solution represents a model of a rotating gravitational singularity. In contrast to the mass parameter, the parameter of the angular momentum may be put into the metric by means of an appropriate normalization. This means that the model depends on *one* irreducible parameter: The properties of rotating gravitational singularities with different masses but the same value of angular momentum emerge from one another by means of scaling transformations.

Problems of stability of these solutions of Einstein's field equations against external perturbations, of the mathematical governing of scattering processes of gravitating or electrically charged particles at the singularity, and of characteristic eigenoscillations of the metric led at the end of the 1960s to the search for differential equations which describe perturbations of the Kerr metric. After preliminary work of several authors (see [9], [22], [28], [123]), it was Saul A. Teukolsky [127] who showed that the perturbations of the Kerr metric are described by two coupled linear ordinary second-order differential equations. These equations are singly confluent Heun equations; they nowadays bear Teukolsky's name. They play a central role in the concept of so-called *quasi-normal modes* (see e.g. [94]): these are quasi-stationary oscillations of the metric which behave as plane waves on asymptotically approaching the hole on the one side and the border of the universe (at infinity) on the other. They transport the energy reflectionlessly into the gravitational singularity and to infinity. To understand this one has to imagine that, at the time $t = 0$, there is a wave packet far from the centre such that, for $t \to +\infty$, it has the above-mentioned boundary behaviour. Such behaviour is possible only for certain values of the frequency of the plane waves. These frequencies are interpreted as eigenfrequencies of the Kerr metric, since they depend only on properties of the metric and not on the kind and the temporal process of the perturbation. They are damped oscillations for which the notion *quasi-normal modes* was coined in the literature. In our notation, the calculation of the quasi-normal modes of a black hole is nothing else than a central two-point connection problem for the singly confluent Heun equation, as discussed in the previous section.

Since the work of Teukolsky was published, there have been efforts to calculate the quasi-normal modes of rotating gravitational singularities by solving the system of the two Teukolsky equations under the above-mentioned boundary conditions [29], [30], [31], [35], [85], [86], [88]. Our investigations are based on a publication of Edvard W. Leaver [87] who, in the second half of the 1980s, has carried out extensive numerical calculations. As a result, he got some of the complex-valued frequencies of the quasi-normal modes.

Besides the mass of the gravitational singularity and the parameter which describes the frequency, there is only one further quantity a contained in the equations which describes the normalized angular momentum of the system. This parameter varies between zero (describing the stationary limiting case) and the maximum value $a = 1/2$ which is a limit given by the theory of general relativity. The solution of the field equations which belongs to vanishing angular momentum is spherically symmetrical. This so-called *Schwarzschild solution* was given by Karl Schwarzschild (see [31] on page 136) in 1915. The opposite limiting case for the maximum value of the angular momentum is called *Kerr limit*.

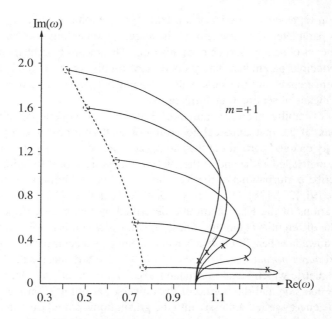

FIG. 4.1. Leaver's quasi-normal mode curves (sketched).

From what has been written above, the calculation of quasi-normal modes requires the minimum solutions of second-order difference equations of Poincaré–Perron type, which may be achieved either by continued-fraction methods or by the Birkhoff-set method. Teukolsky equations are not in a self-adjoint form; thus, for every quasi-normal mode, Leaver [87] gets a curve in the plane of complex-valued frequencies parameters. Moreover, since there is an additional parameter besides the energy, these curves are parametrized by the angular-momentum parameter a. However, while the limiting case of Schwarzschild may be calculated without any problems, the amount of numerical effort increases without bound if the parameter of the angular momentum approaches the upper limiting value. The reason about this is a coalescence of the two finite singularities in one of Teukolsky's equations. Leaver, however, without further dealing with this problem, follows the supposition published ten years before (see [35], [106], [123], [126]) that, in the Kerr limit, the frequencies of the quasi-normal modes degenerate to *real* values and are just the natural numbers after the application of appropriate normalization. It remained an open question whether the metric in this limiting case can ever undergo oscillations with finite amplitudes. If one wants to proceed here, one has first to perform the confluence process at the level of the differential equation, and thereafter solve a central two-point connection problem of the resulting differential equation. This will be done in the following. It should be emphasized that we do not apply parameter asymptotics here, but perform the confluence process, specify the parameters, and then solve a central two-point connection problem for the resulting differential equation. A first investigation of the parameter asymptotic has been given by Abramov *et al.* [3].

4.2.1 Rotating gravitational singularities

The field functions of gravitational perturbation of the Kerr metric may be calculated in Boyer–Lindquist coordinates [21] $[t, r, \theta, \phi]$ by means of the assumed factorization

$$\Phi(t, r, u, \phi) = \exp(-i\omega t)\, \exp(im\phi)\, X_{lm}(u)\, Y_{lm}(r) \qquad (4.2.4)$$

with $u = \cos(\theta)$. Here, the solutions of the Teukolsky equations are denoted X_{lm} und Y_{lm}.

The Teukolsky equation for $X_{lm}(u)$ is

$$\frac{d}{du}\left[(1 - u^2)\frac{dX_{lm}}{du}\right] + \left[a^2\omega^2 u^2 + 4a\omega u - 2 + A_{lm} - \frac{(m - 2u)^2}{1 - u^2}\right] X_{lm} = 0$$

$$(4.2.5)$$

$$\text{for } -1 \le u \le +1, \qquad (4.2.6)$$

while that for $Y_{lm}(r)$ is

$$\Delta\frac{d^2 Y_{lm}}{dr^2} - (2r - 1)\frac{dY_{lm}}{dr} + V(r)\, Y_{lm} = 0 \quad \text{for } 1/2 \le r < \infty, \qquad (4.2.7)$$

where we have used the abbreviation

$$V(r) = \frac{1}{\Delta}[(r^2 + a^2)^2\omega^2 - 2am\omega r + a^2 m^2 - 2i(am(2r - 1) - \omega(r^2 - a^2))]$$

$$+[-4i\omega r - a^2\omega^2 - A_{lm}]. \qquad (4.2.8)$$

The quantity m is a separation constant and is an integer, enumerating the eigenfunctions along with the quantity l. The quantities l, m, and a take on the following values:

$$l = 2, 3, \ldots, \quad l \in \mathbb{N},$$
$$m = -l, \ldots, +l, \quad m \in \mathbb{Z},$$
$$0 \le a \le \frac{1}{2}, \quad a \in \mathbb{R}.$$

One has to look for those values of the separation constant A_{lm} and the complex oscillation frequency ω for which the two differential equations (4.2.5) and (4.2.7) have a solution that meets the boundary conditions. For eqn (4.2.5), this is the regularity at the singularities $u = \pm 1$. For eqn (4.2.7), one has to look for solutions that transport energy towards the gravitational singularity as well as towards the point at infinity.

Equation (4.2.5) as well as eqn (4.2.7) is a singly confluent Heun equation and is a generalization of the spheroidal equation (see [92] page 221). The difference between these two equations consists in the different relevant intervals and in the way that the physical parameter a enters the coefficients.

While the azimuthal equation (4.2.5) in the limiting case $a = 1/2$ does not change its character, the radial equation (4.2.7) becomes a doubly confluent Heun equation by means of a confluence process:

$$\frac{d^2 Y_{lm}}{dz^2} - \frac{2}{z}\frac{dY_{lm}}{dz} + Q(z)\,Y_{lm} = 0 \qquad (4.2.9)$$

with

$$Q(z) = \frac{\frac{1}{4}(m-\omega)^2}{z^4} - \frac{(\omega+2i)(m-\omega)}{z^3}$$
$$+ \frac{\frac{7}{4}\omega^2 - A_{lm}}{z^2} + \frac{2\omega(\omega-2i)}{z} + \omega^2 \qquad (4.2.10)$$

and

$$z = r - \frac{1}{2}. \qquad (4.2.11)$$

If we put $\omega = m$, then eqn (4.2.9) degenerates to a singly confluent hypergeometric equation. This is a remarkable step: If we would like to trace the aforementioned confluence process under the restriction of the eigenconditions, then we should apply asymptotic methods. The result would be logarithmic terms in the series of the small parameter (see [3, 122]).

Treating the central two-point connection problem for this equation, we start with a substitution of the form

$$y = \exp(\nu z)z^\mu (z+1)^\alpha\, w \qquad (4.2.12)$$

for (4.2.9) and require that the factors in front of w should be asymptotic factors of Frobenius solutions and normal solutions, respectively, at the two singularities at $z = 0$ and at infinity. Then the equation for ν is given by

$$\nu^2 + \omega^2 = 0, \qquad (4.2.13)$$

yielding $\nu = \pm i\omega$. Since we are looking for waves which transport energy *into* the point at infinity, we have to choose the plus sign: $\nu = +i\omega$.

μ must be the characteristic exponent of a Frobenius solution at $z = 0$ and thus must obey the equation

$$\mu^2 - 3\mu + \frac{7}{4}\omega^2 - A_{lm} = 0. \qquad (4.2.14)$$

We get the two values

$$\mu = \frac{1}{2}(3 \pm k_{l\omega}) \qquad (4.2.15)$$

with

$$k_{l\omega} = \sqrt{9 - 7\omega^2 + 4A_{l\omega}}. \qquad (4.2.16)$$

Since we look for waves transporting energy *into* the point $z = 0$ we have to choose the minus sign:

$$\mu = \frac{1}{2}(3 - k_{l\omega}). \qquad (4.2.17)$$

The quantity α in (4.2.12) must be chosen in such a way that $z^{\mu+\alpha}$ is the power function of the asymptotic factor of the normal solution for $z \to \infty$ at the singularity located at infinity. This is the case for

$$\alpha = -\frac{D_{-1} + \nu G_{-1}}{G_0 + 2\nu} - \mu. \qquad (4.2.18)$$

If we substitute into the differential equation for w the expansion

$$w(x) = \sum_{n=0}^{\infty} a_n \left(\frac{z}{z+1}\right)^n, \qquad (4.2.19)$$

then the coefficients $\{a_n\}$ $(n \in \mathbb{N}_0)$ obey the following second-order difference equation of Poincaré–Perron type:

a_0 arbitrary,

$\Gamma_1 a_1 + \Delta_1 a_0 = 0$,

$$\left(1 + \frac{\alpha_1}{n} + \frac{\beta_1}{n^2}\right) a_{n+1} + \left(-2 + \frac{\alpha_0}{n} + \frac{\beta_0}{n^2}\right) a_n$$
$$+ \left(1 + \frac{\alpha_{-1}}{n} + \frac{\beta_{-1}}{n^2}\right) a_n = 0 \quad \text{for } n \geq 1. \qquad (4.2.20)$$

The Birkhoff set of this equation is given by

$$s_1(n) = \exp\left(qn^{\frac{1}{2}}\right) n^r \left[1 + \frac{C_{11}}{n^{\frac{1}{2}}} + \frac{C_{12}}{n^{\frac{2}{2}}} + \cdots\right],$$

$$s_2(n) = \exp\left(-qn^{\frac{1}{2}}\right) n^r \left[1 + \frac{C_{21}}{n^{\frac{1}{2}}} + \frac{C_{22}}{n^{\frac{2}{2}}} + \cdots\right], \qquad (4.2.21)$$

with

$$q = 2 \sqrt{-\sum_{i=-2}^{i=0} \alpha_i} = 2\sqrt{-2i\omega}$$

$$r = -\frac{1}{4}(2\alpha_1 - 2\alpha_{-1} - 1) = -\frac{11}{4} - i\omega. \qquad (4.2.22)$$

The general solution of eqn (4.2.20) for $n \to \infty$ may thus be written as

$$a_n \sim L_1 s_1(n) + L_2 s_2(n). \qquad (4.2.23)$$

Here, the coefficients $L_1 = L_1(\omega, A_{l\omega})$ and $L_2 = L_2(\omega, A_{l\omega})$ are independent of n.

Since the azimuthal equation (4.2.5) in the Kerr limit does not undergo a confluence process, it may be handled according to [87]: The expansion

$$S_{lm}(u) = \exp\left(\frac{1}{2}\omega u\right)(1+u)^{\frac{1}{2}|m+2|}(1-u)^{\frac{1}{2}|m-2|}\sum_{n=0}^{\infty}b_n(1+u)^n \qquad (4.2.24)$$

leads to a second-order difference equation of the form

$$b_0 \text{ arbitrary,}$$
$$\alpha_0^\theta b_1 + \beta_0^\theta b_0 = 0,$$
$$\alpha_n^\theta b_{n+1} + \beta_n^\theta b_n + \gamma_n^\theta b_{n-1} = 0 \quad \text{for } n = 1, 2, \ldots, \qquad (4.2.25)$$

with

$$\alpha_n^\theta = -2(n+1)(n+2k_1+1) \quad \text{for } n = 0, 1, 2, \ldots,$$
$$\beta_n^\theta = n(n-1) + 2n(k_1 + k_2 + 1 - \omega)$$
$$\qquad -[\omega(2k_1-1) - (k_1+k_2)(k_1+k_2+1)]$$
$$\qquad -\left[\frac{\omega^2}{4} + 2 + A_{l\omega}\right] \quad \text{for } n = 0, 1, 2, \ldots,$$
$$\gamma_n^\theta = \omega(n+k_1+k_2-2) \quad \text{for } n = 0, 1, 2, \ldots, \qquad (4.2.26)$$

and

$$k_1 = \frac{1}{2}|\omega+2|, \qquad k_2 = \frac{1}{2}|\omega-2|. \qquad (4.2.27)$$

The question of whether quasi-normal modes of rotating gravitational singularities are natural numbers may be answered as follows. One assigns to the parameter ω the value of the natural number in question. With this value, one calculates the parameter $A_{l\omega}$, for which the difference equation (4.2.25) with arbitrary initial conditions b_0 has only minimal solutions; thereby, one has to keep in mind that the only parameters $A_{l\omega}$ that have physical significance are those for which $k_{l\omega}^2$ is a negative real number. (Otherwise eqn (4.2.12) would lose its wave character.) For the pair of values $(\omega, A_{l\omega})$, one has to check by numerical means whether the quantities

$$L_1(\omega, A_{l\omega}) = \lim_{n \to \infty}\left[a_n \exp(-qn^{\frac{1}{2}})n^{-r}\right] \qquad (4.2.28)$$

are zero, where the values a_n $(n = 0, 1, \ldots)$ are calculated recursively from (4.2.20). If this the case, then we have found a quasi-normal mode—otherwise not.

The calculation of the values $A_{l\omega}$ from (4.2.25) may be done—according to Leaver [87]—by means of the continued-fraction condition

$$0 = \beta_0^\theta - \frac{\alpha_0^\theta \gamma_1^\theta}{\beta_1^\theta -} \frac{\alpha_1^\theta \gamma_2^\theta}{\beta_2^\theta -} \frac{\alpha_2^\theta \gamma_3^\theta}{\beta_3^\theta -} \cdots \qquad (4.2.29)$$

Below, we have listed some of the values $A_{l\omega}$ as they result from the continued-fraction condition (4.2.29): In Tables 4.2.1. and 4.2.2, we have listed the corresponding

Table 4.2.1. Values $A_{l\omega}$

	$l = 2$	$l = 3$	$l = 4$
$\omega = 0$	$A_{20} = 4$		
$\omega = 1$	$A_{21} = 3.1828319$	$A_{31} = 9.58$	$A_{41} = 17.71$
$\omega = 2$	$A_{22} = 0.543690$	$A_{32} = 8.25$	$A_{42} = 16.86$
$\omega = 3$		$A_{33} = 5.70$	$A_{43} = 15.39$
$\omega = 4$			$A_{44} = 13.04$

Table 4.2.2. Values $k_{l\omega}^2 = 4A_{l\omega} - 7\omega^2 + 9$

	$l = 2$	$l = 3$	$l = 4$
$\omega = 0$	$k_{20}^2 = 25$		
$\omega = 1$	$k_{21}^2 = 14.72$	$k_{31}^2 = 40.32$	$k_{41}^2 = 72.84$
$\omega = 2$	$k_{22}^2 = -16.84$	$k_{32}^2 = 14$	$k_{42}^2 = 48.44$
$\omega = 3$		$k_{33}^2 = -31.2$	$k_{43}^2 = 7.56$
$\omega = 4$			$k_{44}^2 = -50.84$

quantities $k_{l\omega}^2 = 4A_{l\omega} - 7\omega^2 + 9$. In analogy to the theory of spherical harmonics, one may denote the solutions of the azimuthal equation (4.2.5) for $\omega < l$ as *tesseral* and those for $\omega = l$ as *sectoral* (see e.g. [5] footnote on page 332). As one can see from Table 4.2.2., for $k_{l\omega}^2$, there is a degeneration in the Kerr limit: the physically relevant solutions of eqn (4.2.5) are exclusively of sectoral character.

Concerning the difference equation (4.2.20), we could not find a quasi-normal mode for any of the pairs of values $(\omega = 2, A_{22})$, $(\omega = 3, A_{33})$, $(\omega = 4, A_{44})$, $(\omega = 5, A_{55})$. Thus, according to these calculations, we cannot confirm the above-mentioned conjecture: In the Kerr limit, a gravitational singularity can no longer perform characteristic oscillations—at least not those with a non-vanishing amplitude.

From a general point of view, the example shows that differential equations with complex parameters may well have physical relevance in eigenvalue problems even when the eigenvalues are real numbers. If we compare our result with the eigenvalue curves published by Leaver in [87] on page 295, then one can see that confluence processes of singularities of linear ordinary differential equations may lead to discontinuities of the eigenvalue curves (cf. [76], [77], [84]).

4.3   Dislocation movement in crystalline materials

Plastic deformation of crystalline materials is caused mainly by the migration of dislocations. A crystal dislocation is a one-dimensional misfit of the periodic lattice. It is created under externally applied stresses, moves along so-called glide planes, and may be annihilated by colliding with another dislocation of opposite direction. The movement of a dislocation leads inevitably to an irreversible displacement of the crystal lattice in the glide plane.

In the following, we investigate the elementary migration step of a dislocation in a crystal by means of the so-called line-tension model. This is a one-dimensional nonlinear relativistic field-theoretical formulation of the problem, which is actually a compromise between mathematical solvability and the physical situation.

4.3.1 The line-tension model

We consider an infinitely long straight dislocation in a crystal and regard it as a massive elastic frictionlessly moving string subject to a line tension. The surrounding atoms give rise to a potential (called Peierls potential) along which the dislocation may move. A migration of the dislocation may be caused by an external stress, which in our model results in a slope of the Peierls potential. The dynamical displacement u of such a string is governed by the nonlinear partial differential equation

$$\gamma \frac{\partial^2 u}{\partial z^2} - m \frac{\partial^2 u}{\partial t^2} = \frac{dU}{du}, \quad u = u(z, \tau), \quad U = U\left(\frac{u}{c}\right). \tag{4.3.1}$$

Here c is the distance between the atoms (lattice parameter), γ is the line tension (determined by internal and material properties of the dislocation), and m is the mass density of the string. These three parameters are assumed to be constant. We have

$$U\left(\frac{u}{c}\right) = U_P\left(\frac{u}{c}\right) - b\sigma \frac{u}{c}, \tag{4.3.2}$$

where b is the magnitude of the Burgers vector (i.e. the strength of the dislocation, which may be set at unity without loss of generality) and σ is the magnitude of the component of the externally applied stress acting in the glide plane. U_P, the Peierls potential, is periodic, bounded, and at least differentiable with respect to the displacement u. Our coordinate frame is located such that

$$U(0) = 0, \quad \left.\frac{dU}{du}\right|_{u=0} = 0, \quad \left.\frac{d^2 U}{du^2}\right|_{u=0} \geq 0. \tag{4.3.3}$$

The last of these conditions restricts the external stress σ in the glide plane to a maximum value σ_P (called Peierls stress) at which we have

$$\frac{d^2 U}{du^2} = 0. \tag{4.3.4}$$

By means of appropriate scaling transformations of the dependent and the independent variables, we incorporate the physical parameters into normalized variables and get instead of (4.3.1) the one-dimensional nonlinear wave equation

$$\frac{\partial^2 \chi}{\partial x^2} - \frac{\partial^2 \chi}{\partial \tau^2} = F(\chi) - s \tag{4.3.5}$$

with $F(\chi) = dU/d\chi$. Since the stress σ is confined to the interval $[0, \sigma_P]$, we can normalize s to vary between $s = 0$ and $s = 1$.

It is obvious that

$$\chi \equiv 0 \qquad (4.3.6)$$

is a solution of (4.3.5). Due to the periodicity of $F(\chi)$, we also have trivial solutions of the form

$$\chi \equiv \chi_i \quad (i = 2, 4, 6 \ldots) \qquad (4.3.7)$$

at $\chi = \chi_i$, where

$$F(\chi_i) = 0, \qquad \left. \frac{dF}{d\chi} \right|_{\chi = \chi_i} \geq 0. \qquad (4.3.8)$$

These solutions model a static, straight, and stable configuration of the dislocation.

There are also straight but unstable configurations as trivial solutions of (4.3.5), namely at $\chi = \chi_i$ $(1, 3, 5 \ldots)$, where

$$F(\chi_i) = 0, \qquad \left. \frac{dF}{d\chi} \right|_{\chi = \chi_i} \leq 0. \qquad (4.3.9)$$

It is an important step in the understanding of the line-tension model to realize that, although there are no other stable solutions of (4.3.5) beyond the trivial ones, there are unstable but no longer straight configurations, among which the so-called kinkpair configuration is the important one in the following. It is created from the straight configuration under the thermodynamic action of the vibrating atoms of the crystal. The actual form of this configuration depends upon the value of s. Thus we have to investigate the static solutions of (4.3.5) in general.

The nontrivial static solutions $\chi = \chi_0$ of (4.3.1) are given by the integral

$$x - x_0 = \int^{\chi_0} \frac{d\chi'}{\sqrt{2[U(\chi') + C]}}, \qquad (4.3.10)$$

where x_0 and C are constants of integration. Since $U(\chi)$ is dependent on s, (4.3.10) is a function of the pair of parameters (s, C). We deal with real solutions of (4.3.10) and therefore have to require that the integrand in (4.3.10) remains real. This restricts the values of C such that there exist two functions $C = C_{\min}(s)$ and $C = C_{\max}(s)$ for $0 \leq s \leq 1$ within which C is to be confined.

To get the qualitative behaviour of the static solutions χ_0 coming out of (4.3.10), we have to look at the zeros of $U(\chi) + C$. In the case when this function has a zero of order two, the asymptotic behaviour of $\chi(x)$ for $x \to \pm\infty$ is exponential. This is the case when $C = C_{\max}$ holds. In all the other cases it is algebraic. In particular, if the zero is of order one, we get $\chi_0(x) \sim (x - x_0)^2$ in the neighbourhood of the zero, and therefore a periodic static solution $\chi_0(x)$. Since we are interested in the migration of a dislocation from one straight stable configuration to a neighbouring one through

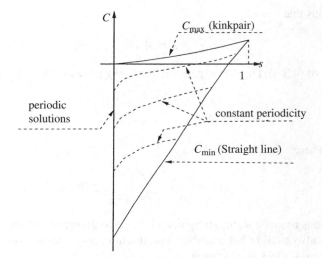

FIG. 4.2. The $C - s$ diagram.

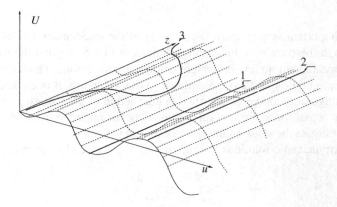

FIG. 4.3. Stable straight configuration (1); unstable configurations: straight line (2) and kinkpair (3).

an unstable nontrivial one, it is clear now that we have to choose $C = C_{\max}$ and have to disregard all the other values. We get a static solution which tends to the trivial stable one exponentially fast as $x \to \pm\infty$ and exhibits a hump in between. It is the above-mentioned kinkpair $\chi = \chi_{kp}(x)$ and represents a saddle ridge or nucleation configuration of the dislocation when migrating from one Peierls valley to a lower-lying neighbouring one.

For not too low temperatures, the vibrating atoms in the neighbourhood of the dislocation act as a thermodynamical heat bath for the nucleation process of the kinkpairs. The main macroscopic quantity in this field is the rate of plastic deformation of the crystal under an applied stress. Under conditions when the line-tension model is applicable, this is dependent only upon the nucleation rate of kinkpairs, which itself is determined by the vibration spectra of the straight stable configuration and the kinkpair configuration. For more physical details see [36], [73], [75], [90], [100]. While the spectrum of the straight configuration is trivial to calculate, this is by no means so for the kinkpair, since it actually is a central two-point connection problem for Heun's differential equation. This will be treated in the following.

In order to calculate the spectrum of the kinkpair, we have to linearize (4.3.5) by means of the separation

$$\chi(x, \tau) = \chi_{kp}(x) + \chi_1(x, \tau). \tag{4.3.11}$$

Assuming $|\chi| \ll 1$, we get

$$\frac{\partial^2 \chi_1}{\partial x^2} - \frac{\partial^2 \chi_1}{\partial \tau^2} = \chi_1 \frac{dF(\chi_{kp})}{d\chi_{kp}}. \tag{4.3.12}$$

By means of the separation

$$\chi_1(x, \tau) = \phi(x) \exp(i\kappa\tau), \tag{4.3.13}$$

eqn (4.3.12) can be reduced to the ordinary differential equation

$$\frac{d^2\phi}{dx^2} + \left[\kappa^2 - \frac{dF(\chi_{kp})}{d\chi_{kp}}\right]\phi = 0. \tag{4.3.14}$$

Since we deal with the elementary migration step of a dislocation from one Peierls valley to the neighbouring one, it is sufficient to consider a double-well potential (e.g. a fourth-order polynomial) instead of a periodic one (called Eshelby potential). In this case, eqn (4.3.14) becomes of Schrödinger type with the following double-well potential (renormalized variables):

$$-V(x) \sim \operatorname{sech}^2(x + x_0) + \operatorname{sech}^2(x - x_0).$$

We look for bounded solutions of (4.3.14) which, for $|x| \to \infty$, either vanish or behave periodically in this limit. The former solutions correspond to discrete

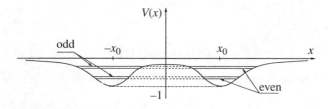

FIG. 4.4. Potential of Schrödinger-type equation in physical coordinates.

eigenvalues κ_i, and the latter ones to a continuous spectrum of a two-parameter eigenvalue problem. Introducing χ_{kp} in (4.3.14) as independent variable transforms it to

$$\frac{d^2\phi}{d\chi_{kp}^2} + \frac{d\ln(d\chi_{kp}/dx)}{d\chi_{kp}} \frac{d\phi}{d\chi_{kp}} + \frac{\kappa^2 + dF(\chi_{kp})/d\chi_{kp}}{(d\chi_{kp}/dx)^2}\phi = 0. \qquad (4.3.15)$$

For the Eshelby potential, (4.3.15) becomes a Fuchsian differential equation with four singularities i.e. Heun's equation, and s plays the role of an accessory parameter.

4.3.2 Differential equations

In the following, we go more deeply into the mathematical treatment, explicitly setting out the calculations that lead to the Schrödinger as well as to the Heun equation.

We start with the dynamical equation of an infinitely long elastic string:

$$\gamma \frac{d^2u}{dz^2} - m \frac{d^2u}{dt^2} = \frac{dU}{du} \qquad (4.3.16)$$

Here, γ is the line tension, m the mass per unit length, u is the displacement of the string, and U the potential; z represents the space coordinate and t the time coordinate.

Suppose that there is a static solution $u_s(z)$ of (4.3.16). Then formally we write

$$u(z, t) = u_s(z) + u_d(z, t),$$

getting the differential equation for the static solution $u_s(z)$:

$$\gamma \frac{d^2u_s}{dz^2} = \frac{dU}{du_s}. \qquad (4.3.17)$$

A Taylor expansion of the potential U about u_s yields

$$\frac{dU}{du} = \frac{dU}{du_s} + u_d \frac{d^2U}{du_s^2},$$

and by means of a linearization we eventually get

$$\gamma \frac{d^2u_d}{dz^2} - m \frac{d^2u_d}{dt^2} = u_d \frac{d^2U}{du_s^2}.$$

In this equation, the independent variables may by separated according to

$$u_d(z, t) = \phi(z)\,\psi(t).$$

This leads to

$$\psi(t) = \exp\left(i\,kt\right)$$

as well as to a Schrödinger-type differential equation for ϕ:

$$\frac{d^2\phi}{dz^2} + \left(\frac{m\,k^2}{\gamma} - \frac{1}{\gamma}\frac{d^2U}{du_s^2}\right)\phi = 0. \qquad (4.3.18)$$

Transformation of the independent variable from z to u_s with the help of

$$\frac{d\phi}{dz} = \frac{d\phi}{du_s}\frac{du_s}{dz}$$

and

$$\frac{d^2\phi}{dz^2} = \frac{d^2\phi}{du_s^2}\left(\frac{du_s}{dz}\right)^2 + \frac{d\phi}{du_s}\frac{d^2u_s}{dz^2},$$

and the integration of the static equation according to

$$\frac{d^2u_s}{dz^2} = \frac{1}{\gamma}\frac{dU}{du_s} \quad \rightarrow \quad \left(\frac{du_s}{dz}\right)^2 = \frac{2}{\gamma}(U(u_s) + C), \tag{4.3.19}$$

leads to

$$\frac{d^2\phi}{du_s^2} + \frac{1}{2}\frac{\frac{dU}{du_s}}{U(u_s) + C}\frac{d\phi}{du_s} + \frac{1}{2}\frac{mk^2 - \frac{d^2U}{du_s^2}}{U(u_s) + C}\phi = 0. \tag{4.3.20}$$

Taking $U(u)$ as an Eshelby potential, i.e.

$$U(u) = P(u) - \sigma u, \tag{4.3.21}$$

with

$$P(u) = 16\,U_0\left(\frac{u}{c}\right)^2\left(\frac{u}{c} - 1\right)^2, \tag{4.3.22}$$

transforms the Schrödinger equation (4.3.18) into (for details see 4.3.4)

$$\frac{d^2\phi}{dx^2} + \frac{1}{4}\beta^2\left\{4\left(\frac{\omega^2}{\beta^2} - 1\right) + 6\,\mathrm{sech}^2\left[\frac{1}{2}\beta(x + x_0)\right]\right.$$
$$\left. + 6\,\mathrm{sech}^2\left[\frac{1}{2}\beta(x - x_0)\right]\right\}\phi = 0 \tag{4.3.23}$$

with the normalizations

$$x = \sqrt{\frac{32\,U_0}{\gamma\,c^2}}\,z,$$

$$\omega^2 = \frac{k^2\,m\,c^2}{32\,U_0},$$

$$\beta = \sqrt{1 - 6\phi^* + 6\phi^{*2}},$$

$$x_0 = \frac{1}{\beta}\ln\left\{\frac{1 - 2\phi^* + \beta}{\sqrt{2\phi^*(1 - \phi^*)}}\right\},$$

$$\phi^* = \frac{1}{2} - \frac{1}{\sqrt{3}}\cos\left\{\frac{1}{3}[\pi - \arccos s]\right\}.$$

Moreover, we have

$$s = \frac{\sigma}{\sigma_{max}} \quad \rightarrow \quad 0 \le s \le 1$$

with σ_{max} obtained from

$$\left. \frac{d^2 P}{du^2} \right|_{u_{max}} = 0,$$

yielding u_{max}, and

$$\left. \frac{dP}{du} \right|_{u_{max}} = \sigma_{max}.$$

Equation (4.3.20) is Fuchsian, and with (4.3.21) and (4.3.22) it has four singularities, i. e. it becomes a Heun equation

$$\frac{d^2 \phi}{dv^2} + \left\{ \frac{1}{v} + \frac{\frac{1}{2}}{v-a} + \frac{\frac{1}{2}}{v-1} \right\} \frac{d\phi}{dv} + \frac{1}{v(v-a)(v-1)}$$
$$\times \left\{ -\frac{\rho^2 a}{v} - 6v + 3(1+a) \right\} \phi = 0$$

with

$$\rho = \sqrt{1 - \frac{\omega^2}{\beta^2}} = \sqrt{-\frac{k^2}{4}},$$

$$v = \frac{\phi - \phi^*}{\tilde{\phi} - \phi^*},$$

$$\tilde{\phi} = 1 - \phi^* + \sqrt{2\phi^*(1 - \phi^*)},$$

$$a = \frac{1 - 2\phi^* - \sqrt{2\phi^*(1 - \phi^*)}}{1 - 2\phi^* + \sqrt{2\phi^*(1 - \phi^*)}} = \mathrm{th}^2 \left(\frac{\beta}{2} x_0 \right), \qquad 0 \le a \le 1.$$

4.3.3 Static solutions

From eqn (4.3.19) we get the static solution, explicitly

$$z - z_0 = \int_{u_0}^{u_s} \frac{du_s}{\sqrt{\frac{2}{\gamma} U(u_s) + C(\sigma)}},$$

which yields

$$\phi_s = \frac{u_s}{c} = \frac{d \, \mathrm{sn}(\alpha z; k) + e}{f \, \mathrm{sn}(\alpha z; k) + g};$$

here d, e, f, g are arbitrary constants, and 'sn' is a Jacobian elliptic function (in the standard notation). Choosing the constant $C(\sigma)$ according to

$$C(\sigma) = -\left[16\,U_0\phi^{*2}(\phi^* - 1)^2 - c\,\sigma\,\phi^*\right]$$

with

$$\phi^* = \frac{u^*}{c},$$

we finally get as the static solution:

$$\phi_s(z) = u^* + \frac{1 - 6\,\phi^* + 6\,\phi^{*2}}{\sqrt{2\,\phi^*(1 - \phi^*)}\,\mathrm{ch}\left[\dfrac{4\,(z - z_0)}{w}\right] + 1 - 2\,\phi^*},$$

where

$$w = \sqrt{\frac{\gamma}{2U_0}}\,\frac{c}{\beta}$$

and z_0 is arbitrary.

4.3.4 Explicit calculations

We have not yet shown the explicit calculations of the Schrödinger equation (4.3.23) from (4.3.18). We give the main lines of this quite lengthy calculation.

Starting with

$$\frac{\mathrm{d}^2 U}{\mathrm{d}u_s^2} = 32\,\frac{U_0}{c^2}\left\{6\left(\frac{u_s}{c}\right)^2 - 6\,\frac{u_s}{c} + 1\right\},$$

normalizing by

$$\phi = \frac{u}{c},$$

and using the abbreviations

$$\phi_s = \phi^* + \frac{\beta^2}{Z},\, u_s = c\phi_s$$

and

$$Z := \sqrt{2\,\phi^*\,(1 - \phi^*)}\,\mathrm{ch}(\beta x) + 1 - 2\phi^*,$$

we get

$$\frac{\mathrm{d}^2\phi}{\mathrm{d}x^2} + \left[\omega^2 - 1 - \beta^2\,V(x)\right]\phi = 0$$

with

$$V(\phi_s) = 6(\phi_s^2 - \phi_s).$$

The crucial step now is to calculate $V(x)$ from $V(\phi_s)$. Omitting the rather long and boring auxiliary calculations, the main steps are

$$\phi_s^2 - \phi_s = \phi^{*2} - \phi^* + \beta^2 \frac{(2\phi^* - 1)Z + \beta^2}{Z^2}$$

$$= \phi^{*2} - \phi^* - \beta^2 \frac{d\,\mathrm{ch}(\beta x) + 1}{\left[\mathrm{ch}(\beta x) + a\right]^2}$$

$$= \phi^{*2} - \phi^* - \beta^2 \frac{1}{2} \frac{\mathrm{ch}^2\left(\frac{1}{2}\beta x_s\right)\mathrm{ch}^2\left(\frac{1}{2}\beta x\right) + \mathrm{sh}^2\left(\frac{1}{2}\beta x_s\right)\mathrm{sh}^2\left(\frac{1}{2}\beta x\right)}{\left[\mathrm{ch}^2\left(\frac{1}{2}\beta x_s\right) + \mathrm{sh}^2\left(\frac{1}{2}\beta x\right)\right]^2}$$

$$= \phi^{*2} - \phi^* - \frac{1}{4}\beta^2 \frac{\mathrm{ch}^2\left(\frac{1}{2}\beta(x - x_s)\right) + \mathrm{ch}^2\left(\frac{1}{2}\beta(x + x_s)\right)}{\mathrm{ch}^2\left(\frac{1}{2}\beta(x - x_s)\right) \cdot \mathrm{ch}^2\left(\frac{1}{2}\beta(x + x_s)\right)}.$$

From this calculation we get

$$V(x) = 6\phi^{*2} - 6\phi^* - \frac{3}{2}\beta^2 \left\{ \mathrm{sech}^2\left[\frac{1}{2}\beta(x + x_0)\right] + \mathrm{sech}^2\left[\frac{1}{2}\beta(x - x_0)\right] \right\}$$

and eventually the final result

$$\omega^2 - 1 + \beta^2 V(x) = \frac{1}{4}\beta^2 \left\{ 4\left[\frac{\omega^2}{\beta^2} - 1\right] + 6\,\mathrm{sech}^2\left[\frac{1}{2}\beta(x + x_0)\right] \right.$$

$$\left. + 6\,\mathrm{sech}^2\left[\frac{1}{2}\beta(x - x_0)\right] \right\}.$$

As a result, the crucial mathematical problem in the theory of plastic deformation of crystals within the line-tension model, namely the calculation of the eigenfrequencies of the saddle-ridge (i.e. kinkpair) configuration of the dislocation, is nothing else than a central two-point connection problem of Heun's differential equation, which is considered in the following. Equation (4.3.15) becomes

$$\frac{d^2\phi}{dv^2} + \left\{ \frac{2}{v} + \frac{1}{v - a} + \frac{1}{v - 1} \right\} \frac{d\phi}{dv}$$

$$+ \frac{1}{v(v - a)(v - 1)} \left\{ -\frac{\rho^2 a}{v} - 6v + 3(1 + a) \right\} \phi = 0 \quad \text{for } v \in \mathbb{C}$$

$$(4.3.24)$$

with $\rho = \rho(\omega^2, s)$, $\omega = \omega(\kappa)$, $v = v(\phi, s)$, and $a = a(s)$, for $0 \le s \le 1$. The physically relevant interval is $0 \le v \le 1$, and a varies between $a = 0$ and $a = 1$

along the real axis when s varies between $s = 1$ and $s = 0$. The generalized Riemann P-symbol of (4.3.24) is given by

$$P = \begin{pmatrix} 0 & a & 1 & \infty & \\ +\rho & 0 & 0 & 3 & v \quad 3(1+a) \\ -\rho & \frac{1}{2} & \frac{1}{2} & -2 & \end{pmatrix}. \tag{4.3.25}$$

As was shown above, the spectrum of (4.3.25) determines the plastic deformation rate of a crystal under an applied stress according to the line-tension model. As we will see below, the spectrum of (4.3.25) consists of discrete eigenvalues and of a non-discrete part. We first display a method to calculate the discrete part.

4.3.5 The discrete spectrum

First we have to note that, according to the symmetry of the kinkpair, eqn (4.3.24) has even and odd eigenvalues with respect to $v = a$: The even ones belong to $\rho = 0$ and the odd ones to $\rho = 1/2$ according to (4.3.25). We are looking for solutions that tend to zero as $|x| \to \infty$, i.e. as $v \to 0$. In terms of eqn (4.3.25), this means that we are looking for functions $\rho = \rho_i(a)$, $(i = 1, 2, \dots)$ such that the corresponding solutions behave like

$$\phi(v) = \sum_{n=0}^{\infty} a_n v^{(n+\rho)} \tag{4.3.26}$$

at $v = 0$, and also like

$$\phi(v) = \sum_{n=0}^{\infty} b_n (v-a)^n \tag{4.3.27}$$

at $v = a$ in the even case or like

$$\phi(v) = \sum_{n=0}^{\infty} b_n (v-a)^{n+1/2} \tag{4.3.28}$$

at $v = a$ in the odd case. This is a two-parameter central two-point connection problem of Heun's differential equation that can be solved easily. First we mention that the even and the odd cases can be treated by the same method. We only have to apply the s-homotopic transformation

$$\phi(v) = (v-a)^{-1/2} \tilde{\phi}(v) \tag{4.3.29}$$

in the odd case to (4.3.25) such that we get an equation for $\tilde{\phi}$ having the P-symbol

$$P = \begin{pmatrix} 0 & a & 1 & \infty & \\ +\rho & -1/2 & 0 & 7/2 & v \quad 3(1+a) \\ -\rho & 0 & \frac{1}{2} & -3/2 & \end{pmatrix}, \tag{4.3.30}$$

and then continue as in the even case, which will be considered now. It is appropriate, instead of an equation related to (4.3.30), to transform to an equation that has a GRS according to

$$P = \begin{pmatrix} 0 & a & 1 & \infty \\ 0 & 0 & 0 & \rho+4 & v & 3(1+a) \\ -2\rho & -\frac{1}{2} & -\frac{1}{2} & \rho-1 \end{pmatrix}. \qquad (4.3.31)$$

Since it is not crucial, whether the singularity at $v = a$ varies between $v = 0$ and $v = 1$ or between $v = 1$ and $v = \infty$, it is elucidating to carry out the transformation

$$\xi = \frac{1-a}{1+a},$$

calculating the discrete eigenvalues in dependence on ξ.

An expansion in the form (4.3.26) leads to a second-order regular difference equation of Poincaré–Perron type for the coefficients a_n:

$$\begin{aligned} a_1 f_1 + a_0 f_0 &= 0, \\ f_1(n)\, a_{n+1} + f_0(n)\, a_n + f_{-1}(n)\, a_{n-1} &= 0 \quad \text{for } n \geq 1, \qquad (4.3.32) \end{aligned}$$

with

$$\begin{aligned} f_1(n) &= \left(a + \frac{\alpha_1}{n} + \frac{\beta_1}{n^2} \right), \\ f_0(n) &= \left((1+a) + \frac{\alpha_0}{n} + \frac{\beta_0}{n^2} \right), \\ f_{-1}(n) &= \left(a + \frac{\alpha_{-1}}{n} + \frac{\beta_{-1}}{n^2} \right), \end{aligned}$$

where α_i and β_i ($i = -1, 0, 1$) are functions of a.

The eigenvalue condition is that the solutions (4.3.26) of (4.3.25) are holomorphic at $z = a$. Since $z = a$ is a regular singularity, this means that the radius of convergence of the series (4.3.26) extend from $\mathcal{R} = a$ in general to $\mathcal{R} = 1$ in the case when ρ becomes an eigenvalue. This causes a sudden change in the asymptotic behaviour of a_n for $n \to \infty$. There is a general procedure to get these special values. For this, we write down the infinite continued fraction

$$F(\rho, a) = f_0 - \cfrac{f_1 f_{-1}(1)}{f_0(1) + \cfrac{f_1(1) f_{-1}(2)}{f_0(2) + \cfrac{f_1(2) f_{-1}(3)}{f_0(3) + \cfrac{f_1(3) f_{-1}(4)}{f_0(4) + \cdots}}}} = 0. \qquad (4.3.33)$$

As can be shown [40], if and only if we have an eigenvalue $\rho = \rho_i(a)$ of the central two-point connection problem is the continued fraction zero and thus $F = 0$.

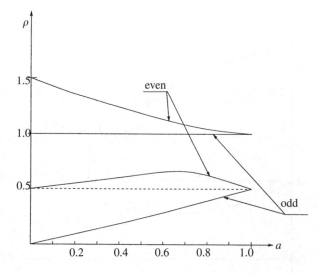

FIG. 4.5. Eigenvalue curves $\rho = \rho(a)$.

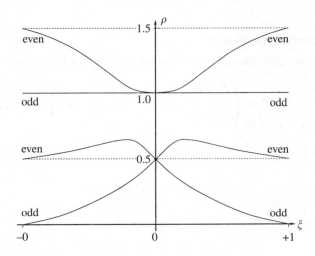

FIG. 4.6. Eigenvalue curves $\rho = \rho(\xi)$.

Numerical algorithms to calculate infinite continued fractions may be seen from [104]. This is a rather comfortable method to get the discrete spectrum of eqn (4.3.24). As is shown in Fig. 4.7, we get four discrete eigenvalues, two belonging to even vibrations and two belonging to odd ones. There is one negative curve which is related to the decay mode of the saddle-ridge configuration and a zero curve that is the Goldstone mode, the possibility to shift the kinkpair along the Peierls potential in a quasi-stationary manner.

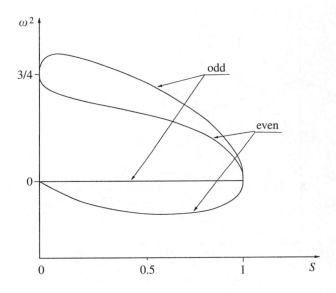

FIG. 4.7.   Eigenvalue curves $\omega = \omega(s)$.

4.3.6   The continuous spectrum

In a physical coordinate x, eqn (4.3.24) represented by the P-symbol (4.3.25)

$$P = \begin{pmatrix} 0 & a & 1 & \infty & & \\ \rho & 0 & 0 & 3 & v & 3(1+a) \\ -\rho & \frac{1}{2} & \frac{1}{2} & -2 & & \end{pmatrix} \qquad (4.3.34)$$

is of Schrödinger type (cf. (4.3.23)):

$$\frac{\mathrm{d}\phi^2}{\mathrm{d}x^2} + \left(E - V(x)\right)\phi = 0 \qquad (4.3.35)$$

the potential of which is

$$-V(x) = \operatorname{sech}^2(x + x_0) + \operatorname{sech}^2(x - x_0). \qquad (4.3.36)$$

Therefore the discrete spectrum consists of only a finite number of eigenvalues for $E < 0$. For $E > 0$, there is a continuous spectrum with generalized (even and odd) eigenfunctions which is characterized by the phase difference δ at $x = \pm\infty$ between the generalized eigenfunctions of (4.3.35)–(4.3.36) and the Schrödinger equation with potential $V(x) \equiv 0$. In both cases, the asymptotic behaviour for $|x| \to \infty$ is a sine for the odd and a cosine for the even eigenfunctions, i.e. a periodic behaviour. In the following we give a procedure how to calculate these phase differences.

In terms of Heun's equation, this is nothing more than a central two-point connection problem which may be stated as follows. Consider the P-symbol (4.3.34): the

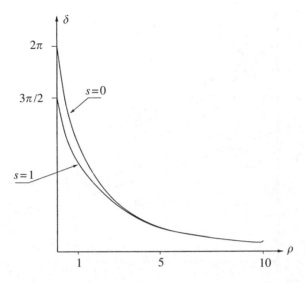

FIG. 4.8. Phase lags $\delta = \delta(\rho)$ for $s = 0$ and $s = 1$.

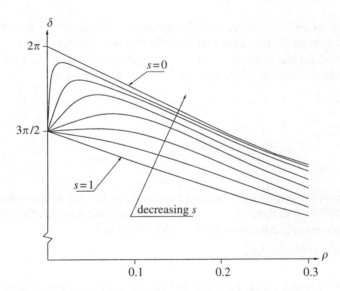

FIG. 4.9. Phase lags $\delta = \delta(\rho)$ for even eigensolutions.

point $v = a$ corresponds to the symmetry point $x = 0$ in the physical system, and the point $v = 0$ corresponds to $x = \pm\infty$. As a consequence, the even solutions with respect to the x-coordinate system are related to the characteristic exponent 0 at $v = a$ and denoted ϕ_e, while the odd ones are related to the characteristic exponent $1/2$ and denoted ϕ_o. The behaviour of the extended states at $x = \pm\infty$ is described by a linear

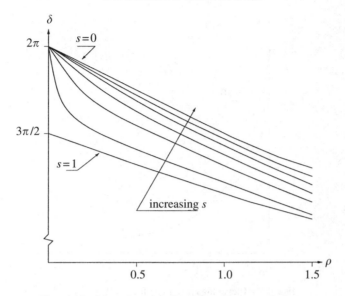

FIG. 4.10. Phase lags $\delta = \delta(\rho)$ for odd eigensolutions.

combination of the two Frobenius solutions at $v = 0$, whereby ρ is to be positive imaginary. In order for this combination of Frobenius solutions to be real on the real axis, one of them has to be the complex conjugate of the other. As a consequence, we may write the general solution of the differential equation as

$$\phi = A\phi_1 + \overline{A\phi_1}$$

with

$$\phi_1 = v^\rho \sum_{n=0}^\infty c_n v^n, \qquad A = A(\delta).$$

The condition for getting the required phase shifts is that, for a fixed value of ρ and a, the coefficient A in dependence of δ is chosen such that ϕ is linearly dependent on $\phi_{e,o}$. A mathematical condition which is continuous in δ is

$$W(\rho, a; \phi)[\phi_{e,o}(z_0); \phi(z_0)] = 0, \qquad (4.3.37)$$

where $W(\rho, a; \phi)[\phi_{e,o}(z_0); \phi(z_0)]$ is the Wronskian of $\phi_{e,o}$ and ϕ, and $0 < z_0 < a$. We have done the numerical calculation of (4.3.37) and give the result in the Figs. 4.8–4.10.

4.3.7 Quantum diffusion of kinks along dislocations

The model described above has taken the potential of the crystal in which the dislocation moves as being periodic in the glide plane only. This is correct as long as the temperature is sufficiently high. If the temperature decreases, then the periodic structure of the crystal becomes important and is modelled in a periodic structure of

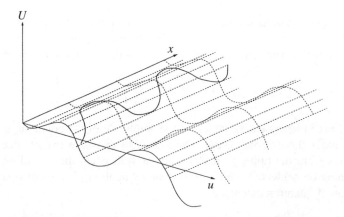

FIG. 4.11. Periodic path of the quantum system for quasi-stationary evaluation in Feynman's path-integral method.

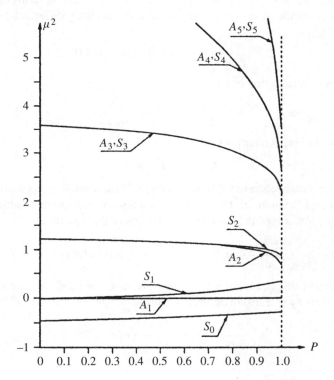

FIG. 4.12. Eigenvalue curves $\mu^2 = \mu^2(P)$ for $s = 0, 5$.

the Peierls potential *perpendicular* to the glide plane of the dislocation and is called a second-order Peierls potential. The kinks are affected by this potential while moving along the Peierls potential of first order. Simultaneously with decreasing temperature, quantum effects appear in this process which are actually tunneling processes of kinks

under the above-mentioned potentials. As is worked out and described extensively
in [42], these quantum effects can be modelled by particles moving in a double-well
potential under the influence of a heat bath. The double well has the form

$$U(u) = 16\,U_0\left[\left(\frac{u}{a}\right)^2\left(1 - \frac{u}{a}\right)^2 - \frac{s}{3\sqrt{3}}\frac{u}{a}\right],$$
(4.3.38)

where u is the independent variable, U_0 the amplitude of the potential, a the lattice
constant, and s the external stress. The aim is to calculate the net decay rate of
the particles from the upper well to the lower well when the tunneling processes
can no longer be neglected. This can be done by applying Feynman's path-integral
formulation of quantum mechanics.

Classical and semiclassical paths
Without going into details, we summarize this method according to our needs. As is
shown in [42], the above-mentioned decay rate can be written in terms of the eigenvalues $\mu = \mu_i$ of the following linear second-order ordinary differential equation

$$\frac{d^2\phi(\rho)}{d\rho^2} + (\mu - \mathcal{V}''(\Phi))\phi(\rho) = 0$$
(4.3.39)

with (Φ is a normalization of u)

$$\mathcal{V} = \frac{1}{2}\Phi^2(1 - \Phi)^2 - \frac{s}{6\sqrt{3}}\Phi$$
(4.3.40)

under the periodicity conditions

$$\phi(\rho + P) = \phi(\rho)$$

where ρ is a normalization of the time variable x, $\mathcal{V}$ the normalized potential (4.3.40),
P is a period of the path (cf. Fig. 4.2), and ϕ a normalized space variable. Equation
(4.3.39) has to be solved in terms of the solutions of the equation

$$\frac{d^2\Phi(\rho)}{d\rho^2} = \mathcal{V}'(\Phi), \qquad \Phi(\rho + P) = \Phi(\rho).$$
(4.3.41)

The solutions of this equation can be written in terms of Jacobian elliptic functions
(see [42]). Writing (4.3.39) in dependence on Φ, we get an equation of the form

$$\frac{d^2\phi}{d\Phi^2} + f(\Phi)\frac{d\phi}{d\Phi} + g(\Phi) = 0$$
(4.3.42)

with

$$f(\Phi) = \frac{\mathcal{V}'}{2(\mathcal{V} + C)}$$
(4.3.43)

where C is from (4.3.10) and Fig. 4.2

$$g(\Phi) = \frac{\mu^2 - \mathcal{V}''}{2(\mathcal{V} + C)}$$

Equation (4.3.42) is Fuchsian, with five singularities, having the P-symbol

$$
\begin{pmatrix}
\frac{1}{2} & \frac{1}{2} & \frac{1}{2} & \frac{1}{2} & 1 & \\
\Phi_1 & \Phi_2 & \Phi_3 & \Phi_4 & \infty & ; \Phi \\
0 & 0 & 0 & 0 & 3 & ; B \\
\frac{1}{2} & \frac{1}{2} & \frac{1}{2} & \frac{1}{2} & -2 &
\end{pmatrix}
$$

with $B = B(\mu)$. The solution of the physical problem is a central two-point connection problem for the Fuchsian equation (4.3.40), (4.3.42) between $\Phi = 0$ and $\Phi = 1$. For the even solutions, one has to look for solutions belonging to the characteristic exponents $\alpha = 0$ at $\Phi = 0$ and $\alpha = 0$ at $\Phi = 1$. For the odd solutions, one has to look for solutions belonging to the characteristic exponents $\alpha = 0$ at $\Phi = 0$ and $\alpha = 1/2$ at $\Phi = 1$. In both cases, an appropriate expansion for the Frobenius solutions in terms of power type series with coefficients a_n, $(n = 0, 1, 2 \dots)$ yields a regular third-order difference equation of Poincaré–Perron type. The characteristic equation of this difference equation has the three solutions

$$
t_1 = 1, \qquad t_2 = 1/b, \qquad t_3 = 1/a
$$

corresponding to the three possible radii of convergence $\mathcal{R} = 1$, $\mathcal{R} = b$, and $\mathcal{R} = a$. Here, the meaning of the t_i is given by

$$
t = \lim_{n \to \infty} \frac{a_{n+1}}{a_n}.
$$

The eigenvalue condition here is given by

$$
\lim_{n \to \infty} \left| \frac{a_{E,n}(\mu_i)}{a_{N,n}(\mu)} \right| = 0,
$$

where $a_{E,n}(\mu_i)$ is a solution of the difference equation that generates an eigensolution and $a_{N,n}(\mu)$ is a solution of the difference equation that does not. These two solutions may be distinguished by linearly independent initial conditions. The result is a two-parameter family of eigenvalues $\mu = \mu_i(s, P)$. Figure 4.12 shows the lowest lying eigenvalues for $s = m^2 = 0.5$. Figure 4.13 shows that $\lim_{m \to \infty} \mu^2 \sim m^2$ where m is the numbering of the energy levels.

4.4 Tunneling in double-well potentials

The studies in the last section may well be considered under the aspect of tunneling processes in double-well potentials. If we start with a Schrödinger equation

$$
\frac{d^2 y}{dx^2} + \left[E - V(x) \right] y(x) = 0,
$$

then the potential $V(x)$ has the following mathematical form:

$$
-V(x) = A \left(\operatorname{sech}^2(x + x_0) + \operatorname{sech}^2(x - x_0) \right), \tag{4.4.1}
$$

where A is a constant and x_0 is a parameter. For this sort of potential, the depth A of its wells keeps constant while the distance $2x_0$ between them can vary. Because there

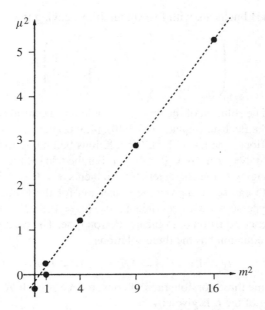

FIG. 4.13. Eigenvalues μ^2 as a function of energy level numbers m^2 for $s = 0.5$.

is a symmetry of (4.4.1) with respect to $x = 0$, the eigenfunctions decay into even and odd ones. We may consider the related eigenvalues as generated by means of a tunneling interaction, as explained in the following. Suppose that we carry out the limiting process $x_0 \to \infty$ so that we keep one well fixed at $x = 0$. The result would yield one well having two localized eigenstates. If we now approach the second well from infinity to the fixed one at zero, these two eigenvalues would split each into two eigenvalues corresponding to even and odd eigensolutions. As is well known, the splitting distance between these even and odd eigenvalues depends exponentially on the distance between the two wells. This reflects the exponentially small tunneling current occurring between the two wells.

Since the wells keep their form while approaching each other, this example of a double-well potential is considered to be a simple one. Moreover, for finite values of x_0, we have

$$\lim_{|x| \to \infty} V(x) = 0.$$

A more complicated example, where these two properties no longer hold, is given by the quartic oscillator. In the following, we calculate the eigenvalues and eigenfunctions of the quartic oscillator as an application of our analysis of the central two-point connection problem of the triconfluent Heun equation. We start with the normal general form of this equation, given by

$$\frac{d^2 y(z)}{dz^2} + \left(E + \sum_{k=1}^{4} D_k z^k \right) y(z) = 0 \quad \text{for } z \in \mathbb{C}, \qquad D_4 \in \mathbb{R}^-. \qquad (4.4.2)$$

The physically relevant interval is the whole real axis. However, the original central two-point connection problem was defined only between $z = 0$ and $z = \infty$. To cope with the present problem, one has to solve it twice: once on the positive and once on the negative real axis (denoted by $+$ and $-$ suffixes in the following). Then these two connection problems have to be matched at $z = 0$ by means of the conditions

$$y^+|_{z=0} = y^-|_{z=0}$$

and

$$\frac{\mathrm{d}y^+}{\mathrm{d}z}\bigg|_{z=0} = \frac{\mathrm{d}y^-}{\mathrm{d}z}\bigg|_{z=0}.$$

Equation (4.4.2) is not an appropriate form for our purposes. We prefer to take the following one:

$$\frac{\mathrm{d}^2}{\mathrm{d}z^2}\Psi(z) + \left(-\frac{p^2}{4}(z^2-1)^2 + p[E-az]\right)\Psi(z) = 0. \qquad (4.4.3)$$

This form clearly shows the three irreducible real parameters E; a, p of the physical problem: E is the accessory parameter playing the role of the spectral parameter, a is the asymmetry parameter of the potential and p is the height of the hump between the two wells. It is important here to restrict p to positive values, since otherwise there would no longer be a double-well potential. (The solution of the single-well quartic oscillator is given in [13].)

Splitting off the asymptotic factor

$$\Psi(z) = \exp\left(\frac{p}{2}(z-z^3)\right)(z+1)^{-a-1}w(z), \qquad (4.4.4)$$

carrying out a Jaffé transformation

$$x := \frac{z}{z+1},$$

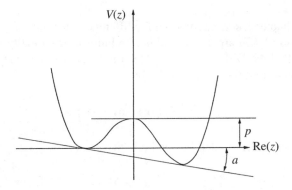

FIG. 4.14. Double-well potential with two parameters.

and applying a power-type series of the form

$$w(x) = \sum_{n=0}^{\infty} a_n x^n \tag{4.4.5}$$

leads to an irregular fourth-order difference equation of Poincaré–Perron type:

a_0 and a_1 arbitrary,

$$f_0(0)a_2 + f_{-1}(0)a_1 + f_{-2}(0)a_0 = 0,$$

$$f_1(1)a_3 + f_0(1)a_2 + f_{-1}(1)a_1 + f_{-2}(1)a_0 = 0,$$

$$f_2(n)a_{n+2} + f_1(n)a_{n+1} + f_0(n)a_n + f_{-1}(n)a_{n-1} + f_{-2}(n)a_{n-2} = 0, \quad \text{for } n \geq 2,$$

$$\tag{4.4.6}$$

where we have, for $n \geq 2$,

$$
\begin{aligned}
f_2(n) &= 1 + \frac{\alpha_2}{n} + \frac{\beta_2}{n}, \\
f_1(n) &= -4 + \frac{\alpha_1}{n} + \frac{\beta_1}{n}, \\
f_0(n) &= 6 + \frac{\alpha_0}{n} + \frac{\beta_0}{n}, \\
f_{-1}(n) &= -4 + \frac{\alpha_{-1}}{n} + \frac{\beta_{-1}}{n}, \\
f_{-2}(n) &= 1 + \frac{\alpha_{-2}}{n} + \frac{\beta_{-2}}{n}.
\end{aligned}
$$

Since we may calculate the solutions of (4.4.6) recursively, the asymptotic behaviour as $n \to \infty$ is of interest. A fundamental system of (4.4.6) in this limit is given by the following Birkhoff solutions ([16, 17, 18])

$$s_l(n) = \exp\left(\sum_{m=1}^{m=3} \gamma_{lm} n^{\frac{4-m}{4}}\right) n^{r_l}\left[1 + \frac{C_{l1}}{n^{\frac{1}{4}}} + \frac{C_{l2}}{n^{\frac{2}{4}}} + \cdots\right] \quad (l = 1, 2, 3, 4).$$

$$\tag{4.4.7}$$

As defined in Chapter 3, normal solutions for difference equations of Poincaré–Perron type in the form (4.4.7) are called Birkhoff solutions. The totality of all Birkhoff solutions is called the Birkhoff set (see [132] p. 274). We thus may represent the general solution of (4.4.6) asymptotically by

$$a_n = \sum_{l=1}^{4} L_l s_l(n) \quad \text{as } n \to \infty. \tag{4.4.8}$$

What is important is that, under the condition that

$$\sum_{i=-2}^{i=+2} \alpha_i \neq 0$$

holds, we have

$$\gamma_{11} = \frac{4}{3}\sqrt[4]{-\sum_{i=-2}^{i=+2}\alpha_i} = \frac{4}{3}\sqrt[4]{2\sqrt{-D_4}} = \frac{4}{3}\sqrt[4]{p},$$

$$\gamma_{21} = -\gamma_{11}, \qquad \gamma_{31} = i\gamma_{11}, \qquad \gamma_{41} = -i\gamma_{11}, \qquad (4.4.9)$$

and

$$\gamma_{12} = -\frac{2\alpha_2 + \alpha_1 - \alpha_{-1} - 2\alpha_{-2}}{\frac{9}{8}\gamma_{11}^2} = -\frac{1}{2}\sqrt{2\sqrt{-D_4}} = -\frac{1}{2}\sqrt{p},$$

$$\gamma_{22} = \gamma_{12}, \qquad \gamma_{32} = -\gamma_{12}, \qquad \gamma_{42} = -\gamma_{12}.$$

The constants γ_{l3} and r_l ($l = 1, \ldots, 4$) may be seen from [79] but are not of significance here.

From (4.4.9) we see that γ_{11} takes a positive real value while γ_{12} takes a negative real one. Therefore, as a result, three of the four Birkhoff solutions (4.4.7) increase exponentially as n tends to infinity, while the fourth one (represented by $s_2(n)$ in (4.4.7)) decreases exponentially in the same limit.

The asymptotic behaviour of the eigensolutions for $z \to \infty$ is given by the asymptotic factor in (4.4.4). If the parameter E is not an eigenvalue, then the asymptotic behaviour of the function $w(z)$ in (4.4.4) represented in (4.4.5) as z tends to infinity (or x tends to 1) is

$$\exp\left(-p(z - z^3)\right),$$

so that the behaviour of the asymptotic factor is affected. As soon as E becomes an eigenvalue, the leading term of the asymptotic behaviour of the function $w(z)$ in (4.4.4), as z tends to infinity, must be

$$\exp f(z), \quad \text{with } f(z) = o(z^3),$$

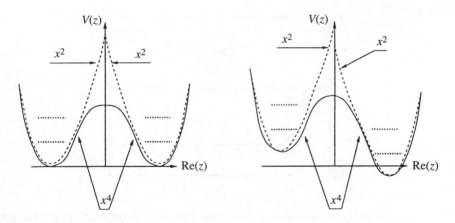

FIG. 4.15. Why avoided crossings must occur.

so that the behaviour of the asymptotic factor in (4.4.4) is not affected. This is the case if and only if the coefficients in front of the increasing solutions in (4.4.8) vanish:

$$L_1(E; a, p) = L_2(E; a, p) = L_3(E; a, p) = 0. \tag{4.4.10}$$

A numerical method of solving (4.4.10) is given in [12]. In Figs 4.16–4.19, we show some of the eigenvalue curves $E = E_i(a)$ for $p = 15$ and some of the corresponding eigenfunctions compared to the approximate ones found by an appropriate Ritz method. As one can see, the eigenvalue curves show up the so-called phenomenon of avoided crossings. It is an exponentially small quantum tunneling effect which we could get since our expansion is exact.

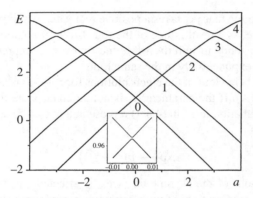

FIG. 4.16. Eigenvalue curves $E = E(a)$ and the effect of avoided crossings (cf. Fig. 3.1 for the asymptotic calculation).

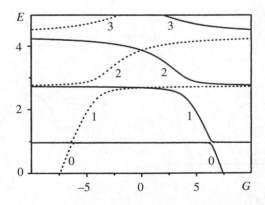

FIG. 4.17. The low-lying eigenvalues of the spectra of the triconfluent Heun equations on the positive and on the negative real axis.

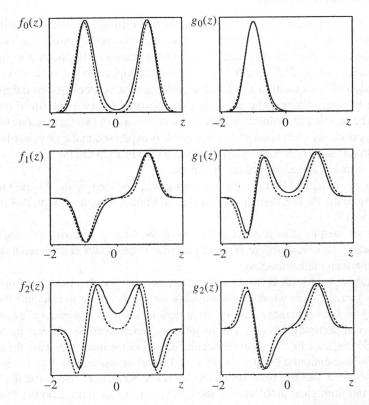

FIG. 4.18. Low-lying eigensolutions for the symmetric (left side) and for the asymmetric (right side; $a = 1$) quantum quartic oscillator.

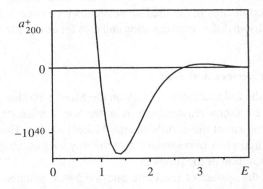

FIG. 4.19. Verifying the eigenvalue condition numerically.

4.5 Hill-type equations

In all of the preceding examples, we have studied differential equations with *polynomial* coefficients and the corresponding central two-point connection problems. Differential equations of Hill type are not of this category but belong to the class of equations with *periodic* coefficients. A physical application of such a situation is the solution of the quantum-mechanical problem of a Schrödinger equation for an electron moving in a periodic potential that consists of only one term of a Fourier series. The fact that the solution of this problem consists of two factors, one of which has the periodicity of the underlying potential, is expressed in the physical literature by the *Bloch theorem* according to Felix Bloch who in 1928 did the calculation in the context of quantum mechanics for the first time.

In the mathematical literature, the same fact is expressed by the *Floquét theorem* according to the French mathematician Gaston Floquét who dealt with this problem as early as 1883.

However, neither Bloch nor Floquét were the first to discover the underlying phenomenon: this was George William Hill who is nowadays considered as the first great American mathematician.

Hill did not get his ideas from purely mathematical considerations but from establishing a method from which it was possible for the first time to calculate the lunar trajectory in such a manner that the numerical labour was reasonably rewarded by the achieved accuracy. He was able to solve the *main problem of lunar theory* (see [49], p. 57) because he could get the secular effects of the motion that are the essence of the above-mentioned Floquét theorem and which reappeared in Bloch's work.

This solution was the basis on which his successors could overcome the mathematical and numerical problems for the landing of human beings on the Moon. As a by-product, Hill discovered so-called infinite determinants, having infinitely many rows, and columns, which nowadays bear his name.

In the following, we give a brief account of Hill's work, emphasizing the physical implications of the main problem of lunar theory and summarizing his considerations about infinite determinants. In the final subsection, we consider a lateral connection problem for a Fuchsian differential equation and thus for an equation with polynomial coefficients.

4.5.1 The lunar perigee and node

It is obvious that the influence of the Earth on the Moon's motion dominates that of the Sun. Therefore it seems reasonable to treat the Sun's influence as a perturbation of a two-centre problem of the Earth's and the Moon's motion. However, what came out after two hundred years of research is that the way the perturbation is formulated is crucial for the solution of the problem.

If we consider the system of the Earth and the Moon situated in the ecliptic of the Sun and the Earth, we have two quantities, the dynamics of which characterize the Keplerian ellipse of the Moon around the Earth: the lunar perigee and the node line. The former is the point of closest approach of the Moon to the Earth. The latter is the line from the point where the Moon intersects the ecliptic upwards, i. e. the ascending node, to the point where it intersects it downwards, i. e. the descending node. If there were no attraction of the Sun on the Moon, then these two quantities would be spatially fixed in a coordinate system the origin of which is located at the centre of the Sun.

The action of the Sun on the Moon can be described by considering the lunar perigee and node line as dynamic variables that vary as follows. While the perigee moves in the same sense as the Moon around the Earth (prograde), and the period of this motion is approximately 9 years, the node line moves retrograde with a period of about 18 years. As a result, the motion of the Moon changes from a periodic one to an almost-periodic one.

While, from a physical point of view, we could keep Kepler's picture by letting the lunar perigee and node line become dynamical variables, the mathematical formulation of the perturbation is by no means as easy. There have been several scientists of the first rank who tried to cope with this problem but failed. The most famous one was the French mathematician Charles Delaunay who after 10 years of calculation in 1860 and 1867 published a two-volume work of more than 900 pages each, the longest formula in which ranges over more than 70 pages. The result was poor: the Moon's motion came out of this calculation with an accuracy that was known already to the ancient Greeks. Hill's analysis of this crucial problem was thus of historic importance, in that it enabled the calculation of the perturbation to the desired degree of accuracy with reasonable effort.

4.5.2 Hill's solution

General remarks
The phase space of the three-body problem consists of 18 dimensions: The common centre of the mass of all three bodies (three degrees of freedom), the vector R from the centre of the Sun to the common centre of mass of the Earth and the Moon (three degrees of freedom), and the vector r from the centre of the Earth to the centre of the Moon (three degrees of freedom) result in nine dimensions. Taking into account the corresponding momenta yields the final number.

Ten constants of motion are obvious: the coordinates of the trajectory and of the momentum of the common mass centre of all the three bodies; the angular momentum of all the three bodies which actually defines the ecliptic; and the total energy of the system.

The result of this reduction procedure is a phase space of eight dimensions. The question which has puzzled scientists since the days of classical antiquity is whether there exist four further constants of the Moon's motion under the influence of the Sun and the Earth. If this was so, then the main problem of lunar theory would be completely integrable. As a consequence, there would be four time-independent frequencies in this problem.

What does observation contribute to this question? Humanity has known of four periods for more than 3000 years. They remain constant within an accuracy of

5 decimal places, and therefore form the basis of the calendars of most human cultures:

(1) the sidereal year (365.257 days) (this is the time for the Sun to reappear under the same fixed star),

(2) the sidereal month (27.32166 days) (this is the time for the Moon to reappear under the same fixed star),

(3) the anomalistic month (27.55455 days) (this is the time for the Moon to reappear at the perigee),

(4) the draconitic or nodical month (27.21222 days) (this is the time for the Moon to reappear at the same (ascending or descending) node).

Incidentally, these periods are the results of the first high-precision observation in physics carried out by the Greeks. From a theoretical point of view, it is still an open question whether these observed quantities are constants of motion.

After Kepler's laws, it was a first-ranking challenge for modern science to give an explanation of the movements of the perigee and of the node line. Eventually this was done by Isaac Newton, who showed by applying mathematical methods that it is the influence of the Sun on the Moon that causes these movements. And it was G.W. Hill who finally gave the correct quantitative results with a reasonable amount of calculation.

Fundamental assumptions
Hill's fundamental observations from which he started are:

(1) Kepler's ellipses are periodic trajectories.

(2) The observed trajectory of the Moon is non-periodic. The periods of the nodal and the perigean motion are non-commensurable with the duration of the sidereal month.

The fundamental new idea Hill took from these observations may be stated as follows. While all the scientists before Hill considered the Moon's motion as a perturbation of the *general* solution of the *two-body problem* Earth–Sun, Hill started with a *specific* periodic trajectory of the *three-body problem* and subsequently solved the main problem of lunar theory as a perturbation with respect to this trajectory.

His basic assumptions are threefold:

(1) The trajectory of the Earth around the Sun is a circle (thus he used the so-called mean anomaly).

(2) The Moon's motion projected on the ecliptic is decoupled from the motion perpendicular to it.

(3) The mathematical solution is to be worked out in a *Cartesian* coordinate system x, y, z that rotates along with the Earth around the Sun while its x axis is pointing to the Sun.

The Hamiltonian function of this system expanded in powers of r/R is given by (see [49])

$$H = \frac{p^2}{2\mu} - \frac{\mu n^2 a^3}{r} - \frac{\mu n'^2 a'^3}{R^5}\left[\frac{3}{2}(\boldsymbol{R},\boldsymbol{r})^2 - \frac{1}{2}r^2 R^2\right] - \cdots . \qquad (4.5.1)$$

Here, a' is the mean radius of the Kepler trajectory of the Earth, a the mean radius of the Kepler trajectory of the Moon, $n' = \frac{2\pi}{T_0}$ is the mean angular velocity of the Earth around the Sun, n is the mean angular velocity of the Moon around the Earth, and $()$ is the scalar product.

The Hamiltonian function (4.5.1) takes into account all observable motions of the Moon. In particular, the influence of the Sun appears in a quadrupole term, which, however, is dependent on R. Therefore, the Hamiltonian function (4.5.1) is time-periodic, the frequency of which is given by n'.

The dynamical problem of Hill's consideration yields a system of three nonlinear ordinary differential equations

$$\frac{\mathrm{d}^2 x}{\mathrm{d}t^2} - 2n'\frac{\mathrm{d}y}{\mathrm{d}t} - n'^2 x = -\frac{n^2 a^3}{r^3}x + 2n'^2 x,$$

$$\frac{\mathrm{d}^2 y}{\mathrm{d}t^2} + 2n'\frac{\mathrm{d}x}{\mathrm{d}t} - n'^2 y = -\frac{n^2 a^3}{r^3}y - 2n'^2 y, \qquad (4.5.2)$$

$$\frac{\mathrm{d}^2 z}{\mathrm{d}t^2} = -\frac{n^2 a^3}{r^3}z - n'^2 z.$$

This system has to be solved under the condition that the solution must have the period of the synodic month (this is the time from full Moon to full Moon, namely 29.53059 days).

The first terms on the left-hand sides describe the acceleration forces, the second ones the Coriolis forces, and the third ones the centrifugal forces. The first terms on the right-hand sides are the gravitational forces, and the second ones the influence of the Sun which—as was already mentioned above—actually is of quadrupolar character.

The solution of (4.5.2) Hill gets by means of the complex-valued Fourier expansion:

$$x_0(\tau) + \mathrm{i}y_0(\tau) = \exp(\mathrm{i}\tau)\sum_{l=-\infty}^{l=+\infty} a_l \exp(\mathrm{i}l\tau), \qquad (4.5.3)$$

where $\tau = (n - n')t$. He recognizes two symmetries being involved in the problem: the symmetry with respect to the x axis makes the coefficients a_l real-valued and the symmetry with respect to the y axis makes the odd coefficients a_{2l+1} vanish.

As a result, Hill's periodic orbits are a singly parametric family of ovals, the centre of each of which is the origin of the coordinate system and which crosses the coordinate axes perpendicularly (see Fig. 4.20). In the first approximation, i. e. taking into account only a_0 in (4.5.3), we get Newton's result

$$a_0 = \frac{a}{(1 + m^2/2)^{1/3}}, \qquad m = \frac{n'}{n}. \qquad (4.5.4)$$

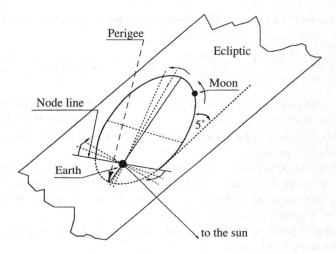

FIG. 4.20. The action of the Sun on the Moon from Kepler's viewpoint.

Formula (4.5.4) shows that, as the Sun pulls the Moon off the Earth, the Moon's response is to lower its mean distance to the Earth, thus compensating the Sun's action.

The linear problem
Given the periodic solution of (4.5.2), Hill introduces perturbations

$$\delta x(t), \ \delta y(t), \ \delta z(t)$$

yielding *linear* differential equations. According to his condition that the motion in the (x, y) plane is decoupled from the motion in the z direction, he gets

$$\frac{d^2 \delta z}{dt^2} + \left(n^2 + \frac{3}{2} n'^2 \right) \delta z = 0$$

as (a simplified form of) the equation governing the motion perpendicular to the ecliptic. Its solution is given by

$$\delta z = \kappa \cos(\omega t + \chi), \qquad \chi \text{ arbitrary,}$$

the angular frequency of which is

$$\omega = n \left(1 + \frac{3 n'^2}{4 n^2} \right). \tag{4.5.5}$$

As we can see from (4.5.5), the vertical motion of the Moon in the vicinity of Hill's periodic orbit has a larger frequency, so that it returns to the ecliptic before it appears under the same fixed star. This is Newton's result of the retrograde-moving node line, which is the lowest-order result of Hill's approach.

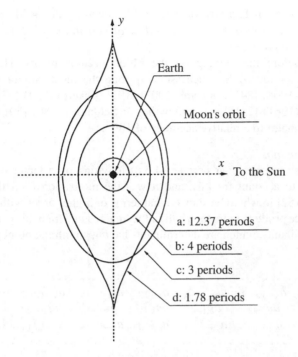

FIG. 4.21. Hill's periodic orbits.

To study the motion in the (x, y) plane, Hill defines the quantity δs, for which he gets

$$\frac{d^2 \delta s}{dt^2} + \Theta(t)\, \delta s = 0. \tag{4.5.6}$$

$\Theta(t)$ is a periodic function having the period $2\pi/(n - n')$. An equation of type (4.5.6) nowadays is called a *Hill equation*. The lowest order of $\Theta(t)$ is given by

$$\Theta(t) = n^2 - \frac{3n'^2}{2}.$$

With (4.5.6) the solution of (4.5.2) is given by

$$\delta s = \epsilon \cos(\Omega t + \psi), \qquad \psi \text{ arbitrary}, \tag{4.5.7}$$

with

$$\Omega = n\left(1 - \frac{3n'^2}{4n^2}\right), \tag{4.5.8}$$

once again a result of Newton, this time for the motion of the perigee, which is wrong by a factor 2. It was Hill's great step to correct this result, which he did by introducing infinite determinants (see below).

The significance of Hill's work may be illustrated by citing Martin Gutzwiller: *This program ... formed the basis for all lunar calculations before the landing of human beings on the Moon.*

However, before the landing on the Moon became reality, Hill's work has had to be elaborated by his successors. Among the most famous of them were Brown (1897–1908) [24], Poincaré (1907) [105], Eckert (1954) [37], and Brower & Clemence (1961) [23]. Later Gutzwiller & Schmidt (1986) [50] calculated the Moon's coordinates to a relative accuracy of 10^{-8}.

Infinite determinants

Newton's result on the lunar perigee (4.5.8) is wrong by a factor of 2. Scientists have puzzled to account for this, and it was the motivation for Hill to write his famous work [52] in which he showed that terms of higher order with respect to the disturbing force produce more than half of the final result. Hill in his paper introduced infinite determinants, which are determinants having an infinite number of rows and columns, e.g.

$$
A = \begin{vmatrix}
\cdots\cdots\cdots\cdots\cdots\cdots\cdots\cdots\cdots\cdots\cdots\cdots\cdots\cdots\cdots \\
\cdots \quad a_{k-1,l-2} \quad a_{k-1,l-1} \quad a_{k-1,l} \quad a_{k-1,l+1} \quad a_{k-1,l+2} \quad \cdots \\
\cdots \quad a_{k,l-2} \quad a_{k,l-1} \quad a_{k,l} \quad a_{k,l+1} \quad a_{k,l+2} \quad \cdots \\
\cdots \quad a_{k+1,l-2} \quad a_{k+1,l-1} \quad a_{k+1,l} \quad a_{k+1,l+1} \quad a_{k+1,l+2} \quad \cdots \\
\cdots\cdots\cdots\cdots\cdots\cdots\cdots\cdots\cdots\cdots\cdots\cdots\cdots\cdots\cdots
\end{vmatrix}
$$

Here, $k, l \in \mathbb{Z}$. In the following we give a brief discussion of how Hill came to this sort of determinant while giving the correct calculation of the motion of the lunar perigee.

In a first step, Hill showed that the motion of the lunar perigee is governed by the following linear homogeneous second-order equation:

$$
\frac{\mathrm{d}^2 w}{\mathrm{d}t^2} + \Theta\, w = 0, \tag{4.5.9}
$$

in which Θ is a periodic function which depends only on the relative position of the Moon with respect to the Sun. It may be developed in a Fourier series of the form

$$
\Theta = \Theta_0 + \Theta_1 \cos(2\tau) + \Theta_2 \cos(4\tau) + \cdots,
$$

where τ denotes the mean angle of the Moon and the Sun (i.e. one point is the eye of the observer on Earth; and the two other points are the Sun and the Moon). Writing

$$
\exp(\mathrm{i}\tau) =: \zeta
$$

and defining the differentiating operator D as

$$
D := -\mathrm{i}\frac{\mathrm{d}}{\mathrm{d}\tau},
$$

(4.5.9) takes the form

$$
D^2 w = \Theta\, w
$$

with

$$\Theta = \sum_{i=-\infty}^{\infty} \Theta_i \zeta^{2i}$$

and the symmetry condition

$$\Theta_{-i} = \Theta_i.$$

The coefficients Θ_i only depend on the quantity

$$m := \frac{n'}{n - n'}.$$

In what follows, Hill assumes the expansion

$$w = \sum_{i=-\infty}^{\infty} b_i \zeta^{c+2i}, \tag{4.5.10}$$

where the exponent c is the ratio of the synodic to the anomalistic month. This quantity, together with the coefficients b_i is given by the equations

$$[c + 2j]^2 b_j - \sum_{i=-\infty}^{\infty} \Theta_{j-i} b_i = 0.$$

Equation (4.5.10) is an infinite number of equations where the number of terms in each equation is also infinite. It may be written as

$$
\begin{array}{ccccccccc}
\cdots & +[-2]b_{-2} & -\Theta_1 b_{-1} & -\Theta_2 b_0 & -\Theta_3 b_1 & -\Theta_4 b_2- & \cdots & = 0 \\
\cdots & -\Theta_1 b_{-2} & +[-1]b_{-1} & -\Theta_1 b_0 & -\Theta_2 b_1 & -\Theta_3 b_2- & \cdots & = 0 \\
\cdots & -\Theta_2 b_{-2} & -\Theta_1 b_{-1} & +[0]b_0 & -\Theta_1 b_1 & -\Theta_2 b_2- & \cdots & = 0 \\
\cdots & -\Theta_3 b_{-2} & -\Theta_2 b_{-1} & -\Theta_1 b_0 & +[1]b_1 & -\Theta_1 b_2- & \cdots & = 0 \\
\cdots & -\Theta_4 b_{-2} & -\Theta_3 b_{-1} & -\Theta_2 b_0 & -\Theta_1 b_1 & +[2]b_2- & \cdots & = 0 \\
\end{array}
$$

with

$$[i] := (c + 2i)^2 - \Theta_0.$$

Then, Hill writes: "If, from this group of equations, infinite in number, and the number of terms in each equation also infinite, we eliminate all the b except one, we get a symmetrical determinant involving c, which, equated to zero, determines this quantity. This equation we will denote thus:—

$$\mathcal{D}(c) = 0." \tag{4.5.11}$$

Subsequently, he shows some properties of c, e. g.

$$\mathcal{D}(-c) = \mathcal{D}(c) \tag{4.5.12}$$

or

$$\mathcal{D}(c + 2v) = \mathcal{D}(c), \quad \text{for } v \in \mathbb{Z}. \tag{4.5.13}$$

As a result, it came out that the quantity c is nothing else than what is known nowadays in the literature as the *Floquét exponent*.

Besides his having understood the role of c, Hill's reputation as a mathematician undoubtedly is based to a great deal on the brilliant way he dealt with infinite determinants, which we will sketch now.

As a first step, he shows that, if the calculation neglects terms higher than m^5, the quantity c is given by

$$c = \sqrt{1 + \sqrt{(\Theta_0 - 1)^2 - \Theta_1^2}}.$$

If the actual numerical values to seven decimal places are taken into account, thus taking

$$\Theta_0 = 1.1588439, \qquad \Theta_1 = -0.0570440,$$

we get

$$c = 1.0715632.$$

From this value, he got the ratio of the motion of the lunar perigee to the sidereal mean motion of the Moon as

$$\frac{1}{n}\frac{d\omega}{dt} = 1 - \frac{c}{1 + m} = 0.008591.$$

Hill proceeded to show that the equation determining c can be put into the form

$$\frac{\sin^2\left(\frac{\pi}{2}c\right)}{\sin^2\left(\frac{\pi}{2}\sqrt{\Theta_0}\right)} = J(0),$$

where J is once again an infinite determinant depending on c which has to be evaluated at $c = 0$:

$$J(0) = \begin{vmatrix}
\cdots & +1 & -\dfrac{\Theta_1}{4^2 - \Theta_0} & -\dfrac{\Theta_2}{4^2 - \Theta_0} & -\dfrac{\Theta_3}{4^2 - \Theta_0} & -\dfrac{\Theta_4}{4^2 - \Theta_0} & \cdots \\
\cdots & -\dfrac{\Theta_1}{2^2 - \Theta_0} & +1 & -\dfrac{\Theta_1}{2^2 - \Theta_0} & -\dfrac{\Theta_2}{2^2 - \Theta_0} & -\dfrac{\Theta_3}{2^2 - \Theta_0} & \cdots \\
\cdots & -\dfrac{\Theta_2}{0 - \Theta_0} & -\dfrac{\Theta_1}{0 - \Theta_0} & +1 & -\dfrac{\Theta_1}{0 - \Theta_0} & -\dfrac{\Theta_2}{0 - \Theta_0} & \cdots \\
\cdots & -\dfrac{\Theta_3}{2^2 - \Theta_0} & -\dfrac{\Theta_2}{2^2 - \Theta_0} & -\dfrac{\Theta_1}{2^2 - \Theta_0} & +1 & -\dfrac{\Theta_1}{2^2 - \Theta_0} & \cdots \\
\cdots & -\dfrac{\Theta_4}{4^2 - \Theta_0} & -\dfrac{\Theta_3}{4^2 - \Theta_0} & -\dfrac{\Theta_2}{4^2 - \Theta_0} & -\dfrac{\Theta_1}{4^2 - \Theta_0} & +1 & \cdots
\end{vmatrix}.$$

$$\tag{4.5.14}$$

Eventually Hill shows how (4.5.14) may be replaced by an infinite series in ascending powers and products of the coefficients $\Theta_1, \Theta_2, \ldots$. We will not give this lengthy formula but will give his final numerical results, the accuracy of which is the central success, as we mentioned in the beginning: Assuming

$$n = 17325594.06085, \qquad n' = 1295977.41516,$$

in units of arc-seconds per year, whence

$$m = 0.080848933808311(6),$$

he got for the coefficient in (4.5.9)

$$\begin{aligned}
\Theta = {} & 1.158843939596583 \\
& - 0.114088037493807 \cos(2\tau) \\
& + 0.000766475995109 \cos(4\tau) \\
& - 0.000018346577790 \cos(6\tau) \\
& + 0.000000108895009 \cos(8\tau) \\
& - 0.000000002098671 \cos(10\tau) \\
& + 0.000000000012103 \cos(12\tau) \\
& - 0.000000000000211 \cos(14\tau).
\end{aligned}$$

Inserting these coefficients into (4.5.14), he gets the value of the infinite determinant as

$$J(0) = 1.00180\,47920\,21011\,2$$

and Floquét's exponent for the motion of the lunar perigee as

$$c = 1.07158\,32774\,16012.$$

The final result of Hill's calculation is the numerical value of the ratio of the motion of the lunar perigee to the sidereal mean motion of the Moon:

$$\frac{1}{n}\frac{d\omega}{dt} = 0.00857\,25730\,04864.$$

This agrees with the observed value of the perigean period of about 9 years, correcting Newton's result by a factor 2.

4.5.3 Floquét solutions and lateral connection problems

Hill's work is a seminal one, in the sense that several ideas have survived and have found their way into scientific works in our days. The most important ones are (4.5.11)–(4.5.13) and

- The infinite determinant $\mathcal{D}(c)$ has a finite Fourier expansion of the form

$$\mathcal{D}(c) = A[\cos(\pi c) - \cos(\pi c_0)],$$

where A is independent of c and c_0.

- As a consequence, in order to calculate Floquét's exponent, it is not necessary to look for the zeros of $\mathcal{D}(c)$ according to (4.5.11), since there is another determinant J having the same zeros as $\mathcal{D}$, for which

$$\frac{\sin^2\left(\frac{\pi}{2}c\right)}{\frac{\pi}{2}\sqrt{\Theta_0}} = J(0)$$

holds. Therefore one has simply to calculate the infinite determinant for a fixed but arbitrary value.

In the following, we will exhibit a special lateral connection problem proposed by D. Schmidt [108] as an example for the use of infinite determinants in the calculation of Floquét's exponents.

We start with the following Fuchsian equation with five singularities:

$$\omega(z)\left(z\frac{d}{dz}\right)^2 y(z) + a(z)\left(z\frac{d}{dz}\right)y(z) + b(z)y(z) = 0 \qquad (4.5.15)$$

with

$$
\begin{aligned}
\omega(z) &= (z - \omega_1)(1 - \omega_{-1}z)(1 - \omega_{-2}z), \\
a(z) &= a_0 + a_1 z + a_2 z^2 + a_3 z^3, \\
b(z) &= b_0 + b_1 z + b_2 z^2 + b_3 z^3.
\end{aligned}
\qquad (4.5.16)
$$

We suppose that the coefficients $\omega_1,\ \omega_{-1},\ \omega_{-2} \in \mathbb{C}$ are constant and that

$$|\omega_1\omega_{-1}| < 1, \qquad |\omega_1\omega_{-2}| < 1.$$

The generalized Riemannian P-symbol is given by

$$
\begin{pmatrix}
1 & 1 & 1 & 1 & & \\
0 & \omega_1 & \frac{1}{\omega_{-1}} & \frac{1}{\omega_{-2}} & z; & p, q \\
\rho_0 & 0 & 0 & 0 & \rho_\infty & \\
\rho_0' & \rho_1' & \rho_{-1}' & \rho_{-2}' & \rho_\infty' &
\end{pmatrix}
$$

with

$$\rho_0 + \rho_0' = \frac{1}{\omega_1}a_0, \qquad \rho_0\rho_0' = -\frac{1}{\omega_1}b_0,$$

$$\rho_1' = 1 - \frac{a(\omega_1)}{\omega_1(1 - \omega_1\omega_{-1})(1 - \omega_1\omega_{-2})},$$

$$\rho'_{-1} = 1 - \frac{\omega_{-1}^2 a\left(\frac{1}{\omega_{-1}}\right)}{(\omega_{-2} - \omega_{-4})(1 - \omega_1\omega_{-1})},$$

$$\rho'_{-2} = 1 - \frac{\omega_{-2}^2 a\left(\frac{1}{\omega_{-2}}\right)}{(\omega_{-1} - \omega_{-2})(1 - \omega_1\omega_{-2})},$$

$$\rho_\infty + \rho'_\infty = \frac{1}{\omega_{-1}\omega_{-2}} a_3, \qquad \rho_\infty\rho'_\infty = \frac{1}{\omega_{-1}\omega_{-2}} b_3.$$

We define the annulus $\mathcal{K}$ by

$$\mathcal{K} = \left\{ z \in \mathbb{C} : |\omega_1| < |z| < \min\left(\frac{1}{|\omega_{-1}|}, \frac{1}{|\omega_{-2}|}\right) \right\}.$$

We are interested in solutions of the differential equation (4.5.15)–(4.5.16) within $\mathcal{K}$. As Schmidt proves for fixed parameters $a_\kappa, b_\kappa \in \mathbb{C}$ ($\kappa = 0, 1, 2, 3$), there is at least one nontrivial Floquét solution of the form

$$y(z; a_0, \ldots, a_3, b_0, \ldots, b_3) = z^\nu h(z; a_0, \ldots, a_3, b_0, \ldots, b_3)$$

with

$$\nu = \nu(z; a_0, \ldots, a_3, b_0, \ldots, b_3) \in \mathbb{C}$$

as the Floquét exponent. The function $h(z)$ is holomorphic in $\mathcal{K}$ and may be represented as a Laurent expansion

$$h(z) = \sum_{n=-\infty}^{n=+\infty} c_n z^n,$$

the coefficients c_n of which obey a third-order regular difference equation of Poincaré–Perron type:

$$\phi_0(\nu + n)c_n - \phi_1(\nu + n - 1)c_{n-1} + \phi_2(\nu + n - 2)c_{n-2} + \phi_3(\nu + n - 3)c_{n-3} = 0$$

with

$$\begin{aligned}
\phi_0(\xi) &= \omega_1\xi^2 - a_0\xi - b_0, \\
\phi_1(\xi) &= (1 + \omega_1\omega_{-1} + \omega_1\omega_{-2})\xi^2 - a_1\xi + b_1, \\
\phi_2(\xi) &= (\omega_{-1} + \omega_{-2} + \omega_1\omega_{-1}\omega_{-2})\xi^2 - a_2\xi - b_2, \\
\phi_3(\xi) &= \omega_{-1}\omega_{-2}\xi^2 - a_3\xi + b_3,
\end{aligned}$$

the characteristic equation of which has the three solutions

$$t_1 = \frac{1}{\omega_1}, \qquad t_2 = \omega_{-1}, \qquad t_3 = \omega_{-2}.$$

As shown by Schmidt, the Floquét exponent $\nu \in \mathbb{C}$ of the differential equation (4.5.15)–(4.5.16) in $\mathcal{K}$ related to the parameters $a_0, \ldots, a_3, b_0, \ldots, b_3$ is given by

$$\cos\left(\pi[2\nu + \tau]\right) = \cos\left(\pi[2\xi + \tau]\right)$$
$$+ 2\pi^2(1 - \omega_1\omega_{-1})(1 - \omega_1\omega_{-2})D(\xi; a_0, \ldots, a_3, b_0, \ldots, b_3)$$

with

$$\tau = \frac{a_1 + a_2\omega_1 + a_3\omega_1^2 + a_0(\omega_{-1} + \omega_{-2} - \omega_1\omega_{-1}\omega_{-2})}{(1 - \omega_1\omega_{-1})(1 - \omega_1\omega_{-2})} \tag{4.5.17}$$

that is, the Floquét exponent ν determines the Wronskian of two linearly independent Floquét solutions of (4.5.15)–(4.5.16) in $\mathcal{K}$. Here, $\xi \in \mathbb{C}$ is an arbitrary value and D is an infinite determinant for which the following statements hold.

- The following limit exists:

$$D(\xi; a_0, \ldots, a_3, b_0, \ldots, b_3) = \lim_{n \to +\infty, m \to -\infty} D_m^n(\xi; a_0, \ldots, a_3, b_0, \ldots, b_3).$$

- D is an entire function with respect to $\xi; a_0, \ldots, a_3, b_0, \ldots, b_3$.
- D_m^n is the following infinite determinant if $n > m - 1$:

$$\begin{vmatrix}
\cdots & \cdots & \cdots & \cdots & \cdots & \cdots \\
\alpha_{1n} & \alpha_{2n-1} & \alpha_{3n-2} & 0 & \cdots & 0 \\
\alpha_{0n} & \alpha_{1n-1} & \alpha_{2n-2} & \alpha_{3n-3} & \cdots & 0 \\
0 & \alpha_{0n-1} & \alpha_{1n-2} & & \cdots & 0 \\
\cdots & \cdots & \cdots & \cdots & \cdots & \cdots \\
0 & \cdots & \alpha_{0m+3} & \alpha_{1m+2} & \alpha_{2m+1} & \alpha_{3m} \\
0 & \cdots & 0 & \alpha_{0m+2} & \alpha_{1m+1} & \alpha_{2m} \\
0 & \cdots & 0 & 0 & \alpha_{0m+1} & \alpha_{1m} \\
\cdots & \cdots & \cdots & \cdots & \cdots & \cdots
\end{vmatrix};$$

otherwise $D_m^n = 0$ if $n < m - 1$, and $D_m^n = 1$ if $n = m - 1$.

- For fixed values of $k \in \mathbb{Z}$, the series $\{D_\kappa^n\}_{n=k-1}^{+\infty}$ ($\kappa \in \mathbb{Z}$, $\kappa \le k+2$) for $n \ge k+2$ are governed by the recursion relation

$$D_\kappa^n - \alpha_{1n}D_\kappa^{n-1} + \alpha_{2n-1}\alpha_{0n}D_\kappa^{n-2} - \alpha_{3n-2}\alpha_{0n-1}\alpha_{0n}D_\kappa^{n-3} = 0. \tag{4.5.18}$$

- For fixed values of $k \in \mathbb{Z}$, the series $\{D_n^\kappa\}_{n=-\infty}^{k+1}$ ($\kappa \in \mathbb{Z}$, $\kappa \ge k-2$) for $n \le k-2$ are governed by the recursion relation

$$D_n^\kappa - \alpha_{1n}D_{n+1}^\kappa + \alpha_{2n}\alpha_{0n+1}D_{n+2}^\kappa - \alpha_{3n}\alpha_{0n+1}\alpha_{0n+2}D_{n+3}^\kappa = 0. \tag{4.5.19}$$

- For $n, k, m \in \mathbb{Z}$ $(n \geq k \geq m)$, the following decompositions hold:

$$D_m^n = D_k^n D_m^{k-1} + D_{k+1}^n D_m^k - \alpha_{1k} D_{k+1}^n D_m^{k-1} + \alpha_{0k+1}\alpha_{0k}\alpha_{3k-1} D_{k+2}^n D_m^{k-2}$$

and

$$D_m^n = D_k^n D_m^{k-1} - \alpha_{0k}\big(\alpha_{2k-1} D_{k+1}^n D_m^{k-2}$$
$$- \alpha_{0k+1}\alpha_{3k-1} D_{k+2}^n D_m^{k-2} - \alpha_{0k-1}\alpha_{3k-2} D_{k+1}^n D_m^{k-3}\big).$$

- For $k \in \mathbb{Z}$ the following limits exist:

$$D_k(\xi; a_0, \ldots, a_3, b_0, \ldots, b_3) = \lim_{n \to +\infty} D_k^n(\xi; a_0, \ldots, a_3, b_0, \ldots, b_3),$$

$$D^k(\xi; a_0, \ldots, a_3, b_0, \ldots, b_3) = \lim_{m \to -\infty} D_m^k(\xi; a_0, \ldots, a_3, b_0, \ldots, b_3).$$

- $$D(\xi; a_0, \ldots, a_3, b_0, \ldots, b_3) = \lim_{n \to +\infty} D^n = \lim_{m \to -\infty} D_m.$$

- For $k \in \mathbb{Z}$, the following decompositions hold:

$$D = D_k D^{k-1} + D_{k+1} D^k - \alpha_{1k} D_{k+1} D^{k-1} + \alpha_{0k+1}\alpha_{0k}\alpha_{3k-1} D_{k+2} D^{k-2}$$

and

$$D = D_k D^{k-1} - \alpha_{0k}\big(\alpha_{2k-1} D_{k+1} D^{k-2} - \alpha_{0k+1}\alpha_{3k-1} D_{k+2} D^{k-2}$$
$$- \alpha_{0k-1}\alpha_{3k-2} D_{k+1} D^{k-3}\big).$$

- For every ξ, we have

$$D(\xi + 1) = D(\xi).$$

- $D(\xi)$ has the Fourier expansion

$$D(\xi; a_0, \ldots, a_3, b_0, \ldots, b_3) = c(\xi; a_0, \ldots, a_3, b_0, \ldots, b_3)$$
$$- \frac{1}{2\pi^2} \frac{\cos(\pi[2\xi + \tau])}{(1 - \omega_1\omega_{-1})(1 - \omega_1\omega_{-1})}$$

whereby c is an entire function with respect to $\xi; a_0, \ldots, a_3, b_0, \ldots, b_3$ and τ is from (4.5.17).

We summarize the result as follows.

> In order to determine Floquét's exponent recursively by applying (4.5.18)–(4.5.19), one has to calculate these two recurrence relations for three linearly independent initial conditions at an arbitrary point ξ.

An actual numerical calculation, however, is not known to the authors.

4.6   An ideal tunneling barrier

4.6.1   Introduction

The doubly confluent Heun equation (DHE) originates from the Heun equation by means of a confluence process when two regular singularities coalesce pairwise into an irregular one [109]. From the analytical and the numerical side, the solutions of the DHE exhibit very specific features that make them useful in some specific physical problems, as for instance in gravitational theory [79]. More precisely, in the case of the DHE, there are no convergent Frobenius solutions, since there are no regular singularities as is normally the case for other special functions. Moreover, the DHE has a rather specific structure of Stokes lines and Stokes domains [43].

In the present book—after a discussion of the differential equation, its generalized Riemann scheme, some basic forms, and simple transformations—we give an asymptotic study resulting in the eigenfunctions and eigenvalues which arise in the central two-point connection problem of the two singularities along the real axis. Thereafter we propose a numeric procedure for computing the eigenvalues. It is based on an extension of Jaffé expansions introduced by W. Lay [79] and on an algorithm and programming code developed by K. Bay *et al.* [12]. Moreover, we emphasize polynomial solutions which appear at certain restrictions. Either of the two presented approaches give numerical results that are in good agreement with one another.

4.6.2   Forms of equations

The doubly confluent Heun equation (DHE) in a canonical form reads (see [116])

$$z^2 \frac{d^2 y(z)}{dz^2} + \left(-z^2 + cz + t\right) \frac{dy(z)}{dx} + \left(-az + \lambda\right) y(z) = 0 \quad \text{for } z \in \mathbb{C}. \quad (4.6.1)$$

Here c and a are local parameters defining the behaviour of the solutions at the irregular singularities located at $z = 0$ and $z = \infty$; t is a scaling parameter defining the location of the turning points. The parameter λ is the so-called accessory parameter.

The behaviour of the solutions at the singularities is exhibited in the corresponding generalized Riemann scheme [117]

$$\begin{pmatrix} 2 & 2 & \\ 0 & \infty & ; z \\ 0 & a & ; \lambda \\ 2-c & c-a & \\ 0 & 0 & \\ t & 1 & \end{pmatrix}. \quad (4.6.2)$$

According to the values of these characteristic exponents, there exist two pairs of local solutions at the singularities of (4.6.1) that behave as

$$y_1(a, c; z = 0, z) = 1(1 + o(1)), \quad y_2(a, c; z = 0, z) = z^{2-c} e^{t/z}(1 + o(1)),$$

as $z \to +0$;

$$y_1(a, c; z = \infty, z) = z^{-a}(1 + o(1)), \quad y_2(a, c; z = \infty, z) = z^{a-c} e^z(1 + o(1)),$$

as $z \to +\infty$. \hfill (4.6.3)

Although eqn (4.6.1) is not written in a self-adjoint form, the corresponding singular boundary–eigenvalue problem, defined for $t > 0$, $z \in [0, \infty[$, can be posed by the boundary conditions

$$|y(0)| < \infty, \qquad e^{-z/2} y(z) \to_{z \to \infty} 0, \qquad (4.6.4)$$

the parameter λ playing the role of the eigenvalue parameter.

From (4.6.3) it is clear that the eigensolutions of the boundary problem (4.6.1–4.6.4) are proportional to $y_1(a, c; z = 0, z)$ at zero and are proportional to $y_1(a, c; z = \infty, z)$ at infinity. It also follows that the necessary condition for an eigensolution to be a polynomial is

$$a = -n, \qquad n = 0, 1, \ldots.$$

Beyond the canonical form, other forms of the doubly confluent equation may be appropriate. First, we carry out a transformation to a more reasonable scaling of the independent variable z when the transition points appear at finite distances:

$$t \mapsto t^2, \qquad z \mapsto tz, \qquad \lambda \mapsto t\lambda \qquad (4.6.5)$$

$$\implies \quad z^2 \frac{d^2 y(z)}{dz^2} + \left(cz - t(z^2 - 1) \right) \frac{dy(z)}{dz} + t(-az + \lambda) y(z) = 0. \qquad (4.6.6)$$

Taking $t(z^2 - 1)$ as the leading term in eqn (4.6.6) at large values of t, one may see that the transition points are located at $z = 0$, $z = -1$, $z = 1$, $z = \infty$. This is even better seen from the normal form of the DHE:

$$z^2 \frac{d^2 w(z)}{dx^2} + \left(t \left[\left(\frac{c}{2} - a \right) z + \left(1 - \frac{c}{2} \right) \frac{1}{z} \right] - t^2 \frac{(z^2 - 1)^2}{4z^2} \right) w + t\tilde{\lambda} w = 0 \qquad (4.6.7)$$

which follows from (4.6.6) by means of the substitution

$$y(z) = \exp \left[\frac{t}{2} \left(z + \frac{1}{z} \right) \right] z^{-\frac{c}{2}} w(z), \qquad \tilde{\lambda} := \lambda - \frac{c(c - 2)}{4t}.$$

For studying boundary–eigenvalue problems, the self-adjoint form of the DHE is conventionally used:

$$\frac{d}{dz} \left(z^2 \frac{dv(z)}{dz} \right) + \left(t \left[\left(\frac{c}{2} - a \right) z + \left(1 - \frac{c}{2} \right) \frac{1}{z} \right] - t^2 \frac{(z^2 - 1)^2}{4z^2} \right) v + t\tilde{\lambda} v = 0, \qquad (4.6.8)$$

where

$$w(z) = zv(z).$$

We introduce new parameters

$$a' := a - 1, \qquad c' := \frac{c}{2} - 1.$$

In terms of these parameters, eqn (4.6.7) is rewritten as

$$z^2 \frac{d^2 w(z)}{dz^2} - \left(t^2 \frac{(z^2 - 1)^2}{4z^2} + t \left(a'z + c' \frac{1 - z^2}{z} \right) \right) w + t\tilde{\lambda} w = 0. \qquad (4.6.9)$$

Equation (4.6.9) does not change under the simultaneous substitutions

$$z \to -z, \qquad a' \to -a', \qquad c' \to -c'.$$

This means that, if we have studied the boundary–eigenvalue problem on the positive real half-axis, then it is no longer necessary to study it on the negative real half-axis, since the corresponding eigenvalues $\tilde{\lambda}^-$ are obtained from the eigenvalues of the boundary–eigenvalue problem on the positive half-axis $\tilde{\lambda}^+$ with the help of the formula

$$\tilde{\lambda}^-(a', c') = \tilde{\lambda}^+(-a', -c'). \qquad (4.6.10)$$

The forms (4.6.7, 4.6.8, 4.6.9) of the DHE are appropriate for studying asymptotic expansions.

Another form of the DHE is needed in order to perform the numeric algorithm. First, we make the term $-atz$ vanish in eqn (4.6.5) by substituting a new dependent variable according to

$$y(z) = (z - 1)^a u(z).$$

The corresponding equation for the function $u(z)$ reads

$$z^2 \frac{d^2 u(z)}{dz^2} + \left(-tz^2 + t + cz - \frac{2az^2}{z + 1} \right) \frac{du(z)}{dz}$$

$$+ \left(-ta - \frac{acz}{z + 1} + \frac{a(a + 1)z^2}{(z + 1)^2} + \tilde{\lambda} \right) u = 0. \qquad (4.6.11)$$

Equation (4.6.11) has the advantage that the required solution has finite limits at both endpoints of the relevant interval $[0, \infty[$.

The next step is the transformation to a new independent variable ξ:

$$\xi := \frac{z-1}{z+1},$$

in which the significant points of the complex z plane convert into the points of the complex ξ plane according to

$$z \mapsto \xi \quad \Rightarrow \quad -1 \mapsto \infty, \; 0 \mapsto -1, \; 1 \mapsto 0, \; \infty \mapsto 1$$

and the original relevant interval $[0, \infty[$ converts into the interval $[-1, 1]$. This transformation leads to the equation

$$\left(1 - \xi^2\right)^2 \frac{d^2 u(\xi)}{d\xi^2} + \left\{ -8t\xi - 2(1 - \xi^2)\left[c + (a+1)(1+\xi)\right] \right\} \frac{du(\xi)}{d\xi}$$
$$+ \left\{ (1+\xi)\left[a(a+1)(1+\xi) - 2ac\right] - 4ta + 4t\tilde{\lambda} \right\} u(\xi) = 0. \qquad (4.6.12)$$

4.6.3 Asymptotic study

Here, methods are used which are valid in the case of 'close' turning points and described in the book [115].

The 'potential' related to equations (4.6.7, 4.6.8, 4.6.9) has the shape of two potential wells separated by an irregular singularity at zero.

The Stokes lines defined by

$$\text{Im} \int_{\pm 1}^{z} \left(\frac{1}{2} - \frac{1}{2z^2} \right) dz = 0$$

comprise the real axis and the unit circle. The anti-Stokes lines defined by

$$\text{Re} \int_{\pm 1}^{z} \left(\frac{1}{2} - \frac{1}{2z^2} \right) dz = 0$$

and represented in polar coordinates (r, φ) on the z plane are

$$r(\varphi) = \frac{1 \pm \sin(\varphi)}{\cos(\varphi)}.$$

Stokes as well as Anti-Stokes lines on the z as well as on the ξ plane are sketched in Figs 4.22 and 4.23.

We look for asymptotic solutions $w_n(z)$ of the posed boundary problem (4.6.9, 4.6.4) in the form

$$w_n(z) = \exp\left(-t \int_{1}^{z} s_n(z, t)\, dz \right)$$

where the new semi-classical variable $s(z, t)$ and the eigenvalues $\tilde{\lambda}_n(t)$ are expanded

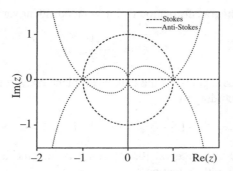

FIG. 4.22. The Stokes set in the complex z plane.

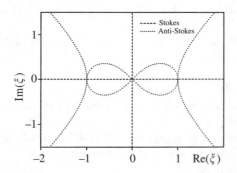

FIG. 4.23. The Stokes set in the complex ξ plane.

in reciprocal powers of t:

$$s_n(z, t) = \sum_{k=0}^{\infty} s_{nk}(z)t^{-k}, \tag{4.6.13}$$

$$\tilde{\lambda}_n(t) = \sum_{k=0}^{\infty} \lambda_{nk}(a, c)t^{-k}.$$

It is important to stress that, in any approximation when a finite number of terms is taken into account, the function on the right-hand side of (4.6.3) is single-valued. This is a consequence of the quantization condition discussed below.

The quantization condition for the low-lying eigenvalues is given by [67]:

$$-t \operatorname{Res}|_{z=1} s_n(t, z) = n, \tag{4.6.14}$$

where $s_n(t, z)$ is obtained from an asymptotic expansion (4.6.13) by recursion processes of

$$z^2 s^2 - \frac{(z^2 - 1)^2}{4z^2} + \frac{1}{t}\left(-\frac{ds(z)}{dz}z^2 - a'z - c'\frac{1 - z^2}{z} + \tilde{\lambda} \right) = 0.$$

The two first terms of the expansion (4.6.13) are

$$s_{n0} = \frac{z^2 - 1}{2z^2}, \qquad s_{n1} = -\frac{c'}{z} + \frac{a' + 1 - \lambda_{n0}}{2(z - 1)} + \frac{a' + 1 + \lambda_{n0}}{2(z + 1)}.$$

The quantization condition (4.6.14) gives

$$\lambda_{n0} = 2n + a' + 1.$$

The computation of s_{n2} is rather tiresome, and we only give the final result for the correction term to the eigenvalues:

$$\lambda_{n1} = \frac{1}{2}n(n + a' + 1) - \frac{1}{4}(a' + 1)(a' + 2) + c'(2 + a' - c').$$

It is important to realize that the representation (4.6.3) is valid all over the complex z plane without adding another exponential term. The proof follows from the general theory of the Stokes phenomenon as given in [43]. In this sense, the eigenfunctions reveal no Stokes phenomenon. This differs from the behaviour of the eigenfunctions of the other confluent Heun equations!

The eigenvalues of the boundary–eigenvalue problem on the negative half-axis are obtained by means of the symmetry relation (4.6.10). In [119], it was found that every confluent case belonging to the Heun class, with exception of the DHE, exhibits the phenomenon of avoided crossings of eigenvalues related to two different potential wells. In the case of the DHE, we have no phenomenon of avoided crossings: the eigenvalue curves cross approximately at integer values of a'. This does not contradict the theory, which forbids degeneration of the eigenfunctions, since these eigenfunctions relate to completely disconnected boundary–eigenvalue problems. A numeric study of this phenomenon is given below.

The values of $\tilde{\lambda}$ and a' at which crossings of eigenvalues occur reveal the following very specific feature of the Stokes phenomenon: at these points the eigenfunctions $y_m^-(z)$ and $y_n^+(z)$, related correspondingly to a central two-point connection problem on the left and on the right half of the real axis, can be taken as an asymptotic basis, and in this basis the Stokes matrix is trivial (i.e. it is diagonal without mixing matrix elements). As far as the authors know, the DHE is the only example of such behaviour.

It is interesting to mention that, at

$$a' = -n - 1,$$

polynomial eigenfunctions appear for which the formula for the eigenvalues simplifies to

$$\tilde{\lambda}_n = n + \frac{1}{t}\left(-\frac{1}{4}n(n - 1) + c'(1 - n - c') \right) + O(t^{-2}).$$

The polynomial solutions cannot arise at crossing points, since this would lead to a reduction of the two linearly independent solutions to a unique solution, and thus to a degeneration of the fundamental system of the differential equation.

4.6.4 Numerical algorithm

The numerical calculation of the eigenvalues and the eigenfunctions within the theory of central two-point connection problems has been extensively covered elsewhere (see [13, 12, 78, 79, 80, 81]). Therefore we may restrict ourselves in the following to a brief account.

The relevant solution of (4.6.12) may be expanded in a convergent series about $\xi = 0$:

$$u(\xi) = \sum_{k=0}^{k=\infty} g_k \xi^k. \tag{4.6.15}$$

The coefficients of (4.6.15) obey a fourth-order difference equation of Poincaré–Perron type with an initial condition:

$$g_{-1} = g_{-2} = 0,$$

g_0, g_1 arbitrary,

$$2g_2 + 2(c - a - 1)g_1 + (4(t\lambda - ta) + a(a + 1) - 2ca)g_0 = 0,$$

$$6g_3 + 4(c - a - 1)g_2$$
$$+ \{-8t - 2(a + 1) + 4(t\lambda - ta) + a(a + 1) - 2ca\}g_1$$
$$+ 2(a(a + 1) - ca)g_0 = 0,$$

$$\left(1 + \frac{\alpha_2}{k} + \frac{\beta_2}{k^2}\right)g_{k+2} + \left(\frac{\alpha_1}{k} + \frac{\beta_1}{k^2}\right)g_{k+1} + \left(-2 + \frac{\alpha_0}{k} + \frac{\beta_0}{k^2}\right)g_k$$
$$+ \left(\frac{\alpha_{-1}}{k} + \frac{\beta_{-1}}{k^2}\right)g_{k-1} + \left(1 + \frac{\alpha_{-2}}{k} + \frac{\beta_{-2}}{k^2}\right)g_{k-2} = 0$$

for $k \geq 2$, \hfill (4.6.16)

The coefficients in (4.6.16) are given by

$$\begin{aligned}
\alpha_2 &:= 3, & \beta_2 &:= 2, \\
\alpha_1 &:= 2(c - a - 1), & \beta_1 &:= 2(c - a - 1), \\
\alpha_0 &:= -8t - 2a, & \beta_0 &:= 4(t\lambda - ta) + a(a + 1) - 2ca, \\
\alpha_{-1} &:= -2(c - a - 1), & \beta_{-1} &:= -2(a - 1)(c - a - 1), \\
\alpha_{-2} &:= 2a - 3, & \beta_{-2} &:= (a - 1)(a - 2),
\end{aligned}$$

and its Birkhoff set ([79, 81]) is

$$s_m(k) = \varrho_m^k \exp\left(\gamma_m k^{\frac{1}{2}}\right) k^{r_m}\left[1 + \frac{C_{m1}}{k^{\frac{1}{2}}} + \frac{C_{m2}}{k^{\frac{2}{2}}} + \cdots\right]$$
$$\text{for } m = 1, 2, 3, 4 \tag{4.6.17}$$

with

$$
\begin{aligned}
\varrho_m &= 1 \quad \text{for } m = 1, 2, \\
\varrho_m &= -1 \quad \text{for } m = 3, 4, \\
\gamma_{m1} &= (-1)^m \sqrt{8t} \quad \text{for } m = 1, 2, 3, 4, \\
r_1 &= r_2 = -1 + a - \frac{c}{2}, \\
r_3 &= r_4 = -2 - \frac{c}{2}.
\end{aligned}
$$

The general solution of (4.6.16) may be put asymptotically in the form

$$
g_k \sim \sum_{m=1}^{4} L_m s_m(k) \tag{4.6.18}
$$

with arbitrary coefficients L_m being dependent on all parameters of the differential equation except the index k.

The series (4.6.15) must converge at $\xi = \pm 1$ when $\tilde{\lambda}$ is an eigenvalue $\tilde{\lambda} = \tilde{\lambda}_n$. Then it necessarily follows that, in this case, the asymptotic behaviour of the coefficients g_n must be described by the exponentially decreasing Birkhoff solutions in (4.6.17)–(4.6.18). This leads to the eigenvalue conditions (see [81])

$$
L_2(\tilde{\lambda}; t, a', c') = L_4(\tilde{\lambda}; t, a', c') = 0. \tag{4.6.19}
$$

A consequence of the two eigenvalue conditions in (4.6.19) is that there must be a set of two eigenvalue parameters $(\tilde{\lambda}; g_1)$. While g_0 in (4.6.16) may be normalized to unity without loss of generality, g_1 in (4.6.15) may be given the role of a second eigenvalue parameter. (It is also possible to take the ratio of g_0 to g_1 as the second eigenvalue parameter.) As a result of our procedure, we have to look for a null-dimensional set $(\tilde{\lambda}_n; g_{1n})$ in a two-parameter space $(\tilde{\lambda}; g_1)$. In the following, we show how to convert this problem into an appropriate numerical procedure.

In a first step, we solve the difference eqn (4.6.16) by *backward recursion*, as outlined in [81]. Using the initial conditions

$$
g_{N-1}^{(1)} = 1, g_N^{(1)} = g_{N+1}^{(1)} = g_{N+2}^{(1)} = 0
$$

for a sufficiently large value N, we calculate the coefficients $g_{-1}^{(1)}$ and $g_{-2}^{(1)}$, representing an exponentially *decreasing* particular solution of (4.6.16) as $k \to \infty$ (this always is achieved independently of the starting values, because of numerical instabilities).

A second linearly independent particular solution $g_k^{(2)}$ ($k = N, N-1, \ldots, 2, 1, 0$) of (4.6.16) is calculated according to the above-mentioned procedure but starting with a linearly independent initial condition:

$$
g_{N-1}^{(2)} = 0, \qquad g_N^{(2)} = 1, \qquad g_{N+1}^{(2)} = g_{N+2}^{(2)} = 0.
$$

The general solution of (4.6.16), in terms of the particular ones $g_k^{(1)}$ and $g_k^{(2)}$, is given by

$$g_{-1} = K_1\, g_{-1}^1 + K_2\, g_{-1}^2 \qquad g_{-2} = K_1\, g_{-2}^1 + K_2\, g_{-2}^2 \,,$$

with two arbitrary and k-independent constants K_1, K_2.

The eigenvalue conditions (4.6.19) are converted into

$$g_{-1} = g_{-2} = 0 \,. \tag{4.6.20}$$

The condition (4.6.20) is then (4.6.20) $\Rightarrow$ (4.6.21)

$$\det A := \begin{vmatrix} g_{-1}^{(1)} & g_{-1}^{(2)} \\ g_{-2}^{(1)} & g_{-2}^{(2)} \end{vmatrix} = 0. \tag{4.6.21}$$

Calculating this determinant by varying $\tilde{\lambda}$ an eigenvalue is given by its zeros. This zero-condition for the eigenvalue may be detected by a Newton algorithm in the numerical calculations.

4.6.5 Results

It is clear that we may consider two relevant intervals of the original equation in z, namely the positive (denoted by $+$) and the negative (denoted by $-$) real half-axis. According to our symmetry considerations, we get two sorts of eigenvalue curves in the $(\tilde{\lambda}, a')$ coordinate system for fixed values of t and c' having the $\tilde{\lambda}$ coordinate as their symmetry axis.

As can be seen from the difference equation (4.6.16), the five-term recurrence relation reduces to a three-term recurrence relation if $c = a + 1$ or $c' = \frac{a'}{2}$, respectively. In this case, there is a decoupling between the even and the odd values g_k of (4.6.16).

If we interpret the differential equation as a Schrödinger equation, then its potential has the form of a double well, the two wells of which are separated by an irregular singularity, and thus is the simplest potential that models the suppression of tunneling fluxes from one well to the neighbouring one. The parameter a in this case governs the asymmetry between the two wells. If the value of the parameter a exceeds a certain threshold (dependent on the other parameters), then eigenvalues lower than the minimum of the higher well appear. It should be mentioned that the corresponding eigenfunctions are generalized polynomials.

In the following, we give some examples of our numerical calculations of the nontrivial eigenvalues and eigenfunctions of the doubly confluent Heun equation, comparing them with the results of the asymptotics.

Figure 4.24 shows the behaviour of the determinant (4.6.21) in dependence on the eigenvalue parameter $\tilde{\lambda}$. The zeros of this curve give rise to the eigenvalues as indicated. In Figs 4.25–4.27 we show the six lowest-lying eigenvalue curves $\tilde{\lambda} - a'$ for $c' = 0$ and for $t = 1$, $t = 3$, and $t = 10$. The ground state is denoted 0, and the excited states are counted according to their number n. The central two-point

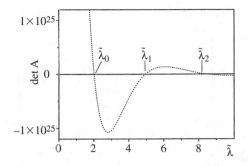

FIG. 4.24. The function $\det A(\tilde{\lambda})$ for $a' = 0.5$, $c' = 0$, $N = 100$.

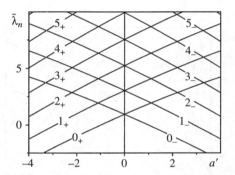

FIG. 4.25. Asymptotic calculation of eigenvalue curves ($t = 10$, $c' = 0$).

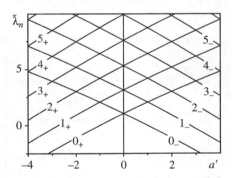

FIG. 4.26. Numeric calculation of eigenvalue curves ($t = 10$, $c' = 0$).

connection problem on the negative half-axis is denoted by $-$, and on the positive half-axis it is denoted by $+$. Figure 4.25 gives the same curves with the same parameters as Fig. 4.26, but as they result from the asymptotic calculation. Thus it should be compared with Fig. 4.26. Figure 4.29 gives a comparison between the asymptotic

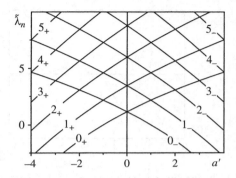

FIG. 4.27. Numeric calculation of eigenvalue curves ($t = 3$, $c' = 0$).

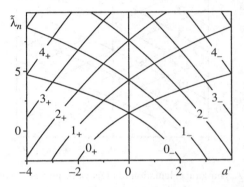

FIG. 4.28. Numeric calculation of eigenvalue curves ($t = 1$, $c' = 0$).

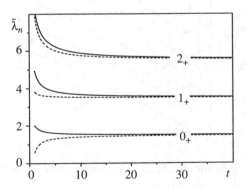

FIG. 4.29. Comparison between asymptotic and numeric calculation in dependence of t for $a' = 0.5$, $c' = 0$.

calculation for large values of t and the numerical ones in dependence on t for fixed values of $a' = 0.5$ and $c' = 0$ for the three lowest-lying eigenvalues.

4.6.6 Conclusion

The doubly confluent Heun equation exhibits several peculiarities: the differential equation has two irregular singularities, the s-ranks of both of which are 2. When being placed at the origin and at infinity, the differential equation becomes symmetrical with respect to inversion at certain restrictions on the parameters. The generalized Jaffé transformation creates an additional regular singularity at infinity. Such a form is appropriate for solving the central two-point connection problems on the positive as well as on the negative half-axis. The coefficients of the Jaffé expansions obey an irregular fourth-order difference equation of Poincaré–Perron type. We have shown that the exact eigenvalue condition for these boundary–eigenvalue problems may be obtained from the Birkhoff set of this difference equation. Moreover, we elaborated a numerical procedure from which we got the eigenvalues in dependence on the parameters. It is well understandable, but still important to stress, that there is no effect of avoided crossing of the eigenvalues in dependence on the asymmetry parameter since—in physical terms—not only the infinite but also the finite singularity is irregular and thus suppresses quantum tunneling fluxes.

The results of the calculations are compared to an asymptotic investigation of the two-point connection problem. The latter is based on a quantization condition that was developed by one of the authors (S. Yu. S.). As is shown graphically, even the lowest asymptotic order is in good agreement with the numerical results for values of the large parameter well below 10. The eigenfunctions of the doubly confluent Heun equation reveal no Stokes phenomenon on the entire complex plane of the argument, in the sense that, at Stokes lines, no other asymptotic solution is added to the existing one. As far as we know, this has not been discovered before, either for this equation or for any other one beyond the hypergeometric class.

Besides the above-mentioned symmetry, we discovered another one that occurs with respect to two of the three parameters. As a result, the eigenvalue curves of the central two-point connection problems on the positive and on the negative half-axis in dependence on the asymmetry parameter become symmetrical with respect to the energy-parameter axis.

Eventually, we discovered a finite set of generalized polynomial solutions for certain combinations of the parameters. These polynomials are not contained within the set of classical orthogonal polynomials and thus do not seem to be known as yet.

4.7 Irradiation-amplified diffusion in crystals

When a crystal is irradiated by electrons, as e. g. is the case in electron microscopy, lattice imperfections as vacancies and interstitials are created in a much higher concentration than in thermodynamical equilibrium. Surfaces in such a situation serve as an ideal sink, which has the consequence that the non-equilibrium distribution will relax to the equilibrium concentration.

Suppose now that the crystal is under constant irradiation such that the creation of lattice imperfections is maintained continually. The surfaces cause a permanent annihilation of vacancies, besides the annihilation processes occurring in the bulk by recombination of vacancies and interstitials. The result is a stationary but space-dependent distribution of vacancies and interstitials for which the stationary condition

$$D_V \, C_V = D_I \, C_I = f(z)$$

holds everywhere (where D is the diffusion coefficient, C is the concentration, z is the distance to the surface; the index V means vacancy and I means interstitial).

For a thin plate, this stationary distribution has been calculated in [114] by

$$f(z) = \lambda^2 \left\{ e + k^2 \, \text{sn}^2 \left(\lambda \sqrt{\frac{\alpha}{6}} z + K(k) | k \right) \right\}, \quad \text{for} \; -\frac{d}{2} \le z \le \frac{d}{2}, \qquad (4.7.1)$$

with

$$f(z) = D_V \, c_V(z), \qquad D_V = \text{const.};$$

here sn is a Jacobian elliptic function: a generalization of the sine function having an additional parameter k, with the help of which the curve may be broadened compared to the sine function. The other parameters are related to k in the following way (cf. [114] p. 162):

$$e = -\frac{1 + k^2}{3}, \quad \text{for} \; \frac{1}{2} < k \le 1,$$

$$\lambda^4 = \frac{9\theta}{\alpha \, (k^4 - k^2 + 1)}, \qquad \theta, \alpha > 0,$$

where α is related to the diffusion coefficients of the vacancies and interstitials (cf. [114] p. 158 and p. 161) and θ is the production rate of vacancies and interstitials by irradiation.

There is the following relation between these parameters:

$$1 + k^2 = 3 \, k^2 \, \text{sn}^2 \left(-\lambda \sqrt{\frac{\alpha}{6}} \frac{d}{2} + K(k) | k \right).$$

The modulus $K(k)$ of the Jacobian elliptic function is defined as

$$K(k) = \int_0^{\frac{\pi}{2}} \left(1 - k^2 \sin^2 v \right)^{-\frac{1}{2}} dv.$$

The distribution is given in Fig. 4.30 .

The stationary distribution for a plate (see Fig. 4.30) can be used for deriving the stationary distribution for a half-space by carrying out a limiting process $k \to 1$ and rearranging the curve such that the surface is located at $z = 0$. The result is

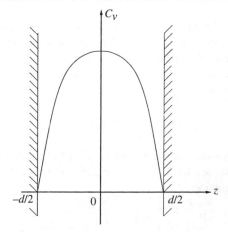

FIG. 4.30.   Space-dependent vacancy distribution in a plate.

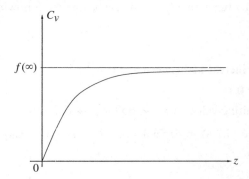

FIG. 4.31.   Space-dependent vacancy distribution in a half plane.

that the Jacobian elliptic function becomes a trigonometric one, namely a hyperbolic tangent:

$$f(z) = 3\sqrt{\frac{\theta}{\alpha}} \left(\tanh^2(\zeta + z_0) - \frac{2}{3} \right) \tag{4.7.2}$$

with

$$\zeta = \left(\frac{\theta\,\alpha}{4} \right)^{\frac{1}{4}} z, \qquad z_0 = \operatorname{artanh}\sqrt{\frac{2}{3}}, \qquad f(\infty) = \sqrt{\frac{\theta}{\alpha}}.$$

This curve is exhibited in Fig. 4.31, below.

Suppose now that there is a tracer in this solid that is diffusing by jumping into the vacancies which are created by irradiation. On its way from the surface of the half-space to the interior of the solid, the tracer will therefore encounter a space-dependent concentration of vacancies resulting in a space-dependent diffusion coefficient. As a consequence, the diffusion equation for the tracer has to take into account this space-dependent diffusion coefficient. The diffusion flux $j_T(z, t)$ of the tracer therefore obeys the relation

$$j_T(z, t) = -D_T \operatorname{grad} c_T(z, t), \tag{4.7.3}$$

where D_T is the diffusion coefficient and $c_T(z, t)$ the concentration distribution of the tracer. On the basis of (4.7.3), the tracer diffusion concentration obeys the linear partial differential equation

$$\frac{\partial}{\partial z}\left(f(z)\,\frac{\partial c_T(z, t)}{\partial z}\right) = \frac{\partial c_T(z, t)}{\partial t}. \tag{4.7.4}$$

Suppose, there is any function $f(z)$ in (4.7.4) having the following properties, as is the case with (4.7.2):

- It is defined on $\mathbb{R}^+$.
- It is monotonely increasing between $z = 0$ and $z = \infty$.
- It is zero for $z = 0$.
- It has a finite limiting value $f(\infty) < \infty$ for $z \to \infty$.

Then we may write (4.7.4) in dependence on f. The necessary formulae for this transformation are

$$\frac{df}{dz} = \sqrt{\frac{2\alpha}{3}}\,\sqrt{f(z) + 2\,f(\infty)}\,(f(\infty) - f(z)),$$

$$\left(\frac{df}{dz}\right)^2 = \frac{2\alpha}{3}\,(f(z) + 2\,f(\infty))\,(f(\infty) - f(z))^2,$$

$$\frac{d^2 f}{dz^2} = -\alpha\,(f(\infty) + f(z))\,(f(\infty) - f(z)),$$

with the help of which we get, after a separation according to

$$c_T(z, t) = C(z)\exp(\varrho\,t),$$

for the function $C(z)$, the ordinary differential equation

$$\frac{2\alpha}{3}\,f[f + 2\,f(\infty)][\,[f(\infty) - f]\,\frac{d^2 C}{df^2}$$
$$+ \left(\frac{2\alpha}{3}\,[f + 2\,f(\infty)]\,[f(\infty) - f]^2 - \alpha\,f[f(\infty)^2 - f^2]^2\right)\frac{dC}{df} - \varrho\,C = 0.$$

The scaling transformation

$$x := \frac{f}{f_\infty}$$

eventually yields the differential equation

$$\frac{d^2 C}{dx^2} + \left(\frac{1}{x-1} + \frac{1}{x} + \frac{1/2}{x+2} \right) \frac{dC}{dx} + \frac{\varrho}{(x-1)^2 \, x \, (x+2)} C = 0. \qquad (4.7.5)$$

This is a Heun equation having the generalized P-symbol

$$P \begin{pmatrix} 1 & 1 & 1 & 1 & \\ -2 & 0 & +1 & \infty & \\ 0 & 0 & -i\sqrt{\frac{\varrho}{3}} & 0 & z \\ \frac{1}{2} & 0 & +i\sqrt{\frac{\varrho}{3}} & \frac{3}{2} & \end{pmatrix}.$$

The physically relevant interval is $[0, 1]$. The boundary condition at the origin is that the solution of (4.7.5) has to be holomorphic there, while at infinity the tracer concentration distribution has to tend to zero fast enough for it to be square-integrable over $\mathbb{R}^+$.

We should mention that, had we used the vacancy distribution (4.7.1) instead of (4.7.2), we would have got also a Fuchsian differential equation—however, not with four but with five singularities and thus no longer an equation of the Heun class.

5

THE PAINLEVÉ CLASS OF EQUATIONS

5.1 Painlevé property

5.1.1 Fixed and movable singular points of a nonlinear ODE

Suppose that an nth-order nonlinear ODE

$$w^{(n)}(t) = G(w^{(n-1)}, \ldots, w, t) \tag{5.1.1}$$

is studied, with the function $G(w^{(n-1)}, \ldots, w, t)$ having good analytical properties in its variables, which we specify later. Solutions of this equation as functions of the independent complex variable t may have singularities determined by the analytical behaviour of the function $G(w^{(n-1)}, \ldots, w, t)$ only (without taking into account initial data for the solution). These singularities are called *fixed* singularities.[1] On the other hand, those singularities of solutions $w(t)$ that cannot be predicted by the coefficients of the equation, and that change their position if the initial data for the solution change, are called *movable* singularities. The movable singularity may be a pole of the solution, an essential singularity, or a branch point (either algebraic or transcendental).

Example: (1) The equation

$$w''(t) + w'^2(t) = 0 \tag{5.1.2}$$

has the solution

$$w(t) = \ln(t + c_1) + c_2.$$

Hence eqn (5.1.2) has a movable singularity $t = -c_1$ which is a transcendent branch point.

(2) The equation

$$w''(t)w(t) - 2w'^2(t) = 0$$

has the solution

$$w(t) = \frac{c_2}{(t + c_1)}$$

characterized by a movable singularity $t = -c_1$, being a simple pole. ☐

Linear equations, which may be considered as a special case of nonlinear equations, have no movable singularities. The classification of their fixed singularities is given in Chapter 1.

---

[1] Several authors also call values of w and t related to a fixed singularity *singular values*.

5.1.2 Painlevé property and Painlevé equations

In a series of articles started by P. Painlevé [97, 98, 99] and continued by his student Gambier [47], the following problem was solved. Consider the second-order nonlinear ODE of the form

$$q''(t) = F(t, q, q'), \tag{5.1.3}$$

where $F(t, q, q')$ is a rational function in its arguments.[2] The reader may ask under what conditions solutions of such an equation have no movable *critical points* (i.e. depending on the initial data). By critical points are meant branch points and essential singularities. In this case, movable singularities may be only poles of a solution. The absence of movable critical points is known as the *Painlevé property*. In the above-mentioned publications, all equations[3] of the form (5.1.1) possessing the Painlevé property have been found. Among these are many equations solvable in terms of elementary or other known functions (e.g. elliptic functions). But several equations could not be simplified to known equations. Nowadays they usually are called *Painlevé equations* and denoted by P^{VI}, P^{V}, P^{IV}, P^{III}, P^{II}, P^{I}. Solutions of these equations are called Painlevé transcendents. Although the method of investigation proposed by Painlevé is fairly simple, the practical calculations involve considering many special cases, and therefore are rather laborious. Since the original studies of the problem have to our mind more historical interest than practical meaning, we give here a list of the Painlevé equations and in addition a draft of the proof that at least one of them possesses the Painlevé property. Standard Painlevé-class equations are the following [53].

$$P^{VI} : q'' - \frac{1}{2}\left(\frac{1}{q} + \frac{1}{q-1} + \frac{1}{q-t}\right)q'^2$$

$$+ \left(\frac{1}{t} + \frac{1}{t-1} + \frac{1}{q-t}\right)q' - \frac{q(q-1)(q-t)}{t^2(t-1)^2}$$

$$\times \left(\alpha + \frac{\beta t}{q^2} + \frac{\gamma(t-1)}{(q-1)^2} + \frac{\delta t(t-1)}{(q-t)^2}\right) = 0, \tag{5.1.4}$$

$$P^{V} : q'' - \left(\frac{1}{2q} + \frac{1}{q-1}\right)q'^2 + \frac{1}{t}q'$$

$$- \frac{(q-1)^2}{t^2}\left(\alpha q + \frac{\beta}{q}\right) - \frac{\gamma q}{t} - \frac{\delta q(q+1)}{q-1} = 0, \tag{5.1.5}$$

$$P^{IV} : q'' - \frac{1}{2q}q'^2 - \frac{3}{2}q^3 - 4tq^2 - 2(t^2 - \alpha)q - \frac{\beta}{q} = 0, \tag{5.1.6}$$

---

[2] More often F is restricted to being a second-order polynomial in q'.

[3] Namely 50 of them (see [54]).

$$P^{III}: \quad q'' - \frac{1}{q}q'^2 + \frac{1}{t}q' - \frac{1}{t}(\alpha q^2 + \beta) - \gamma q^3 - \frac{\delta}{q} = 0, \qquad (5.1.7)$$

$$P^{II}: \quad q'' - q^3 - tq - \alpha = 0, \qquad (5.1.8)$$

$$P^{I}: \quad q'' - 6q^2 - t = 0. \qquad (5.1.9)$$

Above, the conventional notation for Painlevé equations was used. However, for our purposes, we need some additional remarks. First, we perform the transformation from eqn (5.1.7) P^{III} to $\tilde{P}^{III}$

$$\tilde{P}^{III}: \quad q'' - \frac{1}{q}q'^2 + \frac{1}{t}q' - \frac{1}{t^2}(\alpha q^2 + \gamma q^3) - \frac{\beta}{t} - \frac{\delta}{q} = 0 \qquad (5.1.10)$$

with the help of the substitution

$$t \mapsto \sqrt{t}, \qquad q \mapsto \frac{q}{\sqrt{t}}.$$

To our mind $\tilde{P}^{(3)}$ is much better fitted to the confluence process. It possesses, of course, the Painlevé property.

Applying a scaling transformation

$$t \mapsto c_1 t, \qquad q \mapsto c_2 q,$$

with arbitrary complex c_1 and c_2, it is possible to reduce the number of free parameters in eqn (5.1.5) to three and in eqn (5.1.7) (as well as in eqn (5.1.10)) to two.[4]

One of the main advantages of Painlevé equations in comparison with other non-linear equations is that, beyond local solutions, global solutions of these equations can be constructed with prescribed asymptotic behaviour at certain points of the independent variable in the complex plane. This fact is due to the isomonodromic deformation theory, which recently has strongly influenced the theory of special functions [57, 58, 59, 44, 55, 64].

On the other hand, the importance of Painlevé equations for physics lies in the *ARS conjecture* (Ablovitz, Ramani, Segur) [1]. This conjecture has not been proved rigorously yet. However, all known facts in the theory of nonlinear partial differential equations (PDEs) convince us of its correctness.

By a completely integrable nonlinear PDE is meant an equation the solutions of which can be constructed by the inverse-scattering-problem method. The ARS conjecture states that any one-dimensional reduction of such a PDE is a Painlevé equation [1].

Example: Consider the Korteveg–deVries PDE

$$u_t = u_{xxx} - 6uu_x.$$

---

[4]Unless some of the initial coefficients take on the value zero ($\delta = 0$ in the case of P^V, and $\delta = 0$ and $\gamma = 0$ in the case of P^{III}).

Seeking a solution in the form $u(x, t) = f(s)$, where $s = x - ct$ and c is the velocity of propagation, we get an ODE

$$f''' - 6ff' - cf' = 0,$$

or, after integration,

$$f'' - 3f^2 - cf + k = 0.$$

The latter equation coincides with P^I under a scaling transformation of its variables. □

Another remarkable feature of the integrable PDEs is that their solutions may have only poles as singularities on manifolds [129]. This is an extension of the Painlevè property from ODEs.

5.1.3   Proof that movable singularities are poles in the case of P^{II}

Here we sketch the proof that all movable singularities of Painlevé transcendents are poles. This proof has been proposed by N. Joshi and M. Kruskal [60]. Only Painlevé equation P^{II} is considered. All other equations may be studied in an analogous way. Suppose that we have a movable singularity t_0 in the complex t plane. It can always be included in a circle which has t_0 as its center and does not include other singularities. If t_0 is a branch point, then the circle has a straight line cut towards t_0. Suppose that the initial data are posed at the point t_1 which lies on the edge of the circle. If any other point t_2 is taken as an initial point, then the problem reduces to recalculating the initial data or an analytic continuation of the solution $q(t)$ along an appropriate contour on a Riemann surface from t_2 to t_1. The method is proposed in [60] and is based on *integrating the dominant terms* and succeeding iterations of the corresponding integral equation.

The Lipschitz condition (see [32]) fails if $q \to \infty$ in (5.1.8). This is the so-called *singular value* of the solution. Hence the dominant terms in P^{II}, according to (5.1.8), are

$$q'' - q^3.$$

In order to integrate this, terms the integrating factor q' is needed. As a result, we get the integral equation

$$\frac{q'^2}{2} - \frac{q^4}{4} - t\frac{q^2}{2} + \int_{t_1}^t q^2 \, dt - \alpha t + c_1 = 0. \qquad (5.1.11)$$

Here the constant c_1 is fixed by initial conditions and the term integrated by parts. Dividing eqn (5.1.11) by q^4, taking the square root, and integrating yields

$$\frac{1}{q} = \pm \int_{t_1}^t \left(\frac{1}{2} + \frac{t}{q^2} + \frac{2(\alpha t - c_1)}{q^4} - \frac{2}{q^4} \int_{t_1}^t q^2 \, dt \right)^{1/2} dt + c_2 \qquad (5.1.12)$$

as an integral equation related to (5.1.11). If the lower integration limit in (5.1.12) is substituted for t_0 and c_2 is taken to be zero, then we get the solution that satisfies the assumption $q^{-1}|_{t \to t_0} \to 0$.

$$\frac{1}{q} = \pm \int_{t_0}^{t} \left(\frac{1}{2} + \frac{t}{q^2} + \frac{2(\alpha t - c_1)}{q^4} - \frac{2}{q^4} \int_{t_1}^{t} q^2 dt \right)^{1/2} dt. \qquad (5.1.13)$$

Succeeding iterations of eqn (5.1.13) prove that the function $u(t) = 1/q(t)$ is a regular function in a neighbourhood of the point $t = t_0$. Namely

$$u(t) = \sum_{k=1}^{\infty} g_k (t - t_0)^k, \qquad g_1 \neq 0. \qquad (5.1.14)$$

This implies that the point t_0 is a simple pole of the Painlevé transcendent $q(t)$. Similar study of other Painlevé equations shows that Painlevé transcendents either have simple poles or second-order poles as movable singularities.

5.2 The Hamiltonian structure

5.2.1 Heun-class equations and Painlevé equations

Since the publication of Malmqvist [89], it is known that Painlevé equations exhibit Hamiltonian structure, i.e. they can be considered as equations of motion (see also [95, 53]). In 1996, one of the authors of the present book discovered [120] that Painlevé equations are in fact equations of classical motion (Euler–Lagrange equations) for quantum systems described by different types of Heun equations. In this chapter we shall take this fact as a basis for the theory of Painlevé equations.

First, we study the general approach to the Hamiltonian structure for Painlevé equations. Each equation belonging to the Heun class in its canonical form may be presented in a form

$$\frac{1}{f(t)} \left[P_0(z, t) D^2 + P_1(z, t) D + P_2(z, t) \right] y(z) = \lambda y(z). \qquad (5.2.1)$$

In eqn (5.2.1) $P_0(z, t)$, $P_1(z, t)$, $P_2(z, t)$ are polynomials in z of order not higher than three, t is the scaling parameter, and λ is the accessory parameter. If quantum observables $\hat{q}$ and $\hat{p}$, ($\hat{q}$ the coordinate and $\hat{p}$ the momentum) are associated with z and D in eqn (5.2.1), then it gets the Hamiltonian structure and can be rewritten as

$$H(\hat{q}, \hat{p}, t) y = \lambda y. \qquad (5.2.2)$$

In eqn (5.2.2) the function H is the Hamiltonian adiabatically depending on the parameter t, which can be considered as time, and λ is energy. The corresponding Hamiltonian in classical mechanics is quadratic in the classical momentum p:

$$H(q, p, t) = \frac{1}{f(t)} \left[P_0(q, t) p^2 + P_1(q, t) p + P_2(q, t) \right]. \qquad (5.2.3)$$

The Legendre transform can be applied to this Hamiltonian, turning from momentum p to velocity q_t. The corresponding Lagrangian $\mathcal{L}(q, q_t, t)$ reads

$$\mathcal{L}(q, q_t, t) = \frac{f(t)}{4P_0(q, t)}\left(q_t - \frac{P_1(q, t)}{f(t)}\right)^2 - \frac{P_2(q, t)}{f(t)} \tag{5.2.4}$$

taking velocity q_t instead of momentum P as the independent variable. The following Euler–Lagrange equation of motion relates to this Lagrangian:

$$q_{tt} = \frac{1}{2}\frac{\partial}{\partial q}(\ln P_0(q, t))q_t^2 - \left(\frac{\partial}{\partial t}(\ln f(t)) - \frac{\partial}{\partial t}(\ln P_0(q, t))\right)q_t$$
$$+ \frac{P_0(q, t)}{f^2(t)}\left(\frac{\partial}{\partial q}\frac{P_1^2(q, t)}{2P_0(q, t)} + f(t)\frac{\partial}{\partial t}\frac{P_1(q, t)}{P_0(q, t)} - 2\frac{\partial P_2(q, t)}{\partial q}\right). \tag{5.2.5}$$

As it will be shown in the next section, each type of equation belonging to the Heun class is generic for eqn (5.2.5), which coincides with one of the Painlevé equations.

Specially chosen solutions of Painlevé equations constitute the class of special functions related to nonlinear mathematical physics—the so-called Painlevé transcendents. In this sense, they are extensions of special functions of mathematical physics to nonlinear phenomena.

5.2.2 Alternative classification of Painlevé equations

In this subsection the following basic theorem will be proved.

Theorem 5.1 *Each type of equation belonging to the Heun-class is generic for an equation of Painlevé class in the sense that to a Schrödinger-type equation corresponds an equation of motion in classical dynamics. The converse statement holds also: Each type of Painlevé equation may be generated by a corresponding Heun equation.* □

In order to prove this theorem, we need to obtain eqns (5.2.5) for particular cases of Heun-class equations and compare this set with conventional Painlevé equations (5.1.4–5.1.10). For this goal, we will also call particular equations of the form (5.2.5) Painlevé equations but denote them differently from the conventional ones.

The basic classification of types of the Heun-class equations based on confluence has been presented in Chapter 3. If, as a first step, we omit reduced confluent equations, then five types of equation can be distinguished:

(1) the seed Heun equation (HE)—a Fuchsian equation with four regular singularities,

(2) the confluent Heun equation (CHE)—an equation with two regular singularities and one irregular singularity at infinity,

(3) the biconfluent Heun equation (BHE)—an equation with one regular singularity and one irregular singularity at infinity arising as a result of confluence of three regular singularities,

(4) the doubly confluent Heun equation (DHE)—an equation with two regular singularities: one at zero and another at infinity,

(5) the triconfluent Heun equation (THE)—an equation with one irregular singularity at infinity arising as a result of confluence of four regular singularities.

The Hamiltonians corresponding to these types of equations are the following.

HE:

$$H^{(1,1,1;1)}(q, p, t) = \frac{-1}{t(t-1)}\Big[q(q-1)(q-t)p^2$$
$$+ (c(q-1)(q-t) + dq(q-t)$$
$$+ (a+b+1-c-d)q(q-1)p + abq\Big]. \qquad (5.2.6)$$

CHE:

$$H^{(1,1;2)}(q, p, t) = \frac{-1}{t}\Big[q(q-1)p^2$$
$$+ (-tq(q-1) + c(q-1) + dq)\,p - atq\Big]. \qquad (5.2.7)$$

BHE:

$$H^{(1;3)}(q, p, t) = -(qp^2 + (-q(q+t) + c)p - aq). \qquad (5.2.8)$$

DHE:

$$H^{(2;2)}(q, p, t) = \frac{-1}{t}(q^2 p^2 + (-q^2 - t + cq)p - aq). \qquad (5.2.9)$$

THE:

$$H^{(;4)}(q, p, t) = -(p^2 + (-q^2 - t)p - aq). \qquad (5.2.10)$$

Turning to the classical dynamics according to eqns (5.2.3–5.2.5), we obtain five equations which, in the terminology taken for the equations belonging to the Heun-class, can be denoted as Painlevé equation (PE), confluent Painlevé equation (CPE), biconfluent Painlevé equation (BPE), doubly confluent Painlevé equation (DPE), and triconfluent Painlevé equation (TPE). If the same notations for the parameters are used, then the five types of Painlevé equation read as follows.

PE:

$$P^{(1,1,1;1)} : q_{tt} = \frac{1}{2}\left(\frac{1}{q} + \frac{1}{q-1} - \frac{1}{q-t}\right)q_t^2$$
$$- \left(\frac{1}{t} + \frac{1}{t-1} + \frac{1}{q-t}\right)q_t + \frac{q(q-1)(q-t)}{t^2(t-1)^2}$$
$$\times \left(\frac{(a+b+1)^2}{2} - 2ab - \frac{c^2 t}{2q^2}\right.$$
$$+ \frac{d^2(t-1)}{2(q-1)^2} + \left.\frac{(1-(c+d-a-b)^2)t(t-1)}{(q-t)^2}\right). \qquad (5.2.11)$$

CPE:

$$P^{(1,1;2)} : q_{tt} = \frac{1}{2}\left(\frac{1}{q} + \frac{1}{q-1}\right)q_t^2 - \frac{1}{t}q_t$$
$$+ \frac{1}{t^2}\left(\frac{c^2(q-1)}{2q} - \frac{d^2q}{2(q-1)}\right)$$
$$+ \frac{q(q-1)(2q-1)}{2} - \frac{q(q-1)}{t}(1+c+d-2a). \quad (5.2.12)$$

BPE:

$$P^{(1;3)} : q_{tt} = \frac{q_t^2}{2q} + \frac{3}{2}q^3 + 2tq^2 + (\frac{t^2}{2} + 2a - c)q - \frac{c^2}{2q}. \quad (5.2.13)$$

DPE:

$$P^{(2;2)} : q_{tt} = \frac{q_t^2}{q} - \frac{q_t}{t} + \frac{1}{q} + \frac{q^2(2a-c)+q^3}{t^2} + \frac{1-c}{t}. \quad (5.2.14)$$

TPE:

$$P^{(;4)} : \quad q_{tt} = 2q^3 + 2tq + 2a. \quad (5.2.15)$$

The following list of the Hamiltonians includes those which relate to reduced confluent Heun equations with half-integer s-ranks of the irregular singularities (five particular equations).

RCHE:

$$H^{(1,1;3/2)}(q, p, t, c, d) = \frac{-1}{t}\left[q(q-1)p^2 + (c(q-1)+dq)\,p - tq\right]. \quad (5.2.16)$$

RBHE:

$$H^{(1;5/2)}(q, p, t, c) = -(qp^2 + cp - tq). \quad (5.2.17)$$

RDHE:

$$H^{(3/2;2)}(q, p, t, a) = \frac{-1}{t}(q^2p^2 - q^2p - aq - \frac{t}{q}). \quad (5.2.18)$$

DRDHE:

$$H^{(3/2;3/2)}(q, p, t,) = \frac{-1}{t}(q^2p^2 - q - \frac{t}{q}). \quad (5.2.19)$$

RTHE:

$$H^{(;7/2)}(q, p, t,) = -(p^2 + (-q^3 - tq)p). \quad (5.2.20)$$

The following Painlevé equations correspond to these Hamiltonians. The first equation might be called the reduced confluent Painlevé equation—RCPE:

$$P^{(1,1;3/2)} : q_{tt} = \frac{1}{2}\left(\frac{1}{q} + \frac{1}{q-1}\right)q_t^2 - \frac{1}{t}q_t$$

$$+ \frac{1}{t^2}\left(\frac{c^2(q-1)}{2q} - \frac{d^2 q}{2(q-1)}\right) + \frac{2q(q-1)}{t}. \qquad (5.2.21)$$

Then comes the reduced biconfluent Painlevé equation—RBPE:

$$P^{(1;5/2)} : q_{tt} = \frac{q_t^2}{2q} - \frac{c^2}{2q} + 4q^2 - 2tq. \qquad (5.2.22)$$

Next comes the reduced doubly confluent Painlevé equation—RDPE:

$$P^{(3/2;2)} : q_{tt} = \frac{q_t^2}{q} - \frac{q_t}{t} + \frac{2aq^2 + q^3}{t^2} - \frac{2}{t}. \qquad (5.2.23)$$

The next equation might be called doubly reduced doublly confluent Painlevé equation—DRDPE:

$$P^{(3/2;3/2)} : q_{tt} = \frac{q_t^2}{q} - \frac{q_t}{t} + \frac{2q^2}{t^2} - \frac{2}{t}. \qquad (5.2.24)$$

Finally comes the reduced triconfluent Painlevé equation—RTPE:

$$P^{(;7/2)} : q_{tt} = 6q^2 + t. \qquad (5.2.25)$$

Now, as the new list of Painlevé equations has been obtained in order to prove Theorem 5.1, this list should be compared with the conventional Painlevé equations (5.1.4–5.1.10).

Equation $P^{(1,1,1;1)}$ fully coincides with P^{VI} if we admit the following relations between the coefficients:

$$\alpha = \frac{(a+b+1)^2}{2} - 2ab, \qquad \beta = -\frac{c^2}{2},$$

$$\gamma = \frac{d^2}{2}, \qquad \delta = 1 - (a+b-c-d)^2.$$

With Painlevé equation P^V, it is more difficult. First, with the help of the substitution

$$y := \frac{y}{y-1},$$

interchanging the singular values $y = 1$ and $y = \infty$ with one another, we turn from eqn (5.1.5) to a new equation $\tilde{P}^{VI}$:

$$\tilde{P}^V : y_{tt} = \frac{1}{2}\left(\frac{1}{y} + \frac{1}{y-1}\right)y_t^2 - \frac{1}{t}q_t$$

$$- \frac{1}{t^2}\left(\frac{\beta(q-1)}{2q} - \frac{\alpha q}{2(q-1)}\right)$$

$$- \delta q(q-1)(2q-1) - \gamma\frac{q(q-1)}{t}. \qquad (5.2.26)$$

Suppose that δ is not equal to zero. The scaling transformation $t := \kappa t$ leads to a new equation with

$$\alpha \mapsto \alpha, \qquad \beta \mapsto \beta, \qquad \gamma \mapsto \kappa\gamma, \qquad \delta \mapsto \kappa^2\delta.$$

The meaning of this is that, by means of a scaling transformation, the parameter δ (5.2.26) can be settled as unity. In this case, $\tilde{P}^V$ coincides with $P^{(1,1;2)}$ if we take

$$\alpha = \frac{d^2}{2}, \qquad \beta = -\frac{c^2}{2}, \qquad \gamma = 1 + c + d - 2a.$$

On the other hand, if $\delta = 0$, then we can take $\gamma = -2$. In this case $\tilde{P}^V$ coincides with $P^{(1,1;3/2)}$.

Now comes P^{IV}. Under the scaling transformation $q \mapsto \sqrt{2}q$ and $t \mapsto t/\sqrt{2}$, we obtain a new equation from P^{IV}:

$$q_{tt} = \frac{q_t^2}{2q} + \frac{3}{2}q^3 + 2tq^2 + \left(\frac{t^2}{2} - \alpha\right)q + \frac{\beta}{4q}, \qquad (5.2.27)$$

which coincides with $P^{(1;3)}$ by means of $\alpha = c - 2a$, $\beta = -2c^2$. However, what is the relation to $P^{(1;5/2)}$? The answer is given by the following lemma.

Lemma 5.1 *Equation $P^{(1;5/2)}$ can be transformed by means of the contact transform to P^{II}.* $\square$

Proof: Consider the auxiliary linear system

$$y_t = c - 2yz, \qquad z_t = z^2 - 2y + t. \qquad (5.2.28)$$

If the function $z(x)$ is excluded from eqn (5.2.28), then the function $y(x)$ satisfies eqn (5.2.22). On the other hand, the function $z(x)$ satisfies P^{II}, i.e. eqn (5.2.15). $\square$

Considering the Painlevé equation P^{III}, we start already from eqn (5.1.10). Simultaneous scaling transformations $q := hq$ and $t := \kappa t$ lead to the equation

$$q_{tt} = \frac{q_t^2}{q} - \frac{q_t}{t} + \frac{1}{t^2}(h\alpha q^2 + h^2\gamma q^3) + \frac{\kappa\beta}{ht} + \frac{\kappa^2\delta}{h^2q}. \qquad (5.2.29)$$

If $\gamma \neq 0$ and $\delta \neq 0$, by choosing $h^2 = \gamma^{-1}$ and $\kappa = h\delta^{-1/2}$ and by setting $h\alpha = 2a - c$ and $\kappa\beta/h = 1 - c$, we arrive at $P^{(2;2)}$. Equation (5.2.29) is also valid in the case $\delta = 0$. However, under the conditions $h\alpha = 2a$ and $\kappa\beta/h = -2$, it coincides with $P^{(3/2;2)}$. Finally, it is possible to put $\delta = 0$, $\gamma = 0$, $h\alpha = 2$, and $\kappa\beta/h = -2$ in eqn (5.2.29) getting $P^{(3/2;3/2)}$.

In the case of P^{II}, by means of the scaling transformation

$$t := 2^{1/3}t, \qquad q := 2^{-1/3}q,$$

we obtain the Painlevé equation $P^{(;4)}$ from eqn (5.1.8).

Equation $P^{(;7/2)}$ coincides with P^I.

Since we have found a one-to-one correspondence between the two lists of Painlevé equations, the proof of Theorem 5.1 is completed.

5.2.3 Linearization of Painlevé equations

Consider possible linearizations of Painlevé equations. For this object, we additionally assume that $c = 0$. By linearization is meant taking the linear part of the function $F(q, q_t, t)$ in variables q and q_t at the origin:

$$F(q, q_t, t)|_{c=0} \sim \left.\frac{\partial F(q, q_t, t)|_{c=0}}{\partial q_t}\right|_{q=0,q_t=0} q_t + \left.\frac{\partial F(q, q_t, t)|_{c=0}}{\partial q}\right|_{q=0,q_t=0} q.$$
(5.2.30)

Linearization of the seed Painlevé equation $P^{(1,1,1;1)}$ leads to an equation

$$q_{tt} = -\frac{q_t}{t-1} + \left(\frac{(a+b+1)^2 - 4ab}{2(t-1)^2} - \frac{1-(d-a-b)^2}{t^2}\right.$$
$$\left. + \frac{-(a+b+1)^2 + 4ab + d^2 + 2(1-(d-a-b)^2)}{2t(t-1)}\right)q,$$
(5.2.31)

equivalent to the hypergeometric equation. Linearization of the equation $P^{(1,1;2)}$ leads to the equation

$$q_{tt} = -\frac{q_t}{t} + \left(\frac{1}{2} + \frac{1+d-2a}{t} + \frac{d^2}{2t^2}\right)q,$$
(5.2.32)

equivalent to the confluent hypergeometric equation. Linearization of the equation $P^{(1,1;3/2)}$ leads to the equation

$$q_{tt} = -\frac{q_t}{t} + \left(\frac{d^2}{2t^2} + \frac{2}{t}\right)q,$$
(5.2.33)

equivalent to the reduced confluent hypergeometric equation. Linearization of the equation $P^{(1;3)}$ leads to the equation

$$q_{tt} = \left(\frac{t^2}{2} + 2a\right)q,$$
(5.2.34)

equivalent to the biconfluent hypergeometric equation. And the linearization of the equations $P^{(;4)}$ and $P^{(1;5/2)}$ leads to the equation equivalent to the reduced biconfluent hypergeometric equation:

$$q_{tt} = \pm 2tq. \qquad (5.2.35)$$

The Painlevé equations $P^{(2;2)}$, $P^{(3/2;2)}$, $P^{(3/2;3/2)}$, $P^{(;7/2)}$ cannot be linearized in the proposed sense.

The studied relations between the Heun, Painlevé, and hypergeometric classes of equations can be expressed by structure diagrams of the form

$$\{1, 1, 1, 1\} \longrightarrow \{1, 1, 1\}$$
$$\searrow \qquad \nearrow$$
$$P^{(6)}$$

where the horizontal arrow relates to appearance of an additional singularity, the arrow directed down relates to the turning from the Schrödinger equation to the Euler–Lagrange equation and the arrow directed up relates to linearization.

More details about these relations are shown in Figs 5.1 and 5.2.

5.3 Monodromy preserving deformations

Almost from the very beginning of the study of Painlevé equations, they were related not only to the Painlevé property of the solutions but to some linear equations with

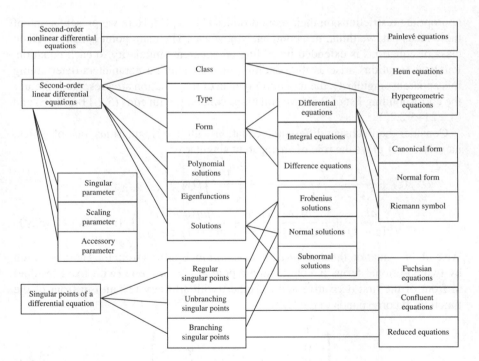

FIG. 5.1. Relations of properties of linear and nonlinear differential equations and their singularities.

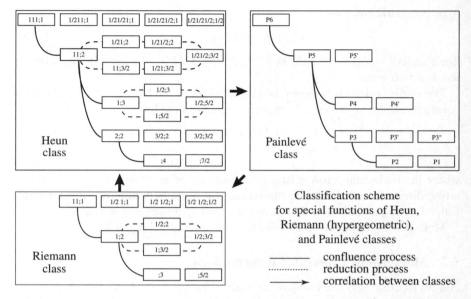

FIG. 5.2. Relations between differential equations of the Heun, Painlevé, and Riemann classes.

appropriate restrictions on their monodromies [112], [46]. Here we present some of these ideas but, we think, in a completely new way. The main point is that the Heun class of equations is extended by adding an apparent singularity to each equation. Painlevé equations arise as the condition that, by altering parameters determining this apparent singularity, the monodromy data of the equation are preserved. Clearly the corresponding list of Painlevé equations coincides with eqns (5.2.11–5.2.13) and (5.2.21–5.2.25).

Consider the Fuchsian differential equation with five singularities, one of which, namely $z = q$ plays the role of an apparent singularity:

$$D^2 y(z) + \left(\frac{c}{z} + \frac{d}{z-1} + \frac{e}{z-t} - \frac{1}{z-q} \right) Dy(z)$$

$$+ \left(\frac{ab}{z(z-1)} + \frac{ht(t-1)}{z(z-1)(z-t)} + \frac{pq(q-1)}{z(z-1)(z-q)} \right) y(z) = 0. \qquad (5.3.1)$$

We call this equation the *deformed Heun equation*. It differs from the Heun equation by two additional terms having a simple pole at $z = q$—one (with fixed residue) in front of the first derivative and the other (with arbitrary residue) in front of the function. It corresponds to the GRS

$$\begin{pmatrix} 1 & 1 & 1 & 1 & 1 \\ 0 & 1 & t & q & \infty \\ 0 & 0 & 0 & 0 & a \\ 1-c & 1-d & 1-e & 2 & b \end{pmatrix}. \qquad (5.3.2)$$

The following Fuchs relation holds:

$$c + d + e = a + b + 3. \tag{5.3.3}$$

The somewhat peculiar introduction of the parameters p and h is justified by the simplicity of the relations

$$h = \text{Res}_{z=t}\frac{ht(t-1)}{z(z-1)(z-t)},$$

$$p = \text{Res}_{z=q}\frac{pq(q-1)}{z(z-1)(z-q)}, \tag{5.3.4}$$

so that h and p are residues of two of the terms in eqn (5.3.1) at the singularities $z = t$ and $z = q$, respectively. The analyticity property at the point $z = q$ holds if eqn (5.3.1) has a local solution in the form of a Taylor expansion

$$y(z) = \sum_{k=0}^{\infty} g_k(z-q)^k, \qquad g_0 = 1, \tag{5.3.5}$$

with arbitrary g_2. For simplicity, we denote the coefficients of eqn (5.3.1) by $P(z)$, $Q(z)$, $f(t)$, and $\rho(z)$:

$$P(z) = \frac{c}{z} + \frac{d}{z-1} + \frac{e}{z-t}, \qquad Q(z) = \frac{ab}{z(z-1)},$$

$$f(t) = t(t-1), \quad \rho(z) = z(z-1)(z-t),$$

so that it reads

$$D^2 y(z) + \left(P(z) - \frac{1}{z-q}\right)Dy(z)$$

$$+ \left(Q(z) + \frac{hf(t)}{\rho(z)} + \frac{pq(q-1)}{z(z-1)(z-q)}\right)y(z) = 0. \tag{5.3.6}$$

Substituting expansion (5.3.5) into eqn (5.3.6), we arrive to recurrence relations for the coefficients g_1 and g_2 of this expansion

$$p_{-1}g_1 + q_{-1} = 0,$$

$$2(1 - p_{-1})g_2 + (p_0 + q_{-1})g_1 + q_0 = 0,$$

$$\dots . \tag{5.3.7}$$

Here p_k and q_k ($k = -1, 0$) are coefficients of the Laurent expansions of the functions $P(z)$ and $Q(z)$ at the point $z = q$ respectively:

$$P(z) = \sum_{k=-1}^{\infty} p_k(z-q)^k, \qquad Q(z) = \sum_{k=-1}^{\infty} q_k(z-q)^k.$$

Their values are directly calculated from eqn (5.3.1):

$$p_{-1} = -1, \qquad q_{-1} = p,$$

$$p_0 = \frac{c}{q} + \frac{d}{q-1} + \frac{e}{q-t},$$

$$q_0 = \frac{ab}{q(q-1)} - p\left(\frac{1}{q} + \frac{1}{q-1}\right) + \frac{ht(t-1)}{q(q-1)(q-t)}. \tag{5.3.8}$$

To clarify the calculations, we rewrite formulae (5.3.8) in terms of eqn (5.3.6):

$$\begin{aligned} p_1 &= -1, \qquad q_{-1} = p, \\ p_0 &= P(q), \\ q_0 &= Q(q) - p\left(\frac{1}{q} + \frac{1}{q-1}\right) + \frac{hf(t)}{\rho(q)}. \end{aligned} \tag{5.3.9}$$

In these terms, eqns (5.3.7) take the form

$$g_1 = p,$$

$$0 \cdot g_2 + (P(q) + p)p + Q(q) - p\left(\frac{1}{q} + \frac{1}{q-1}\right) + \frac{hf(t)}{\rho(q)} = 0,$$

$$\dots \tag{5.3.10}$$

Since the multiplier in front of g_2, due to the value of p_{-1}, is equal to zero, we arrive at a necessary condition for the point $z = q$ to be an apparent singularity. It leads to the following relation between parameters h, p, q, t:

$$h(q, p, t) = -\frac{\rho(q)}{f(t)}\left[p^2 + \left(P(q) - \frac{1}{q} - \frac{1}{q-1}\right)p + Q(q)\right] \tag{5.3.11}$$

or more precisely

$$\begin{aligned} h(q, p, t) = \frac{1}{t(t-1)}[&q(q-1)(q-t)p^2 + ((c-1)(q-1)(q-t) \\ &+ (d-1)q(q-t) + eq(q-1))p + ab(q-t)]. \end{aligned} \tag{5.3.12}$$

It is evident that the function $h(q, p, t)$ coincides with the Hamiltonian related to the Heun equation (see eqn (5.2.6)) up to a shift in the parameters c and d by unity:

$$h(q, p, t, a, b, c, d, e) = H^{(1,1,1;1)}(q, p, t, a, b, c-1, d-1, e).$$

The explanation is that parameters a, b, c, d, e satisfy different Fuchs conditions in case of the Heun equation and in case of the deformed Heun equation.

Suppose that $h(p, q, t)$ is a Hamiltonian for a certain one-dimensional classical dynamical system. The momentum of the particle is denoted by p and the coordinate by q, and t is the time parameter.[5] The classical movement of the particle is determined by the Hamilton equations

$$\frac{dp}{dt} = -\frac{\partial h}{\partial q}, \qquad \frac{dq}{dt} = \frac{\partial h}{\partial p}. \tag{5.3.13}$$

On the other hand, eqns (5.3.11–5.3.13) can be considered as a *preserving-deformation condition*.[6] Clearly, the residue at the point $z = t$ as well as, of course, residues at the points $z = 0$ and $z = 1$ do not depend on the behavior of the functions $p(t)$ and $q(t)$ under conditions (5.3.13).

Equations (5.3.13) are equivalent—under the Legendre transformation from momentum p and coordinate q to velocity $\dot{q}$ and coordinate q—to the Newtonian equation of motion, which turns out to be the Painlevé equation P^{VI}, as proved in the previous section.

The appearance of the shift of the parameters and the somewhat peculiar expression for the deformation term (5.3.4) is due to the Fuchsian structure of Heun's equation.

In the general situation of an arbitrary equation belonging to the Heun class, beyond the Heun equation itself, we write this equation in the form

$$D^2 y(z) + P(z)Dy(z) + \left(Q(z) + \frac{f(t)h}{\rho(z)} \right) y(z) = 0. \tag{5.3.14}$$

The *deformed equation*, with the additional apparent regular singularity at the point $z = q$, reads

$$D^2 y(z) + \left(P(z) - \frac{1}{z-q} \right) Dy(z) + \left(Q(z) + \frac{f(t)h}{\rho(z)} + \frac{p}{z-q} \right) y(z) = 0. \tag{5.3.15}$$

The function $f(t)$ is traced along with a confluence process. Its choice for different equations has already been shown in the previous section.

The condition that eqn (5.3.15) has a local solution of the form (5.3.5) leads to equations for the first two coefficients:

$$-g_1 + p = 0,$$

$$(P(q) + p)p + \left(Q(q) + \frac{f(t)h}{\rho(q)} \right) = 0. \tag{5.3.16}$$

The first of these equations can be interpreted as if the differentiation operator acts like multiplication by classical momentum p. The second of these equations can be interpreted as a substitution of an independent variable z for the classical coordinate

---

[5]In many practical applications, the parameter t has another physical meaning. For instance, in the two-Coulomb-centres problem t is the distance between the centres.

[6]They also are called *isomonodromic conditions*.

q in the Heun-class equation. As a compatibility condition, we come to the following relation between h, p, q, t:

$$h(p, q, t) = -\frac{1}{t}\rho(q)(p^2 + P(q)p + Q(q)).\qquad(5.3.17)$$

This is once again the classical Hamiltonian related to the corresponding Heun-class equation. The preserving deformation conditions have the form of the Hamilton system (5.3.13). In every case, this system is equivalent to the corresponding Newtonian second-order equation which turned out to be one of the Painlevé equations and which already has been dealt with in the previous section. As a result, we get the following theorem:

Theorem 5.2 *Every deformed Heun-class equation with an added apparen singularity is related via a preserving deformation condition to a Painlevé equation. Every Painlevé equation can be considered as a preserving deformation condition for a deformed equation of Heun class.* □

The proposed presentation is inspired by the paper of R. Fuchs [46]. An alternative possibility is to use not a second-order linear ODE but a system of first-order equations. These systems in relation to Painlevé equations were proposed by Schlesinger [112]. In the following, we give a draft of the approach to the Painlevé equations based on the Schlesinger system.

If we take, for instance, P^{VI}, the corresponding Schlesinger system reads

$$\vec{\Psi}_z = A(z, t)\vec{\Psi}\qquad(5.3.18)$$

where $\vec{\Psi}$ is a two dimensional vector and $A(z, t)$ is a matrix having the structure

$$A(z, t) = \frac{A^0}{z} + \frac{A^1}{z-1} + \frac{A^t}{z-t},$$

reflecting the structure of singularities of eqn (5.3.19). Here A^0, A^1, A^t are matrices with constant coefficients. Eigenvalues of these matrices are characteristic exponents at the corresponding singularity (one of them being zero). A problem of explicitly calculating them, however, arises from the parametrization of the matrix elements of the matrices A^0, A^1, A^t in terms of the parameters a, b, c, d, e, t, which is rather cumbersome. It is also supposed that, at infinity, the matrix $A(z, t)$ simplifies to the diagonal matrix A^∞:

$$\lim A(z, t)|_{z\to\infty} = -\frac{A^\infty}{z},$$

$$A^\infty = -(A^0 + A^1 + A^t), \qquad A^\infty = \begin{pmatrix} \kappa_1 & 0 \\ 0 & \kappa_2 \end{pmatrix}.$$

Along with the system (5.3.16), another system is studied:

$$\vec{\Psi}_t = B\vec{\Psi}, \qquad B := \frac{A^t}{t-z}\vec{\Psi}.\qquad(5.3.19)$$

Certain conditions for Ψ should be posed. Namely, as $z \to \infty$, we have

$$\Psi(z, t) \to \begin{pmatrix} z^{\kappa_1} & 0 \\ 0 & z^{\kappa_2} \end{pmatrix}.$$

Then the monodromy-preserving deformation (isomonodromic) condition reads

$$A_t - B_z + [A, B] = 0. \tag{5.3.20}$$

This equation coincides with P^{VI}. The apparent singularity for the second-order equation is 'hidden' in the system (5.3.18). It corresponds to a zero of a nondiagonal matrix element of $A(z, t)$. The matrices A and B are called the *Lax pair* for the corresponding Painlevé equation. In the same way, all other Painlevé equations can be treated. It was not the intention of the authors of the book to derive the asymptotic formulae for Painlevé transcendents as a result of the preserving-deformation condition. The reader can refer to the book [55] and more recent publications.

APPENDIX A: THE GAMMA FUNCTION AND RELATED FUNCTIONS

A.1 The gamma function

In this appendix, we briefly study major properties of the gamma function and introduce some other functions which can be expressed in terms of the gamma function, namely the beta function and the Pochhammer symbol.

The following difference equation may be taken as generic for the gamma function

$$y(z + 1) - zy(z) = 0. \tag{A.1.1}$$

It is studied in the complex plane $\mathbb{C}$. The difference equation for the function $y(z)$ can be transformed to a differential equation for the function $\phi(t)$ by means of a Mellin transformation [10]:

$$y(z) = \int_{\gamma} t^{-z-1} \phi(t) \, dt \tag{A.1.2}$$

where γ is an appropriate contour in $\mathbb{C}$. If we substitute expression (A.1.2) into eqn (A.1.1) and integrate by parts, assuming that the nonintegral terms vanish, we arrive at

$$y(z + 1) - zy(z) = \int_{\gamma} t^{-z} (t^{-2}\phi(t) + \phi'(t)) \, dt = 0.$$

It leads to the following equation for the function $\phi(t)$:

$$\phi'(t) + t^{-2}\phi(t) = 0, \tag{A.1.3}$$

the solution of which is $\phi(t) = C \exp(-t^{-1})$, where C is an arbitrary constant. If the positive half-axis is taken as the contour γ, then the substitution $t := \xi^{-1}$ in the integral (A.1.2) yields the following representation for $y(z)$ as a result:[1]

$$y(z) = C \int_{0}^{\infty} e^{-\xi} \xi^{z-1} \, d\xi.$$

The arbitrary constant in front of the integral can be fixed by the standardization $y(1) = 1$. This leads to the following definition of the function, which is called the gamma function and is denoted $\Gamma(z)$:

$$\Gamma(z) = \int_{0}^{\infty} e^{-\xi} \xi^{z-1} \, d\xi. \tag{A.1.4}$$

This integral was first introduced and studied by Euler.

---

[1] There are other solutions of difference eqn (A.1.1). For instance, $y(z) := y(z)w(z)$ with a periodic function $w(z)$, $w(z + 1) = w(z)$.

Example: Let $z = n + 1$. Then

$$\Gamma(n+1) = \int_0^\infty e^{-\xi} \xi^n \, d\xi = n!. \tag{A.1.5}$$

This means that the gamma function is a generalization of the factorial. It also explains that by definition $0! = \Gamma(1) = 1$. $\square$

Since the integral in the right-hand side of eqn (A.1.4) uniformly converges in the right half plane, the gamma function is holomorphic there. A rigorous proof of this is given in many textbooks on special functions (see [96]).

In order to study the behaviour of the gamma function in the left half plane we split the integral in the right-hand side of (A.1.4) into two parts, namely

$$\Gamma(z) = \int_0^1 e^{-\xi} \xi^{z-1} \, d\xi + \int_1^\infty e^{-\xi} \xi^{z-1} \, d\xi.$$

Denoting the second integral as I_2 and evaluating the first integral as a series generated by the series for the exponential function, we come to a modified presentation of the gamma function:

$$\Gamma(z) = \sum_{n=0}^\infty \frac{(-1)^n}{n!(z+n)} + I_2. \tag{A.1.6}$$

Since the integral I_2 is an entire function in the complex plane $\mathbb{C}$, and the series in (A.1.6) defines a meromorphic function according to Weierstrass's theorem [15], this means that the gamma function is meromorphic in $\mathbb{C}$ with simple poles at nonpositive integers $z = 0, -1, -2 \ldots$, the residues of which are

$$\mathrm{Res}_{z=-n} \Gamma(z) = \frac{(-1)^n}{n!}. \tag{A.1.7}$$

For practical applications, the asymptotic behaviour of the gamma function is often needed. Suppose that z is real and positive. Under substitutions $z \mapsto z+1$ and $\xi := tz$, the integral (A.1.4) transforms to

$$\Gamma(z+1) = z^{z-1} \int_0^\infty e^{z(\ln t - t)} \, dt.$$

The saddle point for the function under the integral sign is found as the root t_s of the equation

$$(\ln t - t)' = 0 \quad \rightarrow \quad t_s = 1.$$

By use of the saddle-point formula [96], we arrive at the Stirling asymptotics for the gamma function:

$$\Gamma(z+1) = \left(\frac{z}{e}\right)^z \sqrt{2\pi z}[1 + O(z^{-1})]. \tag{A.1.8}$$

Although derived for real z, formula (A.1.8) is valid in the sector $|\arg z| < \pi - \epsilon$, where $\epsilon > 0$. Further continuation of (A.1.8) is restricted by the above-mentioned poles of the gamma function.

A.2   The beta function

Several integrals can be expressed in terms of the gamma function. For instance, consider

$$\Gamma(\mu)\Gamma(\nu) = \int_0^\infty \int_0^\infty e^{-\zeta-\eta}\zeta^{\mu-1}\eta^{\nu-1}\,d\zeta\,d\eta.$$

Under substitutions $\zeta := r\cos^2\varphi$ and $\eta := r\sin^2\varphi$, we arrive at

$$\Gamma(\mu)\Gamma(\nu) = 2\int_0^\infty e^{-r}r^{\mu+\nu-1}\,dr\int_0^{\frac{\pi}{2}}\cos^{2\mu-1}\varphi\sin^{2\nu-1}\varphi\,d\varphi.$$

Under one more substitution $\cos^2\varphi := \xi$, the last formula results in

$$\Gamma(\mu)\Gamma(\nu) = \Gamma(\mu+\nu)\int_0^1 \xi^{\mu-1}(1-\xi)^{\nu-1}\,d\xi. \tag{A.2.1}$$

The integral in the right-hand side of eqn (A.2.1) is called *beta function* and is denoted by $B(\mu,\nu)$. Hence

$$B(\mu,\nu) := \int_0^1 \xi^{\mu-1}(1-\xi)^{\nu-1}\,d\xi = \frac{\Gamma(\mu)\Gamma(\nu)}{\Gamma(\mu+\nu)}. \tag{A.2.2}$$

At certain values of arguments μ and ν, this integral can be evaluated explicitly. For instance, let $\nu = 1-\mu$, where $0 < \mu < 1$. Then it follows from eqn (A.2.2) that

$$B(\mu, 1-\mu) = \Gamma(\mu)\Gamma(1-\mu) = \int_0^1 \xi^{\mu-1}(1-\xi)^{-\mu}\,d\xi. \tag{A.2.3}$$

Under the substitution $\xi := \cos^2\varphi$ and further substitution $\cot^2\varphi = t$ in eqn (A.2.3), we obtain

$$\int_0^1 \xi^{\mu-1}(1-\xi)^{-\mu}\,d\xi = 2\int_0^{\frac{\pi}{2}}\cot^{2\mu-1}\varphi\,d\varphi = \int_0^\infty \frac{t^{\mu-1}}{1+t}\,dt := I. \tag{A.2.4}$$

The integral I can be evaluated by means of residue theory. Suppose that a contour ω consists of the line segment $[0, R]$ taken on the upper shore of the cut $[0, \infty]$ of the circle of radius R, and of the line segment $[R, 0]$ taken on the lower shore of the cut. If $R \to \infty$, then the integral over the circle vanishes under the assumed condition on the values of μ, and we obtain

$$\oint_\omega \frac{t^{\mu-1}}{1+t}\,dt = I\left(1 - e^{2\pi\mu i}\right)$$

$$= 2\pi i\,\mathrm{res}_{t=e^{\pi i}}\frac{t^{\mu-1}}{1+t} = -2\pi i\,e^{\pi\mu i}. \tag{A.2.5}$$

As a consequence of eqn (A.2.5) we arrive at a resulting formula which is called the *reflection formula* for the gamma function:

$$\Gamma(z)\Gamma(1 - z) = \frac{\pi}{\sin \pi z}. \tag{A.2.6}$$

Although derived for restricted values of the argument, the reflection formula can be extended to the whole complex plane $\mathbb{C}$, with the exception of integers, by means of analytic continuation.

Example: Let $z = 1/2$. It follows from (A.2.6) that

$$\Gamma^2(1/2) = \pi.$$

Since, for positive z, the gamma function is positive, we have

$$\Gamma\left(\frac{1}{2}\right) = \sqrt{\pi}. \tag{A.2.7}$$

□

Another formula valid for the gamma function, called the *duplication formula*, can be also obtained by use of the beta function. Consider equal arguments in the beta function:

$$B(\mu, \mu) := \int_0^1 \xi^{\mu-1}(1 - \xi)^{\mu-1}\,\mathrm{d}\xi = \frac{\Gamma(\mu)\Gamma(\mu)}{\Gamma(2\mu)}.$$

Under several substitutions of variables, we obtain

$$\int_0^1 \xi^{\mu-1}(1 - \xi)^{\mu-1}\,\mathrm{d}\xi$$

$$= 2\int_0^{\pi/2}(\sin\varphi)^{2\mu-1}(\cos\varphi)^{2\mu-1}\,\mathrm{d}\varphi$$

$$= \frac{1}{2^{2\mu-2}}\int_0^{\pi/2}(\sin 2\varphi)^{2\mu-1}\,\mathrm{d}\varphi = \frac{1}{2^{2\mu-2}}\int_0^{\pi/2}(\sin\omega)^{2\mu-1}\,\mathrm{d}\omega$$

$$= \frac{1}{2^{2\mu-1}}\int_0^1 \xi^{\mu-1}(1 - \xi)^{1/2}\,\mathrm{d}\xi = \frac{1}{2^{2\mu-1}}B(\mu, 1/2)$$

and get the duplication formula for the gamma function as a result:

$$\Gamma(2z) = \frac{2^{2z-1}}{\sqrt{\pi}}\Gamma(z)\Gamma\left(z + \frac{1}{2}\right). \tag{A.2.8}$$

Example: Let $z = n + 1/2$, where n is an integer. Then

$$\Gamma(n + 1/2) = \left(n - \frac{1}{2}\right)\cdots\left(\frac{1}{2}\right)\Gamma(1/2) = \frac{(2n - 1)!!\sqrt{\pi}}{2^n}. \tag{A.2.9}$$

□

A.3   The Pochhammer symbol

The other function which we want to study is related to the gamma function:

$$(z)_n = z(z+1)\dots(z+n-1) = \frac{\Gamma(z+n)}{\Gamma(z)}. \qquad (A.3.1)$$

This function is called *Pochhammer symbol*.[2] A more modern notation for the Pochhammer symbol (see, for instance, [48]) is: $(z)_n = z^{\underline{n}}$. In this book, the notation (A.3.1) is mostly used. The second notation stresses that the function $z^{\underline{n}}$ plays the same role for difference operators as the exponential function z^n plays for differential operators.

Suppose that E is the shift operator: $Ef(z) = f(z+1)$; I is the unity operator: $If(z) = f(z)$; and Δ is the forward difference operator: $\Delta f(z) = (E-I)f(z) = f(z+1) - f(z)$. Then

$$\Delta z^{\underline{n}} = (z+1)z\dots(z-n+2) - z\dots(z-n+1)$$
$$= (z+1-z+n-1)z\dots(z-n+1) = nz^{\underline{n-1}}. \qquad (A.3.2)$$

Further application of the difference operator yields

$$\Delta^n z^{\underline{n}} = n!. \qquad (A.3.3)$$

The function $z^{\underline{n}}$ can be expressed as a polynomial in z with the help of the so-called Stirling numbers C_n^j:

$$z^{\underline{n}} = \sum_{j=0}^{n} C_n^j z^j. \qquad (A.3.4)$$

---

[2]It is also known as the Pochhammer factorial. The notation of (A.3.1) is also used to denote the Jordan factorial $(z)_n = \Gamma(z+1)/\Gamma(z-n+1)$.

APPENDIX B: CTCPS FOR HEUN EQUATIONS IN GENERAL FORM

In Sections 1.6 and 3.6, we have outlined the central two-point connection problem for differential equations with polynomial coefficients of Heun's class. Heun's equations were written in a canonical form. However, when treating such a problem in applications, other forms may occur. Facing this situation, one would like to have solved the CTCPs for all possible forms. This may be achieved by treating the general forms that include all the other forms as special cases. The price one has to pay is that the formulae become more complicated. However, once the calculation is done, the CTCPs are solved definitively. In the following, we exhibit these formulae for use in the reader's own contexts.

B.1 Heun's equation and confluent cases

We will write down Heun's equation and its confluent cases in the natural general form. Heun's differential equation is

$$\frac{d^2y}{dz^2} + \sum_{i=1}^{3} \frac{A_i}{z - z_i} \frac{dy}{dz} + \left(\sum_{i=1}^{3} \frac{C_i}{z - z_i} + \sum_{i=1}^{3} \frac{B_i}{(z - z_i)^2} \right) y = 0 \quad \text{(B.1.1)}$$

with

$$\sum_{i=1}^{3} C_i = 0.$$

This condition tells us that eqn (B.1.1) has a regular singularity at infinity. The singly confluent case is

$$\frac{d^2y_s}{dz^2} + \left[G_{0s} + \frac{A_{1s}}{z - z_{1s}} + \frac{A_{2s}}{z + z_{*s}} \right] \frac{dy_s}{dz}$$

$$+ \left[D_{0s} + \frac{C_{1s}}{z - z_{1s}} + \frac{C_{2s}}{z + z_{*s}} + \frac{B_{1s}}{(z - z_{1s})^2} + \frac{B_{2s}}{(z + z_{*s})^2} \right] y_s = 0. \quad \text{(B.1.2)}$$

The biconfluent case is

$$\frac{d^2y_b}{dz^2} + \left[G_{1b}z + G_{0b} + \frac{A_b}{z - z_{1b}} \right] \frac{dy_b}{dz}$$

$$+ \left[D_{2b}z^2 + D_{1b}z + D_{0b} + \frac{C_b}{z - z_{1b}} + \frac{B_b}{(z - z_{1b})^2} \right] y_b = 0. \quad \text{(B.1.3)}$$

The triconfluent case is

$$\frac{d^2 y_t}{dz^2} + \left[G_{2t} z^2 + G_{1t} z + G_{0t} \right] \frac{dy_t}{dz}$$

$$+ \left[D_{4t} z^4 + D_{3t} z^3 + D_{2t} z^2 + D_{1t} z + D_{0t} \right] y_t = 0. \qquad (B.1.4)$$

The doubly confluent case is

$$\frac{d^2 y_d}{dz^2} + \left[G_{0d} + \frac{G_{-1d}}{z - z_{1d}} + \frac{G_{-2d}}{(z - z_{1d})^2} \right] \frac{dy_d}{dz}$$

$$+ \left[D_{0d} + \frac{D_{-1d}}{z - z_{1d}} + \frac{D_{-2d}}{(z - z_{1d})^2} + \frac{D_{-3d}}{(z - z_{1d})^3} + \frac{D_{-4d}}{(z - z_{1d})^4} \right] y_d = 0.$$

$$(B.1.5)$$

The index s indicates the singly confluent case, the index b indicates the biconfluent case, the index t the triconfluent case, and the index d the doubly confluent case.

The quantities z_i. $(i = *, 1, 2, 3)$ fix the finite singularities. These as well as all the other constants are (complex) parameters of the differential equation.

All the equations (B.1.1)–(B.1.4) have a singularity at infinity, the s-rank of which is at least as large as the s-ranks of each of their other singularities if there are any. Such an equation we call being in *natural* form. Equations (B.1.1)–(B.1.5) therefore are Heun's differential equation and its confluent cases in *natural general form* (cf. Chapter 3).

Without loss of generality, we now put $z_{1s} = z_{1d} = z_{1b} = 0$ in equations (B.1.2)–(B.1.5), getting for the singly confluent case

$$\frac{d^2 y_s}{dz^2} + \left[G_{0s} + \frac{A_{1s}}{z} + \frac{A_{2s}}{z + z_{*s}} \right] \frac{dy_s}{dz}$$

$$+ \left[D_{0s} + \frac{C_{1s}}{z} + \frac{C_{2s}}{z + z_{*s}} + \frac{B_{1s}}{z^2} + \frac{B_{2s}}{(z + z_{*s})^2} \right] y_s = 0, \qquad (B.1.6)$$

biconfluent case

$$\frac{d^2 y_b}{dz^2} + \left[G_{1b} z + G_{0b} + \frac{A_b}{z} \right] \frac{dy_b}{dz}$$

$$+ \left[D_{2b} z^2 + D_{1b} z + D_{0b} + \frac{C_b}{z} + \frac{B_b}{z^2} \right] y_b = 0, \qquad (B.1.7)$$

triconfluent case

$$\frac{d^2 y_t}{dz^2} + \left[G_{2t} z^2 + G_{1t} z + G_{0t} \right] \frac{dy_t}{dz}$$
$$+ \left[D_{4t} z^4 + D_{3t} z^3 + D_{2t} z^2 + D_{1t} z + D_{0t} \right] y_t = 0, \tag{B.1.8}$$

and doubly confluent case

$$\frac{d^2 y_d}{dz^2} + \left[G_{0d} + \frac{G_{-1d}}{z} + \frac{G_{-2d}}{z^2} \right] \frac{dy_d}{dz}$$
$$+ \left[D_{0d} + \frac{D_{-1d}}{z} + \frac{D_{-2d}}{z^2} + \frac{D_{-3d}}{z^3} + \frac{D_{-4d}}{z^4} \right] y_d = 0. \tag{B.1.9}$$

B.2   Asymptotic factors and Jaffé transformations

In the following, we carry out an sg transformation (respectively an s-homotopic transformation in the singly confluent case) for each of the equations (B.1.6)–(B.1.9); each of them contains the asymptotic factors of the singularities of the relevant interval.[1]

Singly confluent case:

$$y_s = \exp(\nu_s z) \, z^{\mu_{1s}} (z + z_{*s})^{\mu_{2s}} \, w_s,$$
$$\nu_s^2 + G_{0s} \nu_s + D_{0s} = 0 \; \rightarrow \; \nu_s = {}_{1,2}\nu_s,$$
$$\mu_{1s}^2 + (A_{1s} - 1)\mu_{1s} + B_{1s} = 0 \; \rightarrow \; \mu_{1s} = {}_{1,2}\mu_{1s},$$
$$\mu_{2s} = -\mu_{1s} - \frac{(A_{1s} + A_{2s}) \nu_s + C_{1s} + C_{2s}}{G_{0s} + 2\nu_s}. \tag{B.2.1}$$

Biconfluent case

$$y_b = \exp\left(\nu_b z + \frac{\kappa_b z^2}{2} \right) z^{\mu_b} (z + z_{*b})^{\alpha_b} \, w_b,$$
$$\kappa_b^2 + G_{1b} \kappa_b + D_{2b} = 0 \; \rightarrow \; \kappa_b = {}_{1,2}\kappa_b,$$
$$\nu_b = -\frac{D_{1b} + G_{0b} \kappa_b}{G_{1b} + 2\kappa_b},$$
$$\mu_b^2 + (A_b - 1) \mu_b + B_b = 0 \; \rightarrow \; \mu_b = {}_{1,2}\mu_b,$$
$$\alpha_b = -\frac{\nu_b^2 + G_{0b} \nu_b + D_{0b} + \kappa_b (A_b + 1)}{G_{1b} + 2\kappa_b} - \mu_b. \tag{B.2.1a}$$

---

[1] For the sake of simplicity, the z_* are not the same as the ones in Sections 1.6 and 3.6 but have the opposite sign!

Triconfluent case

$$y_t = \exp\left(v_t z + \frac{\kappa_t z^2}{2} + \frac{\eta_t z^3}{3}\right)(z + z_{*t})^{\alpha_t} w_t,$$

$$\eta_t^2 + G_{2t}\,\eta_t + D_{4t} = 0 \;\rightarrow\; \eta_t = {}_{1,2}\eta_t,$$

$$\kappa_t = -\frac{D_{3t} + \eta_t\,G_{1t}}{G_{2t} + 2\eta_t},$$

$$v_t = -\frac{D_{2t} + \eta\,G_{0t} + \kappa_t\,(G_{1t} + \kappa_t)}{G_{2t} + 2\eta_t},$$

$$\alpha_t = -\frac{D_{1t} + v_t\,G_{1t} + \kappa_t\,G_{0t} + 2\kappa_t v_t + 2\eta_t}{G_{2t} + 2\eta_t}.$$

$$(B.2.2)$$

Doubly confluent case:

$$y_d = \exp\left(v_d z - \frac{\psi_d}{z}\right) z^{\mu_d}(z + z_{*d})^{\alpha_d} w_d,$$

$$\psi_d^2 + G_{-2d}\,\psi_d + D_{-4d} = 0 \;\rightarrow\; \psi_d = {}_{1,2}\psi_d,$$

$$\mu_d = -\frac{D_{-3d} + \psi_d\,(G_{-1d} - 2)}{G_{-2d} + 2\psi_d},$$

$$v_d^2 + G_{0d}\,v_d + D_{0d} = 0 \;\rightarrow\; v_d = {}_{1,2}v_d,$$

$$\alpha_d = \frac{-D_{-1d} - G_{0d}\,\mu_d - v_d\,(G_{-1d} + 2\mu_d)}{G_{0d} + 2v_d}.$$

$$(B.2.3)$$

The quantities z_{*b}, z_{*t}, and z_{*d} determine the locations of the generated singularities.

As one can see, at each singularity of the differential equations of Heun's class, there exist in general two qualitatively different behaviours of its solutions while approaching the singularity radially. This is expressed by the quadratic equations in (B.2.1)–(B.2.3).

In each of the confluent cases, via the sg (s-homotopic), respectively transformation, we get a differential equation of the form

$$\Xi(z)\frac{d^2 w}{dz^2} + \left[g_2 z^2 + g_1 z + g_0 + \frac{g_{-1}}{z + z_*}\right]\frac{dw}{dz}$$

$$+ \left[d_0 + \frac{d_{-1}}{z + z_*} + \frac{d_{-2}}{(z + z_*)^2}\right] w = 0, \qquad (B.2.4)$$

in which the function $\Xi(z)$ and the parameters g_i, d_j ($i = 2, 1, 0, -1$; $j = 0, -1, -2$) are, in the singly confluent case,

$$\Xi =: P_{2s}(z) = z^2,$$

$$g_2 =: g_{2s} = G_{0s} + 2v_s,$$

$$g_1 =: g_{1s} = A_{1s} + A_{2s} + 2(\mu_{1s} + \mu_{2s}),$$

$$g_0 =: g_{0s} = -(A_{2s} + 2\mu_{2s})z_{*s},$$

$$g_{-1} =: g_{-1s} = (A_{2s} + 2\mu_{2s})z_{*s}^2,$$

$$d_0 =: d_{0s} = A_{1s}\mu_{2s} + A_{2s}(\mu_{1s} + \mu_{2s} - \nu_s z_{*s})$$
$$+ B_{2s} - C_{2s}z_{*s} - G_{0s}\mu_{2s}z_{*s}$$
$$+ 2\mu_{1s}\mu_{2s} + \mu_{2s}^2 - \mu_{2s}(2\nu_s z_{*s} + 1),$$

$$d_{-1} =: d_{-1s} = -z_{*s}[A_{1s}\mu_{2s} + A_{2s}(\mu_{1s} + 2\mu_{2s} - \nu_s z_{*s}) + 2B_{2s}$$
$$- C_{2s}z_{*s} - \mu_{2s}\{G_{0s}z_{*s} - 2(\mu_{1s} + \mu_{2s} - \nu_s z_{*s} - 1)\}],$$

$$d_{-2} =: d_{-2s} = \{B_{2s} + \mu_{2s}(A_{2s} + \mu_{2s} - 1)\}z_{*s}^2,$$

$$\text{(B.2.5)}$$

in the biconfluent case,

$$\Xi =: P_{1b}(z) = z,$$

$$g_2 =: g_{2b} = G_{1b} + 2\kappa_b,$$

$$g_1 =: g_{1b} = G_{0b} + 2\nu_b,$$

$$g_0 =: g_{0b} = A_b + 2(\mu_b + \alpha_b),$$

$$g_{-1} =: g_{-1b} = -2\alpha_b z_{*b},$$

$$d_0 =: d_{0b} = C_b + \nu_b A_b + \mu_b G_{0b} + 4\mu_b \nu_b$$
$$+ \alpha_b\{G_{0b} + 2\nu_b - (G_{1b} + 2\kappa_b)z_{*b}\} + 2\alpha_b\frac{\mu_b}{z_{*b}},$$

$$d_{-1} =: d_{-1b} = \alpha_b\{(G_{1b} + 2\kappa_b)z_{*b}^2 - (G_{0b} + 2\nu_b)z_{*b} + A_b + 2\mu_b + \alpha_b - 1\},$$

$$d_{-2} =: d_{-2b} = -\alpha_b(\alpha_b - 1)z_{*b},$$

$$\text{(B.2.6)}$$

in the triconfluent case,

$$\Xi =: P_{0t}(z) \equiv 1,$$

$$g_2 =: g_{2t} = G_{2t} + 2\eta_t,$$

$$g_1 =: g_{1t} = G_{1t} + 2\kappa_t,$$

$$g_0 =: g_{0t} = G_{0t} + 2\nu_t,$$

$$g_{-1} =: g_{-1t} = 2\alpha_t,$$

$$d_0 =: d_{0t} = D_{0t} + G_{0t}\nu_t + \nu_t^2 + \kappa_t - \alpha_t\{(G_{2t} + 2\eta_t)z_{*t} - (G_{1t} + 2\kappa_t)\},$$

$$d_{-1} =: d_{-1t} = \alpha_t[(G_{2t} + 2\eta_t)z_{*t}^2 - \{(G_{1t} + 2\kappa_t)z_{*t} - (G_{0t} + 2\nu_t)\}],$$

$$d_{-2} =: d_{-2t} = \alpha_t(\alpha_t - 1),$$

$$\text{(B.2.7)}$$

and in the doubly confluent case,

$$\Xi =: P_{2d}(z) = z^2,$$

$$g_2 =: g_{2d} = G_{0d} + 2\nu_d,$$

$$g_1 =: g_{1d} = G_{-1d} + 2(\mu_d + \alpha_d),$$

$$g_0 =: g_0 = G_{-2d} + 2(\psi_d - \alpha_d z_{*d}),$$

$$g_{-1} =: g_{-1d} = 2\alpha_d z_{*d}^2,$$

$$d_0 =: d_{0d} = D_{-2d} + \nu_d G_{-2d} + \psi_d G_{0d} + 2\nu_d \psi_d + \mu_d G_{-1d}$$
$$+ \mu_d^2 - \mu_d + \alpha_d\{G_{-1d} - (G_{0d} + 2\nu_d)z_{*d} + 2\mu_d + \alpha_d - 1\},$$

$$d_{-1} =: d_{-1d} = \alpha_d\{(G_{0d} + 2\nu_d)z_{*d}^2 - (G_{-1d} + 2\mu_d + 2\alpha_d - 2)z_{*d}$$
$$+ G_{-2d} + 2\psi_d\},$$

$$d_{-2} =: d_{-2d} = \alpha_d(\alpha_d - 1)z_{*d}^2.$$

$$(B.2.8)$$

B.3 Jaffé expansions and difference equations

In the following, for each confluent case we carry out a Möbius transformation [78]

$$x = \frac{z - z_{**}}{z + z_{*}}.$$

$$(B.3.1)$$

The constant quantities z_* and z_{**} are given in the following table:

Name of equation	z_{**}	z_*
Singly confluent case	0	z_{*s}
Biconfluent case	0	z_{*b}
Triconfluent case	0	z_{*t}
Doubly confluent case	z_{*d}	z_{*d}

The differential equation (B.2.4) transformed by (B.3.1) has the form

$$P_4(x)\frac{d^2 w}{dx^2} + \sum_{i=0}^{3}\Gamma_i x^i \frac{dw}{dx} + \sum_{j=0}^{2}\Delta_j x^j w = 0.$$

$$(B.3.2)$$

$P_4(x)$ is a polynomial of order four and Γ_i $(i = 0, 1, 2, 3)$ and Δ_j $(j = 0, 1, 2)$ are parameters, which are exhibited explicitly in the following.

Singly confluent case:

$$P_4(x) =: P_{4s}(x) = x^4 - 2x^3 + x^2 = x^2(x - 1)^2,$$

$$\Gamma_3 =: \Gamma_{3s} = 2 - \frac{g_{-1s}}{z_{*s}^2},$$

$$\Gamma_2 =: \Gamma_{2s} = g_{2s}z_{*s} - g_{1s} - 2 + \frac{g_{0s}}{z_{*s}} + \frac{3g_{-1s}}{z_{0s}^2},$$

$$\Gamma_1 =: \Gamma_{1s} = g_{1s} - \frac{2g_{0s}}{z_{*s}} - \frac{3g_{-1s}}{z_{0s}^2},$$

$$\Gamma_0 =: \Gamma_{0s} = \frac{g_{0s}}{z_{*s}} + \frac{g_{-1s}}{z_{0s}^2} \equiv 0,$$

$$\Delta_2 =: \Delta_{2s} = \frac{d_{-2s}}{z_{*s}^2},$$

$$\Delta_1 =: \Delta_{1s} = \frac{-2d_{-2s}}{z_{*s}^2} - \frac{d_{-1s}}{z_{*s}},$$

$$\Delta_0 =: \Delta_{0s} = \frac{d_{-2s}}{z_{*s}^2} + \frac{d_{-1s}}{z_{*s}} + d_{0s} \equiv 0.$$

$$(\text{B.3.3})$$

Biconfluent case:

$$P_4(x) =: P_{4b}(x) = x^4 - 3x^3 + 3x^2 - x = x\,(x-1)^3,$$

$$\Gamma_3 =: \Gamma_{3b} = 2 + \frac{g_{-1b}}{z_{*b}},$$

$$\Gamma_2 =: \Gamma_{2b} = -4 - z_{*b}(g_{2b}z_{*b} - g_{1b}) - g_{0b} - 3\frac{g_{-1b}}{z_{*b}},$$

$$\Gamma_1 =: \Gamma_{1b} = 2 - g_{1b}z_{*b} + 2g_{0b} + 3\frac{g_{-1b}}{z_{*b}},$$

$$\Gamma_0 =: \Gamma_{0b} = -(g_{0b} + \frac{g_{-1b}}{z_{*b}}),$$

$$\Delta_2 =: \Delta_{2b} = -\frac{d_{-2b}}{z_{*b}},$$

$$\Delta_1 =: \Delta_{1b} = \frac{2d_{-2b}}{z_{*b}} + d_{-1b},$$

$$\Delta_0 =: \Delta_{0b} = -\frac{d_{-2b}}{z_{*b}} - d_{-1b} - d_{0b}z_{*b}.$$

$$(\text{B.3.4})$$

Triconfluent case:

$$P_4(x) =: P_{4t}(x) = x^4 - 4x^3 + 6x^2 - 4x + 1 = (x-1)^4,$$

$$\Gamma_3 =: \Gamma_{3t} = 2 - g_{-1t},$$

$$\Gamma_2 =: \Gamma_{2t} = -6 + z_{*t}\{g_{2t}z_{*t}^2 - g_{1t}z_{*t} + g_{0t}\} + 3g_{-1t},$$

$$\Gamma_1 =: \Gamma_{1t} = 6 + z_{*t}\{g_{1t}z_{*t} - 2g_{0t}\} - 3g_{-1t},$$

$$\Gamma_0 =: \Gamma_{0t} = -2 + z_{*t}g_{0t} + g_{-1t},$$

$$\Delta_2 =: \Delta_{2t} = d_{-2t},$$
$$\Delta_1 =: \Delta_{1t} = -2d_{-2t} - d_{-1t}z_{*t},$$
$$\Delta_0 =: \Delta_{0t} = d_{-2t} + d_{-1t}z_{*t} + d_{0t}z_{*t}^2.$$

$$(B.3.5)$$

Doubly confluent case:

$$P_4(x) =: P_{4d}(x) = x^4 - 2x^2 + 1 = (x^2 - 1)^2,$$

$$\Gamma_3 =: \Gamma_{3d} = -\frac{g_{-1d}}{z_{*d}^2} + 2,$$

$$\Gamma_2 =: \Gamma_{2d} = 2g_{2d}z_{*d} - 2g_{1d} + 2\frac{g_{0d}}{z_{*d}} + 3\frac{g_{-1d}}{z_{*d}^2} + 2,$$

$$\Gamma_1 =: \Gamma_{1d} = 4g_{2d}z_{*d} - 4\frac{g_{0d}}{z_{*d}} - 3\frac{g_{-1d}}{z_{*d}^2} - 2,$$

$$\Gamma_0 =: \Gamma_{0d} = 2g_{2d}z_{*d} + 2g_{1d} + 2\frac{g_{0d}}{z_{*d}} + \frac{g_{-1d}}{z_{*d}^2} - 2,$$

$$\Delta_2 =: \Delta_{2d} = \frac{d_{-2d}}{z_{*d}^2},$$

$$\Delta_1 =: \Delta_{1d} = -2\frac{d_{-2d}}{z_{*d}^2} - 2\frac{d_{-1d}}{z_{*d}},$$

$$\Delta_0 =: \Delta_{0d} = \frac{d_{-2}}{z_{*d}^2} + 2\frac{d_{-1d}}{z_{*d}} + 4d_{0d}.$$

In the singly confluent, biconfluent, and triconfluent cases, the results of the formulae (B.3.3)–(B.3.5) may be arranged more clearly. To do this, we define the quantities $\tilde{\Gamma}_i$ ($i = 0, 1, 2, 3$) and $\tilde{\Delta}_i$ ($i = 0, 1, 2$) according to

$$\tilde{\Gamma}_0 = g_0 z_* - g_{-1},$$
$$\tilde{\Gamma}_1 = g_1 z_*^2 - 2g_0 z_* - 3g_{-1},$$
$$\tilde{\Gamma}_2 = g_2 z_*^3 - g_1 z_*^2 + g_0 z_* + 3g_{-1},$$
$$\tilde{\Gamma}_3 = -g_{-1},$$
$$\tilde{\Delta}_0 = d_0 z_*^2 + d_{-1} z_* + d_{-2},$$
$$\tilde{\Delta}_1 = -2d_{-2} - d_{-1} z_*,$$
$$\tilde{\Delta}_2 = d_{-2}.$$

Then we have, in the triconfluent case,

$$\Gamma_{0t} = \tilde{\Gamma}_0 - 2, \qquad \Gamma_{1t} = \tilde{\Gamma}_1 + 6, \qquad \Gamma_{2t} = \tilde{\Gamma}_2 - 6, \qquad \Gamma_{3t} = \tilde{\Gamma}_3 + 2,$$
$$\Delta_{it} = \tilde{\Delta}_i \quad (i = 0, 1, 2),$$

in the biconfluent case,

$$\Gamma_{0b} = -\frac{1}{z_{*b}}\tilde{\Gamma}_0, \quad \Gamma_{1b} = -\frac{1}{z_{*b}}\tilde{\Gamma}_1 - 2, \quad \Gamma_{2b} = -\frac{1}{z_{*b}}\tilde{\Gamma}_2 + 4, \quad \Gamma_{3b} = -\frac{1}{z_{*b}}\tilde{\Gamma}_3 - 2,$$

$$\Delta_{ib} = -\frac{1}{z_{*b}}\tilde{\Delta}_i \quad (i = 0, 1, 2),$$

and, in the singly confluent case,

$$\Gamma_{0s} = \frac{1}{z_{*s}^2}\tilde{\Gamma}_0, \qquad \Gamma_{1s} = \frac{1}{z_{*s}^2}\tilde{\Gamma}_1, \qquad \Gamma_{2s} = \frac{1}{z_{*s}^2}\tilde{\Gamma}_2 - 2, \qquad \Gamma_{3s} = \frac{1}{z_{*s}^2}\tilde{\Gamma}_3 + 2,$$

$$\Delta_{is} = \frac{1}{z_{*s}^2}\tilde{\Delta}_i \quad (i = 0, 1, 2).$$

The sg transformations (B.2.1)–(B.2.3) have the property that the equations (B.3.2) resulting from them possess in each of the four confluent cases a particular solution which may be expanded in a power series about $x = 0$ according to

$$w(x) = \sum_{n=0}^{\infty} a_{n,.} x^n,$$ (B.3.6)

which converges within the unit disc. Along with (B.2.1)–(B.2.3), the expansion (B.3.6) meets the boundary conditions at $x = 0$.

The coefficients a_n of the series (B.3.6) all are solutions of irregular difference equations of Poincaré–Perron type, the order of which is precisely the sum of the s-ranks of the *irregular* singularities of the according differential equation. Thus (B.3.6) may be seen as a mapping of the differential equations (B.1.6)–(B.1.9) onto the difference equations (B.3.7)–(B.3.10) written below. In the following, we exhibit these difference equations for the confluent cases of Heun's equation.

Singly confluent case:

$$a_{0,s} \text{ arbitrary,}$$

$$\Gamma_{1s} a_{1,s} + \Delta_{1s} a_{0,s} = 0,$$

$$\left(1 + \frac{\alpha_{1s}}{n} + \frac{\beta_{1s}}{n^2}\right) a_{n+1,s} + \left(-2 + \frac{\alpha_{0s}}{n} + \frac{\beta_{0s}}{n^2}\right) a_{n,s}$$

$$+ \left(1 + \frac{\alpha_{-1s}}{n} + \frac{\beta_{-1s}}{n^2}\right) a_{n-1,s} = 0 \quad \text{for } n \geq 1,$$

$$\begin{aligned}
\alpha_{1s} &:= \Gamma_{1s} + 1, & \beta_{1s} &:= \Gamma_{1s}, \\
\alpha_{0s} &:= \Gamma_{2s} + 2, & \beta_{0s} &:= \Delta_{1s}, \\
\alpha_{-1s} &:= \Gamma_{3s} - 3, & \beta_{-1s} &:= \Delta_{2s} - \Gamma_{3s} + 2.
\end{aligned}$$

(B.3.7)

Biconfluent case:

$a_{0,b}$ arbitrary,

$\Gamma_{0b} a_{1,b} + \Delta_{0b} a_{0,b} = 0,$

$2 (\Gamma_{0b} - 1) a_{2,b} + (\Gamma_{1b} + \Delta_{0b}) a_{1,b} + (\Delta_{1b} - \Gamma_{2b}) a_{0,b} = 0,$

$$\left(-1 + \frac{\alpha_{1b}}{n} + \frac{\beta_{1b}}{n^2} \right) a_{n+1,b} + \left(3 + \frac{\alpha_{0b}}{n} + \frac{\beta_{0b}}{n^2} \right) a_{n,b}$$

$$+ \left(-3 + \frac{\alpha_{-1b}}{n} + \frac{\beta_{-1b}}{n^2} \right) a_{n-1,b} + \left(1 + \frac{\alpha_{-2b}}{n} + \frac{\beta_{-2b}}{n^2} \right) a_{n-2,b} = 0$$

for $n \geq 2$,

$$
\begin{aligned}
\alpha_{1b} &:= \Gamma_{0b} - 1, & \beta_{1b} &:= \Gamma_{0b}, \\
\alpha_{0b} &:= \Gamma_{1b} - 3, & \beta_{0b} &:= \Delta_{0b}, \\
\alpha_{-1b} &:= \Gamma_{2b} + 9, & \beta_{-1b} &:= \Delta_{1b} - \Gamma_{2b} - 6, \\
\alpha_{-2b} &:= \Gamma_{3b} - 5, & \beta_{-2b} &:= -2\Gamma_{3b} + \Delta_{2b} + 6.
\end{aligned}
$$

(B.3.8)

Triconfluent case:

$a_{0,t}, a_{1,t}$ arbitrary,

$2 a_{2,t} + \Gamma_{0t} a_{1,t} + \Delta_{0t} a_{0,t} = 0,$

$6 a_{3,t} + (2\Gamma_{0t} - 8) a_{2,t} + (\Gamma_{1t} + \Delta_{0t}) a_{1,t} + \Delta_{1t} a_{0,t} = 0,$

$$\left(1 + \frac{\alpha_{2t}}{n} + \frac{\beta_{2t}}{n^2} \right) a_{n+2,t} + \left(-4 + \frac{\alpha_{1t}}{n} + \frac{\beta_{1t}}{n^2} \right) a_{n+1,t} + \left(6 + \frac{\alpha_{0t}}{n} + \frac{\beta_{0t}}{n^2} \right) a_{n,t}$$

$$+ \left(-4 + \frac{\alpha_{-1t}}{n} + \frac{\beta_{-1t}}{n^2} \right) a_{n-1,t} + \left(1 + \frac{\alpha_{-2t}}{n} + \frac{\beta_{-2t}}{n^2} \right) a_{n-2,t} = 0$$

for $n \geq 2$,

$$
\begin{aligned}
\alpha_{2t} &:= 3, & \beta_{2t} &:= 2, \\
\alpha_{1t} &:= \Gamma_{0t} - 4, & \beta_{1t} &:= \Gamma_{0t}, \\
\alpha_{0t} &:= \Gamma_{1t} - 6, & \beta_{0t} &:= \Delta_{0t}, \\
\alpha_{-1t} &:= \Gamma_{2t} + 12, & \beta_{-1t} &:= \Delta_{1t} - \Gamma_{2t} - 8, \\
\alpha_{-2t} &:= \Gamma_{3t} - 5, & \beta_{-2t} &:= \Delta_{2t} - 2\Gamma_{3t} + 6.
\end{aligned}
$$

(B.3.9)

Doubly confluent case:

$a_{0,d}, a_{1,d}$ arbitrary,

$$2\,a_{2,d} + \Gamma_{0d}\,a_{1,d} + \Delta_{0d}\,a_{0,d} = 0,$$

$$6a_{3,d} + 2\Gamma_{0d}\,a_{2,d} + (\Gamma_{1d} + \Delta_{0d})\,a_{1,d} + \Delta_{1d}\,a_{0,d} = 0,$$

$$\left(1 + \frac{\alpha_{2d}}{n} + \frac{\beta_{2d}}{n^2}\right) a_{n+2,d} + \left(\frac{\alpha_{1d}}{n} + \frac{\beta_{1d}}{n^2}\right) a_{n+1,d} + \left(-2 + \frac{\alpha_{0d}}{n} + \frac{\beta_{0d}}{n^2}\right) a_{n,d}$$

$$+ \left(\frac{\alpha_{-1d}}{n} + \frac{\beta_{-1d}}{n^2}\right) a_{n-1,d} + \left(1 + \frac{\alpha_{-2d}}{n} + \frac{\beta_{-2d}}{n^2}\right) a_{n-2,d} = 0 \quad \text{for } n \geq 2,$$

$$
\begin{aligned}
\alpha_{2d} &:= 3, & \beta_{2d} &:= 2, \\
\alpha_{1d} &:= \beta_{1d} := \Gamma_{0d}, & & \\
\alpha_{0d} &:= \Gamma_{1d} + 2, & \beta_{0d} &:= \Delta_{0d}, \\
\alpha_{-1d} &:= \Gamma_{2d}, & \beta_{-1d} &:= -\Gamma_{2d} + \Delta_{1d}, \\
\alpha_{-2d} &:= \Gamma_{3d} - 5, & \beta_{-2d} &:= 6 - 2\Gamma_{3d} + \Delta_{2d}.
\end{aligned}
\tag{B.3.10}
$$

The solutions of these difference equations may be computed recursively: To do this in the singly confluent and biconfluent cases, one has to fix the first term $a_{0,.}$. In the triconfluent and doubly confluent cases, in addition to the first term, also the second terms $a_{1,t}$ and respectively $a_{1,d}$ have to be determined.

All the coefficients in front of the $a_{n,.}$ of the four difference equations written above converge to a finite value for $n \to \infty$. Thus, we have difference equations exclusively of irregular *Poincaré–Perron type*.

B.4 Characteristic equations and Birkhoff sets

Since the zeroth-order characteristic equations (3.6.19)–(3.6.22) do not reveal the asymptotic factors in Birkhoff solutions, we had to introduce algebraic equations of one-higher order, namely up to $1/n$, in order to get a characteristic equation (i.e. of algebraic type) with the help of which one can distinguish the different particular solutions of the difference equations. In the following, these are written for the confluent cases of Heun's equation.

Singly confluent case:

$$\left(1 + \frac{\alpha_{1s}}{n}\right) t_s^2(n) + \left(-2 + \frac{\alpha_{0s}}{n}\right) t_s(n) + \left(1 + \frac{\alpha_{-1s}}{n}\right) = 0. \tag{B.4.1}$$

Biconfluent case:

$$\left(-1 + \frac{\alpha_{1b}}{n}\right) t_b^3(n) + \left(3 + \frac{\alpha_{0b}}{n}\right) t_b^2(n) + \left(-3 + \frac{\alpha_{-1b}}{n}\right) t_b(n)$$

$$+ \left(1 + \frac{\alpha_{-2b}}{n}\right) = 0. \tag{B.4.2}$$

Triconfluent case:

$$\left(1 + \frac{\alpha_{2t}}{n}\right) t_t^4(n) + \left(-4 + \frac{\alpha_{1t}}{n}\right) t_t^3(n) + \left(6 + \frac{\alpha_{0t}}{n}\right) t_t^2(n)$$

$$+ \left(-4 + \frac{\alpha_{-1t}}{n}\right) t_t(n) + \left(1 + \frac{\alpha_{-2t}}{n}\right) = 0. \qquad (B.4.3)$$

Doubly confluent case:

$$\left(1 + \frac{\alpha_{2d}}{n}\right) t_d^4(n) + \frac{\alpha_{1t}}{n} t_d^3(n) + \left(-2 + \frac{\alpha_{0d}}{n}\right) t_d^2(n) + \frac{\alpha_{-1d}}{n} t_d(n)$$

$$+ \left(1 + \frac{\alpha_{-2d}}{n}\right) = 0. \qquad (B.4.4)$$

The solutions of these four equations are in turn

$$t_{s,1,2}(n) = 1 \pm \sqrt{\frac{-\sum_{i=-1}^{i=1} \alpha_{is}}{n}} + \mathcal{O}(1/n);$$

$$t_{b,1}(n) = 1 - \sqrt[3]{\frac{-\sum_{i=-2}^{i=1} \alpha_{ib}}{n}} + \mathcal{O}\left(1/n^{\frac{2}{3}}\right),$$

$$t_{b,2}(n) = 1 + \frac{1}{2}\sqrt[3]{\frac{-\sum_{i=-2}^{i=1} \alpha_{ib}}{n}}(1 - \sqrt{3}i) + \mathcal{O}\left(1/n^{\frac{2}{3}}\right),$$

$$t_{b,3}(n) = 1 + \frac{1}{2}\sqrt[3]{\frac{-\sum_{i=-2}^{i=1} \alpha_{ib}}{n}}(1 + \sqrt{3}i) + \mathcal{O}\left(1/n^{\frac{2}{3}}\right);$$

$$t_{t,1,2}(n) = 1 \pm \sqrt[4]{\frac{-\sum_{i=-2}^{i=+2} \alpha_{it}}{n}} + \mathcal{O}\left(1/n^{\frac{1}{2}}\right),$$

$$t_{t,3,4}(n) = 1 \pm i\sqrt[4]{\frac{-\sum_{i=-2}^{i=+2} \alpha_{it}}{n}} + \mathcal{O}\left(1/n^{\frac{1}{2}}\right);$$

$$t_{d,1,2}(n) = 1 \pm \frac{1}{2}\sqrt{\frac{-\sum_{i=-2}^{i=2} \alpha_{id}}{n}} + \mathcal{O}(1/n),$$

$$t_{d,3,4}(n) = -1 \pm \frac{1}{2}\sqrt{\frac{-\sum_{i=-2}^{i=2}(-1)^i \alpha_{id}}{n}} + \mathcal{O}(1/n).$$

One may get the results exhibited in (B.4.1)–(B.4.4) also by considering the ratio of two consecutive terms $a_{n+1,.}/a_{n,.}$ and assuming the existence of $\lim_{n\to\infty} a_{n+1,.}/a_{n,.}$.

but writing down the coefficients only up to the order $O(1/n)$; or if, in each of the confluent cases, one formally calculates the expressions $s^{(m.)}(n+1)/s^{(m.)}(n)$ for each Birkhoff solution according to (B.4.5)–(B.4.8) (see below) asymptotically for $n \to \infty$. The solutions of the equations (B.4.1)–(B.4.4) are of practical use in so far as one can get from them without heavy computation an overview of which ratios of two consecutive terms in the solutions of the difference equations for $n \to \infty$ may occur.

The Birkhoff sets are given as follows.

Singly confluent case:

$$s^{(1s)}(n) = \varrho_s^n \exp\left(\gamma_{1s}n^{\frac{1}{2}}\right) n^{r_{1s}} \left[1 + \frac{C_{11s}}{n^{\frac{1}{2}}} + \frac{C_{12s}}{n^{\frac{2}{2}}} + \cdots\right],$$

$$s^{(2s)}(n) = \varrho_s^n \exp\left(\gamma_{2s}n^{\frac{1}{2}}\right) n^{r_{2s}} \left[1 + \frac{C_{21s}}{n^{\frac{1}{2}}} + \frac{C_{22s}}{n^{\frac{2}{2}}} + \cdots\right]. \tag{B.4.5}$$

Biconfluent case:

$$s^{(1b)}(n) = \varrho_b^n \exp\left(\gamma_{11b}n^{\frac{2}{3}} + \gamma_{12b}n^{\frac{1}{3}}\right) n^{r_{1b}} \left[1 + \frac{C_{11b}}{n^{\frac{1}{3}}} + \frac{C_{12b}}{n^{\frac{2}{3}}} + \cdots\right],$$

$$s^{(2b)}(n) = \varrho_b^n \exp\left(\gamma_{21b}n^{\frac{2}{3}} + \gamma_{22b}n^{\frac{1}{3}}\right) n^{r_{2b}} \left[1 + \frac{C_{21b}}{n^{\frac{1}{3}}} + \frac{C_{22b}}{n^{\frac{2}{3}}} + \cdots\right],$$

$$s^{(3b)}(n) = \varrho_b^n \exp\left(\gamma_{31b}n^{\frac{2}{3}} + \gamma_{32b}n^{\frac{1}{3}}\right) n^{r_{3b}} \left[1 + \frac{C_{31b}}{n^{\frac{1}{3}}} + \frac{C_{32b}}{n^{\frac{2}{3}}} + \cdots\right]. \tag{B.4.6}$$

Triconfluent case:

$$s^{(1t)}(n) = \varrho_t^n \exp\left(\gamma_{11t}n^{\frac{3}{4}} + \gamma_{12t}n^{\frac{2}{4}} + \gamma_{13t}n^{\frac{1}{4}}\right) n^{r_{1t}} \left[1 + \frac{C_{11t}}{n^{\frac{1}{4}}} + \frac{C_{12t}}{n^{\frac{2}{4}}} + \cdots\right],$$

$$s^{(2t)}(n) = \varrho_t^n \exp\left(\gamma_{21t}n^{\frac{3}{4}} + \gamma_{22t}n^{\frac{2}{4}} + \gamma_{23t}n^{\frac{1}{4}}\right) n^{r_{2t}} \left[1 + \frac{C_{21t}}{n^{\frac{1}{4}}} + \frac{C_{22t}}{n^{\frac{2}{4}}} + \cdots\right],$$

$$s^{(3t)}(n) = \varrho_t^n \exp\left(\gamma_{31t}n^{\frac{3}{4}} + \gamma_{32t}n^{\frac{2}{4}} + \gamma_{33t}n^{\frac{1}{4}}\right) n^{r_{3t}} \left[1 + \frac{C_{31t}}{n^{\frac{1}{4}}} + \frac{C_{32t}}{n^{\frac{2}{4}}} + \cdots\right],$$

$$s^{(4t)}(n) = \varrho_t^n \exp\left(\gamma_{41t}n^{\frac{3}{4}} + \gamma_{42t}n^{\frac{2}{4}} + \gamma_{43t}n^{\frac{1}{4}}\right) n^{r_{4t}} \left[1 + \frac{C_{41t}}{n^{\frac{1}{4}}} + \frac{C_{42t}}{n^{\frac{2}{4}}} + \cdots\right]. \tag{B.4.7}$$

Doubly confluent case:

$$s^{(1d)}(n) = \varrho_{1d}^n \exp\left(\gamma_{1d} n^{\frac{1}{2}}\right) n^{r_{1d}} \left[1 + \frac{C_{11d}}{n^{\frac{1}{2}}} + \frac{C_{12d}}{n^{\frac{2}{2}}} + \cdots\right],$$

$$s^{(2d)}(n) = \varrho_{1d}^n \exp\left(\gamma_{2d} n^{\frac{1}{2}}\right) n^{r_{2d}} \left[1 + \frac{C_{21d}}{n^{\frac{1}{2}}} + \frac{C_{22d}}{n^{\frac{2}{2}}} + \cdots\right],$$

$$s^{(3d)}(n) = \varrho_{2d}^n \exp\left(\gamma_{3d} n^{\frac{1}{2}}\right) n^{r_{3d}} \left[1 + \frac{C_{31d}}{n^{\frac{1}{2}}} + \frac{C_{32d}}{n^{\frac{2}{2}}} + \cdots\right],$$

$$s^{(4d)}(n) = \varrho_{2d}^n \exp\left(\gamma_{4d} n^{\frac{1}{2}}\right) n^{r_{4d}} \left[1 + \frac{C_{41d}}{n^{\frac{1}{2}}} + \frac{C_{42d}}{n^{\frac{2}{2}}} + \cdots\right]. \quad (B.4.8)$$

In the following are written the coefficients of the asymptotic factors. It is not necessary to calculate the coefficients $C_{...}$ of the series in our context.

Singly confluent case:

$$\varrho_s = 1,$$

$$\gamma_{1s} = 2\sqrt{-\sum_{i=-1}^{i=1} \alpha_{is}},$$

$$\gamma_{2s} = -\gamma_{1s},$$

$$r_{1s} = r_{2s} = -\frac{1}{4}(2\alpha_{1s} - 2\alpha_{-1s} - 1) =: r_s. \quad (B.4.9)$$

In order for r_s in (B.4.9) to admit the given value, we have to suppose $\gamma_{1s} \neq 0$. A similar statement holds in the subsequent cases.

Biconfluent case:

$$\varrho_b = 1,$$

$$\gamma_{11b} = -\frac{3}{2}\sqrt[3]{-\sum_{i=-2}^{i=+1} \alpha_{ib}} \neq 0,$$

$$\gamma_{21b} = \frac{3}{4}\sqrt[3]{-\sum_{i=-2}^{i=+1} \alpha_{ib}}(1 - \sqrt{3}i), \qquad i = \sqrt{-1},$$

$$\gamma_{31b} = \frac{3}{4}\sqrt[3]{-\sum_{i=-2}^{i=+1} \alpha_{ib}}(1 + \sqrt{3}i),$$

$$\gamma_{12b} = -\frac{3\alpha_{1b} + \alpha_{0b} - \alpha_{-1b} - 3\alpha_{-2b}}{2\sqrt[3]{-\sum_{i=-2}^{i=+1} \alpha_{ib}}},$$

$$\gamma_{22b} = \frac{3\alpha_{1b} + \alpha_{0b} - \alpha_{-1b} - 3\alpha_{-2b}}{4\sqrt[3]{-\sum_{i=-2}^{i=+1}\alpha_{ib}}}(1 + i\sqrt{3}),$$

$$\gamma_{32b} = \frac{3\alpha_{1b} + \alpha_{0b} - \alpha_{-1b} - 3\alpha_{-2b}}{4\sqrt[3]{-\sum_{i=-2}^{i=+1}\alpha_{ib}}}(1 - i\sqrt{3}),$$

$$r_{1b} = \frac{1}{3}(\alpha_{1b} + \alpha_{-2b} + 1),$$

$$r_{2b} = r_{3b} = r_{1b} =: r_b. \tag{B.4.10}$$

Triconfluent case:

$$\varrho_t = 1,$$

$$\gamma_{11t} = \frac{4}{3}\sqrt[4]{-\sum_{i=-2}^{i=+2}\alpha_{it}} \neq 0,$$

$$\gamma_{21t} = -\gamma_{11t},$$

$$\gamma_{31t} = \frac{4}{3}i\sqrt[4]{-\sum_{i=-2}^{i=+2}\alpha_{it}},$$

$$\gamma_{41t} = -\gamma_{31t},$$

$$\gamma_{12t} = -\frac{2\alpha_{2t} + \alpha_{1t} - \alpha_{-1t} - 2\alpha_{-2t}}{2\sqrt{-\sum_{i=-2}^{i=+2}\alpha_{it}}},$$

$$\gamma_{22t} = \gamma_{12t},$$

$$\gamma_{32t} = -\gamma_{12t},$$

$$\gamma_{42t} = -\gamma_{12t},$$

$$\gamma_{13t} = -\frac{1}{24(-\sum_{i=-2}^{i=+2}\alpha_{it})^{5/4}}[-32\alpha_{2t}^2 - 16\alpha_{2t}(4\alpha_{-1t} + 7\alpha_{-2t})$$
$$-5\alpha_{1t}^2 - 2\alpha_{1t}(20\alpha_{2t} + 11\alpha_{-1t} + 32\alpha_{-2t})$$
$$+4\alpha_{0t}^2 - 4\alpha_{0t}(\alpha_{1t} + 10\alpha_{2t} + \alpha_{-1t} + 10\alpha_{-2t})$$
$$-5\alpha_{-1t}^2 - 40\alpha_{-1t}\alpha_{-2t} - 32\alpha_{-2t}^2],$$

$$\gamma_{23t} = -\gamma_{13t},$$

$$\gamma_{33t} = -i\gamma_{13t},$$

$$\gamma_{43t} = i\gamma_{13t},$$

$$r_{1t} = -\frac{1}{8}(2\alpha_{2t} - 2\alpha_{-2t} - 3),$$

$$r_{2t} = r_{3t} = r_{4t} = r_{1t} =: r_t. \tag{B.4.11}$$

Doubly confluent case:

$$\varrho_{1d} = 1,$$

$$\gamma_{1d} = \sqrt{-\sum_{i=-2}^{i=+2} \alpha_{id}} \neq 0 \,,$$

$$\gamma_{2d} = -\gamma_{1d} \,,$$

$$r_{1d} = r_{2d} = -\frac{1}{8}(2\alpha_{2d} + \alpha_{1d} - \alpha_{-1d} - 2\alpha_{-2d} - 2),$$

$$\varrho_{2d} = -1,$$

$$\gamma_{3d} = \sqrt{-\sum_{i=-2}^{i=+2} (-1)^i \alpha_{id}} \neq 0 \,,$$

$$\gamma_{4d} = -\gamma_{3d},$$

$$r_{3d} = r_{4d} = -\frac{1}{8}(2\alpha_{2d} - \alpha_{1d} + \alpha_{-1d} - 2\alpha_{-2d} - 2). \tag{B.4.12}$$

The fact that the $\varrho_{.}$ are $+1$ (respectively $+1$ and -1 in the doubly confluent case) is a consequence of the fact that the relevant irregular singularities were placed at unity (respectively $x = +1$ and $x = -1$ in the doubly confluent case) by the applied Möbius transformations.

As can be seen from equations (B.4.9)–(B.4.12), the asymptotic factors depend exclusively on the quantities

$$-\sum_i \alpha_{i.}.$$

These quantities themselves depend upon the scaling parameters of the related differential equation:

$$-\sum_{i=-1}^{i=+1} \alpha_{is} = -\sum_{i=0}^{i=+3} \Gamma_{is} = -g_{2s} z_{*s},$$

$$-\sum_{i=-2}^{i=+1} \alpha_{ib} = -\sum_{i=0}^{i=+3} \Gamma_{ib} = +g_{2b} z_{*b}^2,$$

$$-\sum_{i=-2}^{i=+2} \alpha_{it} = -\sum_{i=0}^{i=+3} \Gamma_{it} = -g_{2t} z_{*t}^3,$$

$$-\sum_{i=-2}^{i=+2} \alpha_{id} = -\sum_{i=0}^{i=+3} \Gamma_{id} = -8 g_{2d} z_{*d},$$

$$-\sum_{i=-2}^{i=+2} (-1)^i \alpha_{id} = -\sum_{i=0}^{i=+3} (-1)^i \Gamma_{id} = \frac{8}{z_{*d}}\left(\frac{g_{-1d}}{z_{*d}} + g_{0d}\right).$$

Recalling (B.2.5)–(B.2.8), and according to the quadratic equations for determining the quantities ν_s, κ_b, η_t, and ψ_d in (B.2.1)–(B.2.3), we have

$$g_{2s} = \pm\sqrt{G_{0s}^2 - 4D_{0s}},$$

$$g_{2b} = \pm\sqrt{G_{1b}^2 - 4D_{2b}},$$

$$g_{2t} = \pm\sqrt{G_{2t}^2 - 4D_{4t}},$$

$$g_{2d} = \pm\sqrt{G_{-2d}^2 - 4D_{-4d}},$$

$$g_{0d} + \frac{g_{-1d}}{z_{*d}} = \pm\sqrt{G_{-2d}^2 - 4D_{-4d}}.$$

Eventually we obtain the useful formula

$$\gamma_{12t} = -\frac{1}{2}\sqrt{-g_{2t}z_{*t}^3} - \frac{g_{1t}z_{*t}^2}{2\sqrt{-g_{2t}z_{*t}^3}}.$$

APPENDIX C: MULTIPOLE MATRIX ELEMENTS

C.1 Introduction

Let $\psi_n(z)$ and λ_n be an eigenfunction and corresponding eigenvalue of a spectral problem on the whole axis for a 1D Schrödinger equation

$$(-L + \lambda)\psi = \frac{d^2}{dz^2}\psi(z) + (\lambda - q(z))\psi(z) = 0, \qquad (C.1.1)$$

$$\lim \psi(z)_{z \to -\infty} = 0, \qquad \lim \psi(z)_{z \to \infty} = 0.$$

The eigenfunctions obey the orthogonality condition

$$\int_{-\infty}^{\infty} \psi_n(z)\psi_m(z)\,dz = \delta_{nm},$$

where δ_{nm} is the Kronecker symbol. The function $q(z)$—playing the role of a potential in the Schrödinger equation—is assumed to be a polynomial of even order satisfying $q(z)|_{z \to \pm\infty} \to \infty$. By multipole matrix elements we mean the integrals

$$V_{nm}^{(k)} = \int_{-\infty}^{\infty} z^k \psi_n(z)\psi_m(z)\,dz \qquad (C.1.2)$$

with positive integers k. In the case $n = m$, the matrix elements are called diagonal; otherwise they are nondiagonal. Our goal is to obtain recursive relations for matrix elements for different k and fixed values of n and of m.

Many equations studied in this book were written in a more general self-adjoint form with the first derivative included, namely

$$(-L + \lambda)\psi = \frac{d}{dz}\left(r(z)\frac{d}{dz}\psi(z)\right) + (\lambda - q(z))\psi(z) = 0. \qquad (C.1.3)$$

Here, once again, the matrix elements corresponding to eigenfunctions for eqn (C.1.3) can be studied. Below, the calculations are given only for the more simple case of eqn (C.1.1).

C.2 Auxiliary differential equations

The method is based on equations for the square of the eigenfunctions and the product of two eigenfunctions related to different eigenvalues. First, several lemmas are proved.

Lemma C.1 *The function defined as a square $w_{nn}(z) = (\psi_n(z))^2$ of an eigenfunction of (C.1.1) is a solution of the third-order equation*

$$w_{nn}'''(z) + 4(\lambda_n - q)w_{nn}'(z) - 2q'w_{nn}(z) = 0. \qquad (C.2.1)$$

$\square$

Proof: The derivatives of $w(z)$ (the indexes are omitted for brevity) are

$$w' = 2\psi\psi', \qquad w'' = 2(\psi\psi'' + (\psi')^2), \qquad w''' = 2(\psi\psi''' + 3\psi'\psi'').$$

On the other hand, we obtain from eqn (C.1.1) that

$$2\psi''\psi' = (q-\lambda)w', \qquad \psi''\psi = (q-\lambda)w, \qquad 2\psi'''\psi + 3(\lambda-q)w' = w'''.$$

Excluding the function ψ from these equalities, we arrive at (C.2.1). $\square$

Consider two equations

$$y''(z) - f(z)y(z) = 0, \qquad u''(z) - g(z)u(z) = 0. \tag{C.2.2}$$

The function $v(z)$ is defined as a product of these equations, i.e. $v := yu$. According to these definitions, the following relations hold:

$$v' = y'u + yu', \qquad v'' = y''u + yu'' + 2y'u',$$

$$v''' = y'''u + yu''' + 3(y''u' + y'u''),$$
$$v'''' = y''''u + yu'''' + 4(y'''u' + y'u''') + 6y''u''.$$

On the other hand, it follows from eqn (C.2.2) that

$$y''u = fv, \qquad yu'' = gv, \qquad y''u'' = fgv, \qquad 2y'u' = v'' - (f+g)v,$$
$$y'''u + y''u' = f'v + fv', \qquad yu''' + y'u'' = g'v + gv',$$
$$y''''u + 2y'''u' + y''u'' = f''v + 2f'v' + fv'',$$
$$yu'''' + 2y'u''' + y''u'' = g''v + 2g'v' + gv''.$$

Excluding from these equalities the functions y and u, we arrive at

$$v''''(z) - 2y'''u' - 2y'u''' - (f+g)v''(z)$$
$$- 2(f'+g')v'(z) - 4fgv(z) - (f''+g'')v(z) = 0. \tag{C.2.3}$$

We apply some more equalities:

$$y''u' = fyu', \qquad y'u'' = gy'u, \qquad y'''u' = fy'u' + f'yu',$$
$$y'u''' = gy'u' + g'y'u.$$

Let $f = q - \lambda_n$ and $g = q - \lambda_m$, resulting in $f' = g'$ and $f'' = g''$. This leads to the following conjecture.

Lemma C.2 *The product of the wave eigenfunctions* $v_{nm}(z) = (\psi_n(z))\psi_m(z)$ *in (C.1.1) satisfies the fourth-order equation*

$$v''''_{nm}(z) + 2(\lambda_n + \lambda_m - 2q)v''_{nm}(z) - 6q'v'_n(z)$$
$$+ ((\lambda_n - \lambda_m)^2 - 2q'')v_{nm}(z) = 0. \tag{C.2.4}$$

$\square$

It should be noticed that, by differentiating eqn (C.2.1), we arrive at eqn (C.2.4) for $n = m$. In other terms, eqns (C.2.1, C.2.4) can be symbolically written as

$$T_4(z, D)v_{nm}(z) = 0, \qquad T_3(z, D)w_{nn}(z) = 0, \qquad D = \frac{\mathrm{d}}{\mathrm{d}z}, \qquad (C.2.5)$$

where $T_m(z, D)$ is a polynomial in both variables z and D, being either of the third or of the fourth order in D.

Consider now solutions of eqn (C.1.3). The following conjecture is a generalization of Lemma C.1.

Lemma C.3 *The square $w_{nn}(z) = (\psi_n(z))^2$ of the wave eigenfunction related to eqn (C.1.3) satisfies the third-order equation*

$$w_{nn}'''(z) + 3r'w_{nn}''(z) + \left(4(\lambda_n - q) + r'' + \frac{r'^2}{r}\right)w_{nn}'(z)$$

$$+ \left(-2q' + \frac{2(\lambda - q)r'}{r}\right)w_{nn}(z) = 0, \qquad (C.2.6)$$

while the product of the eigenfunctions $v_{nm}(z) = (\psi_n(z)\psi_m(z))^2$ satisfies the fourth-order equation (the analogue of eqn (C.2.4))

$$rv_{nm}'''' + 5r'v_{nm}''' + \left(2\Sigma\lambda - 4q + 4r'' + \frac{4r'^2}{r}\right)v_{nm}''$$

$$\left(-6q' - \frac{3(2q - \Sigma\lambda)r'}{r} + r''' + \frac{3r'r''}{r}\right)v_{nm}'$$

$$+ \left(\frac{(\delta\lambda)^2}{r} - 2q'' - \frac{4q'r'}{r} - \frac{(2q - \Sigma\lambda)r''}{r}\right)v_{nm} = 0. \qquad (C.2.7)$$

□

Derivation of eqns (C.2.6)–(C.2.7) needs calculations similar to that in Lemmas C.1 and C.2 but more cumbersome. Both equations can be easily transformed to the form of eqns (C.2.5).

C.3 The integral transform

On the functions $w_{nn}(z)$ and $v_{nm}(z)$ an analogue of the Laplace transformation can be defined:

$$w_{nn}(z), v_{nm}(z) \mapsto u_{nm}(p),$$

$$u_{nm}(p) = \int_{-\infty}^{\infty} \exp(-pz)v_{nm}(z)\,\mathrm{d}z,$$

$$u_{nn}(p) = \int_{-\infty}^{\infty} \exp(-pz)w_{nn}(z)\,\mathrm{d}z. \qquad (C.3.1)$$

Convergence of the integrals is guaranteed by the behavior of the eigenfunctions $\psi_n(z)$ at infinity. On the other hand, according to the properties of the Laplace transformation which are also valid in our case, the function $u_{nm}(p)$ should satisfy the equation

$$T(D_p, -p)u_{nm}(p) = 0. \tag{C.3.2}$$

Equation (C.3.2) has a regular singularity at $p = 0$. Its solution appropriate to our goals is holomorphic at this point. However, these facts are not in contradiction, as will be clarified below.

The coefficients of the Taylor expansion of the function $u_{nm}(p)$ are, in fact, the sought matrix elements according to

$$u_{nm}(p) = \sum_0^\infty V_{nm}^{(k)}(-p)^k/k! := \sum_0^\infty g_k p^k. \tag{C.3.3}$$

Our main result can be formulated as follows.

Theorem C.1 *Matrix elements on wave eigenfunctions $V_{nn}^{(k)}$ related to the 1D Schrödinger eqn (C.1.1) are determined up to the trivial multipliers $(-1)^k k!$ by coefficients of the Taylor expansion of a proper regular solution of eqn (C.3.2) at zero.* □

The proof is given above.

The formulated theorem being a general guide to actions, however, leaves many questions. Indeed the function $u_{nm}(p)$ is holomorphic at zero, but it may have a high-order root at this point. This will influence the calculation of matrix elements. The other practical problem is: what are the initial conditions for the solution of the recurrent system. In order to answer this question, the case of the 'general situation' should be analysed as well as concrete exclusive examples of the potential $q(z)$. In this section we study the case of the 'general situation'.

Lemma C.4 *Equation (C.3.2) in the case $n \neq m$ reads*

$$\left(p^2 q(D_p) + \frac{1}{2} pq'(D_p) - \frac{1}{4}\left((\delta\lambda)^2 + 2\Sigma\lambda p^2 + p^4 \right) \right) u_{nm}(p) = 0. \tag{C.3.4}$$

□

Proof: Transforming the known eqn (C.2.3) to eqn (C.3.2) and applying the equalities

$$q(D_p)p^2 u = p^2 q(D_p)u + 2pq'(D_p)u + q''(D_p)u,$$
$$q'(D_p)pu = pq'(D_p)u + q''(D_p)u,$$

we come to eqn (C.3.4). □

Equation (C.3.4) is an equation of $2N$th order identical to the order of the polynomial $q(z)$. The point $p = 0$ is a regular singularity of eqn (C.3.4) and is characterized by the characteristic exponents $\rho = 0, 1 \ldots, 2N - 1$. However, only the solution that is holomorphic at zero fits to our goals. Moreover, its Taylor expansion should start from the term linear in p:

$$u_{nm}(p) = \sum_{0}^{\infty} g_k p^{k+1}. \tag{C.3.5}$$

Let the function $q(D)$ be

$$q'(D) = \sum_{0}^{2N-1} a_j D^j. \tag{C.3.6}$$

Then the initial equation starting the recurrent procedure of calculating the coefficients g_k, i.e.

$$\sum_{1}^{2N-1} a_j j! g_{j-1} - \frac{1}{4}(\delta\lambda)^2 g_0 = 0, \tag{C.3.7}$$

linearly binds $g_0, \ldots, g_{2N-2}$. Hence, at fixed m and n, knowing numerically or asymptotically the eigenvalues λ_n and λ_m and several 'basic' matrix elements (in our case of the 'general situation' $g_0, \ldots, g_{2N-3}$), it is possible to calculate the other matrix elements according to recurrent procedure from 'the beginning'. On the other hand, knowing the asymptotic decrease of g_k at large values of k, it is possible to proceed with the calculations from 'the end'. In this case, we need only eigenvalues.

The analogue of eqn (C.3.4) at $n = m$ is the equation

$$\left(pq(D_p) + \frac{1}{2}q'(D_p) - \lambda p + p^3/4\right) u_{nn}(p) = 0. \tag{C.3.8}$$

Here the first term of the Taylor expansion of the function $u_{nn}(p)$ is always fixed, since the starting matrix element is the normalization of the function $\psi(z)$:

$$u_{nn}(p) = \sum_{0}^{\infty} g_k p^k. \tag{C.3.9}$$

The study of the recurrent procedure for g_k is similar to the case of eqn (C.3.4).

C.4 The harmonic oscillator

The first example is associated with the biconfluent hypergeometric equation. The problem of the harmonic oscillator is studied on $]-\infty, \infty[$:

$$\psi''(z) + (\lambda - z^2)\psi(z) = 0, \tag{C.4.1}$$

$$|\psi(-\infty)| < \infty, \qquad |\psi(\infty)| < \infty.$$

The wave eigenfunctions ψ_n and the corresponding eigenvalues (energies) λ_n are known to be

$$\psi_n(z) = e^{-z^2/2} H_n(z), \qquad \lambda_n = 2n + 1.$$

Here $H_n(z)$ are the Hermite polynomials. The third-order equation for the square of the eigenfunction reads

$$w'''_{nn}(z) + 4(\lambda - z^2)w'_{nn}(z) - 4z w_{nn}(z) = 0, \qquad \text{(C.4.2)}$$

while the fourth-order equation for the products of eigenfunctions reads

$$v''''_{nm}(z) + 2(\lambda_n + \lambda_m - 2z^2)v''_{nm}(z) - 12z v'_{nm}(z) + ((\lambda_n - \lambda_m)^2 - 4)v_{nm}(z) = 0. \qquad \text{(C.4.3)}$$

Matrix elements for the case of eqn (C.4.2) were studied in publication [118]. Here we study eqn (C.4.3). Let $n - m = l$. Transform to the equation for the function u_{nm}:

$$p^2 u''_{nm}(p) + p u'_{nm}(p) - l^2 u_{nm}(p) - ((n + m + 1)p^2 + p^4/4)u_{nm}(p) = 0. \qquad \text{(C.4.4)}$$

This is a second-order equation with a regular singularity at zero and an irregular singularity at infinity. This equation is a specialization of the biconfluent equation [107]. The characteristic exponents at zero are $\pm l$. The solution regular at zero is sought in the form

$$u_{nm}(p) = \sum_{k=0}^{\infty} c_k^{(nm)} p^{l+k}. \qquad \text{(C.4.5)}$$

As a result of substitution of (C.4.5) into eqn (C.4.4), the three-term recurrence relation for the coefficients c_k follows as

$$4(l + 1)c_2 - (n + m + 1)c_0 = 0,$$

$$k(k + 2l)c_k - (n + m + 1)c_{k-2} - \frac{1}{4}c_{k-4} = 0. \qquad \text{(C.4.6)}$$

The recurrence relation (C.4.6) shows that all matrix elements for fixed n and m are expressed in terms of the first nonzero matrix element.

C.5   The anharmonic oscillator

Here the basic equation is a specialization of the triconfluent Heun equation. Consider one of the possible models of the anharmonic oscillator on $]-\infty, \infty[$:

$$\psi''(z) + (\lambda - t^2(z^2 + 1)^2)\psi(z) = 0, \qquad \text{(C.5.1)}$$

$$|\psi(-\infty)| < \infty, \qquad |\psi(\infty)| < \infty.$$

Equation (C.5.1) is a specialization of the triconfluent Heun equation (see Section 3.1). By a scaling transformation of the independent variable $z \mapsto zt^{-1/2}$, eqn (C.5.1) can be transformed to the form more customary for physicists

$$\psi''(z) + (\tilde{\lambda} - (2z^2 + t^{-1}z^4))\psi(z) = 0, \tag{C.5.2}$$

where $\tilde{\lambda} = \lambda t^{-1} - t$. Let $l = n - m$. Then the parity of the function under the integral sign in the definition of the matrix elements coincides with that of $k + l$. If $k + l$ is an odd integer, then the corresponding matrix element is equal to zero. Equation (C.2.4) in the studied case reads

$$v_{nm}''''(z) + 2(\Sigma\lambda - 2t^2(z^2 + 1)^2)v_{nm}''(z) - t^2 z(z^2 + 1)v_{nm}'(z)$$
$$+ ((\delta\lambda)^2 - t^2 z^2 - 8t^2)v_{nm}(z) = 0 \tag{C.5.3}$$

with the notations

$$\Sigma\lambda = \lambda_n + \lambda_m, \qquad \delta\lambda = \lambda_n - \lambda_m.$$

After a Laplace transformation, we arrive at the equation

$$p^2 u_{nm}''''(p) + 2p u_{nm}'''(p) + 2p^2 u_{nm}''(p) + 2p u_{nm}'(p)$$
$$+ \left(-\frac{(\delta\lambda)^2}{4t^2} + \left(1 - \frac{\Sigma\lambda}{2t^2}\right)p^2 - \frac{p^4}{4t^2} \right) u_{nm}(t) = 0. \tag{C.5.4}$$

The point $p = 0$ is a regular singularity of eqn (C.5.4). The characteristic exponents at this point are $\rho_1 = 0$, $\rho_{2,3} = 1$, $\rho_4 = 1$. One of the solutions related to the characteristic exponent $\rho_{2,3} = 1$ and the solution related to the characteristic exponent $\rho_4 = 1$ are functions holomorphic at zero and thus can be presented in the form of Taylor expansions:

$$u_{nm}(p) = \sum_0^\infty c_{k+1} p^{k+1}, \tag{C.5.5}$$

$$u_{nm}(p) = \sum_0^\infty d_{k+2} p^{k+2}. \tag{C.5.6}$$

Substitution of these expansions into the equation leads to the following recurrent relations for the coefficients c_k:

$$12c_3 + \left(-2 - \frac{(\delta\lambda)^2}{4t^2}\right)c_1 = 0,$$

$$a.2.240c_5 + \left(6 - \frac{(\delta\lambda)^2}{4t^2}\right)c_3 + \left(1 - \frac{\Sigma\lambda}{2t^2}\right)c_1 = 0,$$

$$(k+6)^2(k+7)(k+5)c_{k+7} + \left(2(k+5)(k+3) - \frac{(\delta\lambda)^2}{4t^2}\right)c_{k+5}$$

$$+ \left(1 - \frac{\Sigma\lambda}{2t^2}\right)c_{k+3} - \frac{1}{4t^2}c_{k+1} = 0 \quad (k \text{ even, } k \leq 0), \tag{C.5.7}$$

and for the coefficients d_k

$$72d_4 - \left(\frac{(\delta\lambda)^2}{4t^2}\right)d_2 = 0,$$

$$600d_6 + \left(16 - \frac{(\delta\lambda)^2}{4t^2}\right)d_4 + \left(1 - \frac{\Sigma\lambda}{2t^2}\right)d_2 = 0,$$

$$(k+7)^2(k+8)(k+6)d_{k+8} + \left(2(k+6)(k+4) - \frac{(\delta\lambda)^2}{4t^2}\right)d_{k+6}$$

$$+ \left(1 - \frac{\Sigma\lambda}{2t^2}\right)d_{k+4} - \frac{1}{4t^2}d_{k+2} = 0 \quad (k \text{ even, } k \geq 0). \tag{C.5.8}$$

Finally, at $n = m$, i.e. in the case of diagonal matrix elements, the solution regular at zero takes the form

$$u_{nn}(p) = \sum_0^\infty g_k p^k \tag{C.5.9}$$

which leads to the following recurrence relations for the coefficients g_k:

$$g_4 + 6g_2 + (1 - \lambda_n/t^2)g_0 = 0,$$

$$\left[(k+4)(k+3)(k+2)^2 g_{k+4} + (k+2)(2k+3)g_{k+2}\right.$$

$$\left. + (1 - \lambda_n/t^2)g_k + \frac{1}{4t^2}\right]g_{k-2} = 0. \tag{C.5.10}$$

All even matrix elements are expressed in terms of normalization integrals and the quadrupole matrix element.

APPENDIX D: SFTOOLS—DATABASE OF THE SPECIAL FUNCTIONS

D.1 Introduction

In the following we describe a knowledge-based computer project on special functions of mathematical physics, which uses the unified structural approach presented in the book. A program package called *SFTools* has been created by A. Akopyan, A. Piroznikov, V. Zhigunov, and A. Yatzik under the supervision of S. Slavyanov. Special functions addressed by the system include hypergeometric-type functions, Heun functions, Painlevé transcendents, and orthogonal polynomials. For these functions, basic differential equations are given with emphasis on singularities, integral relations, series expansions, Riemann P-symbols (Riemann scheme), and other important characteristics. In order to structure the information, the processes of confluence and reduction of equations are widely used. This means that the underlying principles of the book are preserved in the software package.

The unified structural approach to the classification of special functions in conjunction with modern information technology allow the grouping of all knowledge so that it can be used with less effort than in the traditional 'bookshelf' approach.

In addition to the formulas presented in the book, the package SFTools gives more detailed information about polynomial solutions of differential and difference equations. The classical orthogonal polynomials, regarded as specializations of the hypergeometric function, are considered in Section 2.5.

More generally, polynomials arise as a specialization of the generalized hypergeometric function

$$_pF_q(a_1, \ldots, a_p, c_1, \ldots, c_q, x) = \sum \frac{(a_1)_n, \ldots, (a_p)_n}{(c_1)_n, \ldots, (c_q)_n} \frac{x^n}{n!},$$

where $(k)_n$ is the Pochhammer symbol. To classify these polynomials, a modified version of the Askey–Wilson scheme [7] was chosen. For each of the classical polynomials, a difference equation can be written [66]. By means of a Mellin transformation with a proper integration contour

$$\int_\gamma t^{-z-1}\phi(t)\, dt,$$

this equation can be transformed into a differential one.

The *SFTools* program covers two classes of linear ODEs, namely the hypergeometric (Riemann) and the Heun, one class of nonlinear ODEs—Painlevé transcendents—and the class of classical orthogonal polynomials.

For the Heun and Riemann classes, each equation is presented in the scheme by its s-rank multisymbol, which is a two-dimensional structure consisting of horizontally placed clusters. Inside the cluster, all equations could be obtained from the first (left) equation in the cluster by means of the sequence of reduction processes. The inter-cluster (vertical) connection is one of confluence.

The schemes for the Painlevé class and the class of classical orthogonal polynomials were built to reflect their correspondence with the equations of Heun and Riemann classes.

At the present moment, two versions of the system exist. The one is a stand-alone dialogue program currently developed for running under the Microsoft Windows'95, Windows'98, or Windows NT 4.0 or later versions. The second version coded in Java was designed bearing in mind modern trends in information technology, and is designed to work in the Internet.

D.2 Stand-alone version of the program

The current version of the stand-alone program was developed in the Borland Delphi 3.0 development environment. A full installation of the program requires about 11 Mbyte of hard disk space. This number is likely to change slightly with further system development. A graphic user interface supports all widespread screen modes. Since SFTools uses external databases to work with scientific data, the Borland Database Engine (BDE) should be installed.

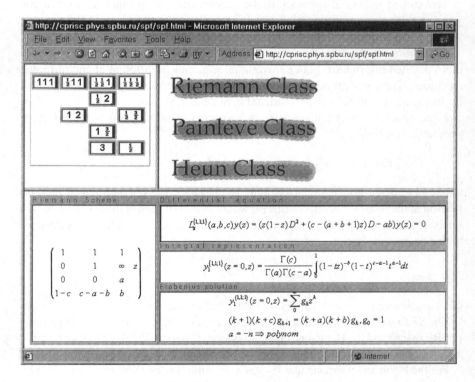

FIG. D.1. The starting window of SFT.

All information on the four above-mentioned classes of special functions is organized in one tab-set window with four pages. Classification schemes are constructed of button-like elements. Each element corresponds to one particular equation. To show this correspondence, each button carries the image of an s-rank multisymbol (for Heun, Riemann, and Painlevé classes) or an abbreviation of the polynomial name. Once pushed, an element becomes highlighted by red, and all the information available for the equation appears in the information windows. The confluent equations of the currently chosen one are highlighted in green. If the mouse-guided pointer is moved to such a green-coloured element, then a new information window containing analytical formulas for the corresponding transformation appears. Information which can be found in the current version of SFTools is the following:

for the Painlevé class:	corresponding equation,
for the Riemann class:	corresponding equation,
	generalized Riemann scheme (GRS),
	integral representation of solutions,
	series solutions,
for the Heun class:	corresponding equation,
	generalized Riemann scheme (GRS),
for classical polynomials:	corresponding differential equation,
	orthogonality condition,
	recurrence relations,
	corresponding difference equation.

Help information presenting the mathematical theory is stored in a standard MS Windows 32-bit help file and can be accessed by direct opening as well as during SFTools session.

We have satisfied four main conditions to make SFTools powerful and flexible at the same time:

- Data should be stored in a way which can provide easy access from different instrument tools such as editors with mathematical support, symbolic algebra software, etc.

- There should be an easy way to add new data.

- Adding new and editing old data should minimally affect the development of other parts of the software product such as GUI, the computational part, etc.

- The process of installation of the program and necessary files (i.e. BDE) should be automated and require the least effort from the user.

To achieve these goals, it was decided to use several external PARADOX databases for holding all scientific data. All the necessary functionality for operating these databases is provided by Borland Database Engine (BDE), which certainly adds some requirements to the system. On the other hand, this software makes the process of interaction between programmers and scientists more convenient when dealing with a large amount of information. Our own experience shows that this facility is

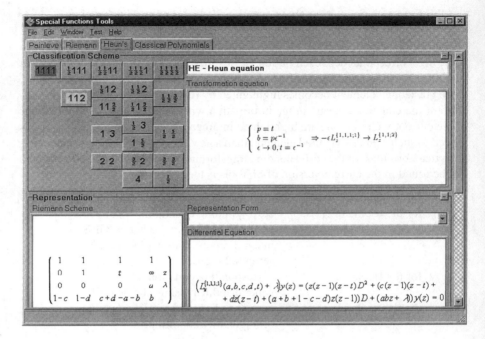

FIG. D.2. A typical window of SFT.

essential for projects of this kind. The graphic interface of SFTools was built to reflect as many of the properties of the classification scheme described above as possible.

The process of program installation with necessary system files modifications and tuning of BDE is automated by means of the Install Shield Light 1.1 utility, which also generates all necessary codes and data.

D.3 Internet version

The Internet version of SFTools program was developed on the base of the stand-alone version. It supports nearly all the features of the stand-alone version and also allows the rapid modification and renewal of information. The program is supposed to work with any browser that supports HTML 2.0, a graphical user interface, and frames.

The major problem that arises in nearly any publication for Internet is the representation of scientific data—especially formulas involving special mathematical symbols and abbreviations. At the present moment, there is no universal standard for such a representation in plain HTML format. Under these circumstances, we see three general ways to represent scientific data:

- By using a Java application embedded into the HTML document. This application should draw a formula in a graphical window of the browser based on its text representation (for example in TEX-like format). We have found some shareware Java applets performing this task, and are now testing them within the project.

- By transferring a representation of a text formula held on the server through a CGI request. The server draws the formula and supplies it in the form of an ordinary graphical image (for example GIF file). The necessary set of tools for this method of data processing is also under construction now.
- By the use of XML capabilities.

Each of these ways has its own advantages and drawbacks. The Java applet usually is larger than an image, but needs to be downloaded only once, while the latter must be reconstructed after any change in the coefficients of the formula. Also, not all browsers support Java capabilities and XML. So it was decided to move in both directions.

D.4   Possible future development

At the present moment, SFTools provides static information only. Such an information system is still valuable, since much information earlier dispersed in many books is concentrated in one place. Moreover, SFTools provides the user with an easy way to obtain all the necessary relations between the equations. The authors see that the system needs to be developed further both in its code and in the theory. Here are some of the directions of possible future development:

- The database could be integrated with some computer algebra system that can perform all necessary symbolic, as well as numeric, computations.
- All the scientific information that is now included in the system as images, and thus is static, will be changed to its proper text representation in the database, with the possibility of drawing formulas on the screen.
- Correspondences between orthogonal polynomials and Heun-type functions, which are of great theoretical interest, could be included.
- Asymptotic relations for the Heun and hypergeometric classes should be included into the system.
- The class of Painlevé transcendents needs further investigation.

To provide SFTools with these features is our next goal, which will help to turn our product into a very efficient scientific instrument.

D.5   How to order SFTools

The copies of SFTools on a CD-ROM will be distributed starting from summer 2000. This software package is primarily regarded as a shareware. Hence, only basic costs—costs of manufacturing the CD-ROM and delivery costs—should be covered when ordering it. In Europe, the reader can order the software at the company: Con. Science Dr. Michael Zaiser & Partner. Address of the company: Torgasse 9, D-71672 Marbach/Neckar, Germany. Further information can be obtained directly from one of the authors: S. Slavyanov, E-mail: slav@slav.usr.pu.ru

BIBLIOGRAPHY

[1] Ablowitz, M.J., Ramani, A., and Segur, H., J. Math. Phys. **21**, 1980, 715–721, 1006–1015.

[2] Abramov, D.I., Sov. J. Vestnik LGU, 1984.

[3] Abramov, D.I., Slavyanov, and S.Yu., J. Phys. B **11**(13), 1978, 2229–2241.

[4] Abramov, D.I. and Komarov I.V., Sov. J. Theor. Math. Phys. **29**, N 2, 1976, 235–243.

[5] Abramowitz, M. and Stegun, I., *Handbook of Mathematical Functions.* 9th edition, Dover Publications New York 1970.

[6] Adams, C.R., Trans. Amer. Math. Soc. **30**, 1928, 507–541.

[7] Askey, R. and Ismail, M., Amer. Math. Soc. **5**, 1984, pp. 108.

[8] Balser, W., *Formal Power Series and Linear Systems of Meromorphic Ordinary Differential Equations*, Springer Berlin, Heidelberg, New York, 1999.

[9] Bardeen, J.M., Press, W.H., and Teukolsky, S.A., The Astrophysical Journal **178**, 1972, 347–369.

[10] Barkatou, M.A. and Jung, F., Programmirovanie (a journal of the Russian Academy of Science), 1997.

[11] Bates, D.R., Ledsham, K., and Stewart, A.L., Phil. Trans. Roy. Soc. Lond. A **246**, 1953/54, 215–240.

[12] Bay, K. Lay, W., and Akopyan, A., Journal of Physics A **30**, 1997, 3057–3067.

[13] Bay, K. and Lay, W., Journal of Mathematical Physics **38**(5), 1997, 2127–2131.

[14] Berry, M.V., Proc. R. Soc. Lond. A **430**, 1990, 653–667.

[15] Bieberbach, L., *Theorie der gewöhnlichen Differential gleichungen* Springer, Berlin, Heidelberg, New York, 1965, (second edition).

[16] Birkhoff, G.D., Trans. Am. Math. Soc. **17**, 1911, 243–284.

[17] Birkhoff, G.D., Acta Math. **54**, 1930, 205–246.

[18] Birkhoff, G.D. and Trjitzinsky, W.J., Acta Math. **60**, 1932, 1–89.

[19] Bolibrukh, A.A., *The 21st Hilbert Problem for Linear Fuchsian Systems.* Proceedings of the Steklov Institute of Mathematics **206**(5), 1995 (translated from Russian by AMS).

[20] Born, M. and Oppenheimer, J., Ann. d. Physik **84**, 1927, pp. 457.

[21] Boyer, R.H. and Lindquist, R.W., J. Math. Phys. **8**, 1967, 265–281.

[22] Brill, D.R., Chrzanowski, P.L., Pereira, C.M., Fackerell, E.D., and J.R. Ipser, Phys. Rev. D **5**, 1972, 1913–1915.

[23] Brower, D. and Clemence, G.M., *Methods of Celestial Mechanics.* Academic Press, New York, 1961.

[24] Brown, E.W., *An Introductory Treatise on the Lunar Theory.* Cambridge University Press Cambridge 1896. Reprinted by Dover Publications, New York, 1960.

[25] de Bruijn, N.G., *Asymptotic Methods in Analysis.* North-Holland Publishing Co., Amsterdam 1961, (second edition).

[26] Bühring, W., Proc. Am. Math. Soc. **118**, 1993, 801–812.

[27] Bühring, W., Meth. Appl. Analysis **1**, 1994, 348–370.

[28] Carter, B., Comm. Math. Phys. **10**, 1968, 280–310.

[29] Chandrasekhar, S., Proc. Roy. Soc. Lond. A. **343**, 1975, 289–298.

[30] Chandrasekhar, S., and Detweiler, S., Proc. Roy. Soc. Lond. A. **344**, 1975, 441–452.

[31] Chandrasekhar, S., *The Mathematical Theory of Black Holes*. The International Series of Monographs on Physics 69, Clarendon Press, Oxford, Oxford University Press, New York, 1983.

[32] Coddington, E.A. and Levinson N., *Theory of Ordinary Differential Equations*. McGraw-Hill, New York, 1955.

[33] Decarreau, A., Dumont-Lepage, M.C., Maroni, P., Robert, A., and Ronveaux, A., Ann. Soc. Sc. de Bruxelle **92**, 1978, 53–78.

[34] Decarreau, A., Maroni, P., and Robert, A., Ann. Soc. Sc. de Bruxelle **92**, 1978, 151–189.

[35] Detweiler, S., The Astrophysical Journal **239**, 1980, 292–295.

[36] Döttling, R., Esslinger, J., Lay, W., and Seeger, A., in: Lecture Notes in Physics **353**, 1990, 193–201. Nonlinear Coherent Structures, Proceedings of the 6th Interdisciplinary Workshop on Nonlinear Coherent Structures in Physics, Mechanics, and Biological Systems. Held at Montpellier, France, June 21–23, 1989. Eds.: M Barthes, J. Léon.

[37] Eckert, W.J., Jones, R., and Clark, H.K., *Construction of the Lunar Ephemeris*. In: Improved Lunar Ephemeris 1952–1959. Washington: US Government Printing Office: 242–363.

[38] Epstein, P.S., Phys. Rev. **28**, 1926, pp. 695.

[39] Erdelyi A., Quart. J. Math. (Oxford) **13**, 1942, 107–112.

[40] Erdélyi, A., Magnus, W., Oberhettinger, F., and Tricomi, F.G., *Higher Transcendental Functions*. Volume III, McGraw-Hill, New York, Toronto, London, 1955.

[41] Erdélyi, A., *Asymptotic Expansions*. Dover Publications Inc., 1956.

[42] Esslinger, J., Dissertation thesis, Universität Stuttgart, 1990.

[43] Fedoryuk, M.V., *Asymptotic Analysis*. Springer, Berlin, Heidelberg, New York, 1993.

[44] Flashka, H. and Newell, A.C., Commun. Math. Phys. **76**, 1980, 65–116.

[45] Fuchs, L., Journal für Reine und Angewandte Mathematik **66**, 1866, 121–160.

[46] Fuchs, R., Math. Ann. **63**, 1907, 301–321.

[47] Gambier, B., Acta Math. **33**, 1909, 1–55.

[48] Graham, R., Knuth, D.E. and Patashnik, O., *Concrete Mathematics*. Addison-Wesley, 1994.

[49] Gutzwiller, M.C., *Chaos in Classical and Quantum Mechanics*. Springer, New York, Berlin, Heidelberg, 1990.

[50] Gutzwiller, M.C. and Schmidt, D., *The Motion of the Moon as Computed by the Method of Hill, Brown, and Eckert*. Astronomical Papers prepared for the use of the American Ephemeris and Nautical Almanac. Volume XXIII Part I. Washington: US Naval Observatory, 1986, 1–272.

[51] Heun, K., Math. Annalen **33**, 1889, 161–179.

[52] Hill, G.W., Acta Mathematica **8**, 1886, 1–36.

[53] Iwasaki, K., Kimura, H., Shimomura, S., and Ioshida, M., *From Gauss to Painleve: a Modern Theory of Special Functions*. Vieweg, Braunschweig, 1991.

[54] Ince, E.L., *Ordinary Differential Equations*. Dover Publications, New York, 1956.

[55] Its, A.R. and Novokshenov, V.Yu., *The Isomonodromic Deformation Method in the Theory of Painlevé Equations*, in "Lecture Notes in Mathematics", v. 1191, Springer Verlag, Berlin–Heidelberg, New York–Tokyo, 1986.

[56] Jaffé, G., Zeitschrift für Physik **87**, 1933, 535–544.

[57] Jimbo, M., Miva, T., and Ueno K., Physica D **2**, 1981, 306–352.

[58] Jimbo, M. and Miwa, T., Physica D **2**, 1981, 407–448.

[59] Jimbo, M. and Miwa, T., Physica D **4**, 1981, 26–46.

[60] Joshi, N. and Kruskal, M., Studies in Appl. Math. **93**, 1994, 187–207.

[61] Kazakov, A.Ya. and Slavyanov, S.Yu., Theor. Math. Phys., **105**, N 6, 1996.

[62] Kazakov, A.Ya. and Slavyanov, S.Yu., Methods and Applic. of Anal., **3**, N 4, 1996, 447–456.

[63] Kerr, R.P., Phys. Rev. Lett. **11**, 1963, 237–238.

[64] Kitaev, A.V., Sfb 288 Preprint N 272, 1997.

[65] Knopp, K., *Theorie und Anwendung der unendlichen Reihen.* Springer Berlin, Heidelberg, New York, 1964 (5th edition).

[66] Koekoek, R. and Swarttouw, R. *The Askey-scheme of hypergeometric orthogonal polynomials and its q-analogue.* Delft Univ. of Technology. Netherlands, 1994.

[67] Kohen, I. and Slavyanov, S.Yu., St. Petersburg. Math. J. **3**(2), 1992, 355–361.

[68] Komarov, I.V. and Slavyanov, S.Yu., Journal of Physics B **1**, 1968, 1066–1072.

[69] Komarov, I.V. and Ponomarev, L.J., and Slavyanov, S.Yu., *Spheroidal and Coulomb Spheroidal Functions,* (in Russian). Nauka Moskow, 1976.

[70] Kovalevskii, M.A., J. Soviet. Math. **50**, 1990.

[71] Lamé, G. *Leçons sur les fonctions inverses des transcendantes et les surfaces isothermes.* Paris, 1857.

[72] Landau, L.D. and Lifshitz, E.M. *Quantum mechanics.*

[73] Lay, W., Seeger, A., and Esslinger, J., Journal de Physique, Colloque **50**(3), 1989, 107–112.

[74] Lay, W., Diploma thesis, Universität Stuttgart, 1983.

[75] Lay, W., in: *Proceedings of the Adriatico Research Conference on Quantum Fluctuations in Mesoscopic and Macroscopic Systems.* Miramare, Trieste, Italy 3–6 July 1990. Ed: H.A. Cerdeira, F.G. Lopez, U. Weiss. World Scientific Publishing Co. Pte. Ltd. 1991, 97–120.

[76] Lay, W., in: *Centennial Workshop on Heun's Equation—Theory and Applications.* Sept. 3–8, 1989. Schloš Ringberg (Rottach-Egern). Edited by A. Seeger and W. Lay. Max-Planck-Institut für Metallforschung, Institut für Physik. Stuttgart, 1990.

[77] Lay, W., Dissertation thesis, Universität Stuttgart, 1987.

[78] Lay, W., Theor. Math. Phys. **101**(3), 1994, 1413–1418.

[79] Lay, W., *Das zentrale Zweipunkt-Verbindungsproblem für die Klasse der Heunschen Differentialgleichungen.* Habilitation thesis, Stuttgart, 1995.

[80] Lay, W., ZAMM **78**, Supplement 3 (GAMM 97, Annual Meeting), 1998, S991–S992.

[81] Lay, W. and Slavyanov, S.Yu., Journal of Physics A **31**, 1998, 4249–4261.

[82] Lay, W. and Slavyanov, S.Yu., Proc. Roy. Soc. A., to appear in Dec. '99.

[83] Lay, W. and Slavyanov, S.Yu., Journal of Physics A **31**, 1998.

[84] Lay, W. and Slavyanov, S.Yu., Proceedings of the Royal Society of London A **455**, 1999, 4347–4361.

[85] Leaver, E.W., J. Math. Phys. **27**(5), 1986, 1238–1265.

[86] Leaver, E.W., Phys. Rev. D **41**(10), 1990, 2986–2997.

[87] Leaver, E.W., Proc. R. Soc. Lond. A **402**, 1985, 285–298.

[88] Leins, M., Dissertation thesis, Universität Tübingen, 1994.

[89] Malmquist, J., Arkiv. Math. Astr. Fys. **17**, 1922–23, 1–89.

[90] Mann, E., Phys. Stat. Sol. (b) **111**, 1982, 541–553.

[91] Maroni, P., Ann. Inst. Henri Poincare **30**, N 4, 1979, 315–332.

[92] Meixner, J. and Schäfke, F.W., *Mathieusche Funktionen und Sphäroidfunktionen.* Springer Berlin, Göttingen, Heidelberg, 1954.

[93] Messiah, A., *Quantum Mechanics, I-II.* North Holland Publishing House, Amsterdam, Oxford, 1962.

[94] Nollert, H.-P., Dissertation thesis, Universität Tübingen, 1990.

[95] Okamoto, K., Ann. Mat. Pura Appl. **146**, 1986, 337–381.

[96] Olver, F.W.J., *Asymptotics and Special Functions.* Academic Press, New York, London, 1974.

[97] Painlevé, P., Bull. C. R. Acad. Sci. Paris **126**, 1898, 1697–1700.

[98] Painlevé, P., Bull. Soc. Math. France **28**, 1900, 201–261.

[99] Painlevé, P., Acta Math. **25**, 1902, 1–85.

[100] Pantleon, W., Diploma thesis, Universität Stuttgart, 1991.

[101] Perron, O., Journal f. reine und angew. Math. **136**(1), 1909, 17–37.

[102] Perron, O., Journal f. reine und angew. Math. **137**(1), 1909, 6–64.

[103] Perron, O., Acta math. **34**, 1910, 109–137.

[104] Perron, O., Die Lehre von den Kettenbrüchen.

[105] Poincaré, H., *Sur les petits diviseurs dans la theéorie de la Lune.* Bull. Astr. **25**, 321–360.

[106] Press, W.H. and Teukolsky, S.A., The Astrophysical Journal **185**, 1973, 649–673.

[107] *Heun's Differential Equations.* Ed: A. Ronveaux. Oxford University Press, Oxford, New York, Tokyo, 1995.

[108] Schmidt, D., Dissertation thesis, Köln, 1970.

[109] Schmidt, D. and Wolf, G., *Double confluent Heun equation* in: *Heun's Differential Equations*, Ed.: A. Ronveaux, Oxford University Press, Oxford, New York, Tokyo, 1995.

[110] Schmidt, D., J. Reine Angew. Math. **309**, 1979, 127–148.

[111] Schmidt, D. and Wolf, G., SIAM J. Math. Anal. **10**, 1979, 823–838.

[112] Schlesinger, L., J. Reine u Angew. Math. **141**, 1912, 96–145.

[113] *Centennial Workshop on Heun's Equation—Theory and Applications.* Sept. 3–8, 1989 Schloss Ringberg (Rottach-Egern). Ed: A. Seeger und W. Lay. Max-Planck-Institut für Metallforschung, Institut für Physik, Stuttgart, 1990.

[114] Seeger, A., Phys. Stat. Sol.(a) **28**, 1975, 157–168.

[115] Slavyanov, S.Yu., *Asymptotic Solutions of the One-dimensional Schrödinger Equation.* Leningrad University Press, Leningrad, 1991, (in Russian). Translation into English: S.Yu. Slavyanov, *Asymptotic Solutions of the One-dimensional Schrödinger Equation.* Providence, Rhode Island, Amer. Math. Soc.,
Trans. of Math. Monographs; Volume 151, 1996.

[116] Slavyanov, S.Yu., Lay, W., and Seeger, A., *Classification* in: *Heun's Differential Equations*, 289–323. Ed: A. Ronveaux, Oxford University Press, Oxford, New York, Tokyo, 1995.

[117] Slavyanov, S.Yu., Lay, W., and Seeger, A., International Journal of Mathematical Education in Science and Technology **28**, 1997, 641–660.

[118] Slavyanov, S.Yu., J. Phys. A: Math. Gen. **32** , 1773–1778, 1999.

[119] Slavyanov, S.Yu. and Veshev, N.A., J. Phys. A: Math. Gen. **30**, 1997, 673–687.

[120] Slavyanov, S.Yu., J. Phys. A.: Math. Gen. **29**, 1996, 7329–7335.

[121] Sleeman, B.D., *Multiparameter spectral theory in Hilbert space.* Pitman, 1978.

[122] Solov'ev, E.A., Sov. Phys. JETP **54**(5), 1981, 893–898.

[123] Starobinskii, A.A. and Churilov, S.M., Sov. Phys.—JETP, **38**, 1974, 1–5.

[124] Sternin, B. and Schatalov, V., Borel–Laplace Transform and Asymptotic Theory: Introduction to Resurgent Analysis, CRC-Press, 1995.

[125] Svartholm, N., Math. Ann. **116**, 1939, 413–421.

[126] Teukolsky, S.A. and Press, W.H., The Astronomical Journal **193**, 1974, 443–461.

[127] Teukolsky, S.A., Phys. Rev. Lett. **29**, 1972, 1114–1118.

[128] Voros, A., J. Phys. A: Math. Gen. **32**, 1999, 5993–6007.

[129] Weiss, J., J. Math. Phys. **24**, 1983, 1405–1413.

[130] Wentzel, G., Z. Phys. **33**, 1926.

[131] Wilson, A.H., Proc. Roy. Soc. A **118**, 1928, 617–647.

[132] Wimp, J., *Computations with Recurrence Relations.* Pitman Advanced Publishing Program, Boston, London, Melbourne, 1984.

[133] Wong, R. and Li, H., Journal of Computational and Applied Mathematics **41**(1-2), 1992, 65–94.

INDEX